WHERE THE SUN NEVER SHINES

WHERE THE SUN NEVER SHINES

A HISTORY OF AMERICA'S BLOODY COAL INDUSTRY

PRISCILLA LONG

PARAGON HOUSE
NEW YORK

FIRST PAPERBACK EDITION, 1991
PUBLISHED IN THE UNITED STATES BY
PARAGON HOUSE
90 FIFTH AVENUE
NEW YORK, NY 10011

LIBRARY OF CONGRESS CATALOGING-IN-PUBLICATION DATA

LONG, PRISCILLA.
   WHERE THE SUN NEVER SHINES: A HISTORY OF
AMERICA'S BLOODY COAL INDUSTRY
   PRISCILLA LONG.
      P.   CM.
   BIBLIOGRAPHY: P.
   INCLUDES INDEX.
   ISBN 1-55778-465-5
   1. COAL MINERS—UNITED STATES—HISTORY.
2. TRADE-UNIONS—COAL MINERS—UNITED STATES—
HISTORY.   3. STRIKES AND LOCKOUTS—COAL
MINING—UNITED STATES—HISTORY.   I. TITLE.
HD8039.M62U6444   1989
338.2'724'0973—DC19   89-3157
                                       CIP

DESIGNED BY KATHY KIKKERT

MANUFACTURED IN THE UNITED STATES OF AMERICA

THE PAPER USED IN THIS PUBLICATION MEETS THE
MINIMUM REQUIREMENTS OF
AMERICAN NATIONAL STANDARD FOR INFORMATION
SCIENCES—PERMANENCE OF PAPER
FOR PRINTED LIBRARY MATERIALS, ANSI Z39.48-1984.

*It's dark as a dungeon and damp as the
  dew,
Where the danger is double and pleasures
  are few,
Where the rain never falls and the sun
  never shines,
It's dark as a dungeon way down in the
  mine.*

—*Merle Travis,* "Dark as a Dungeon"[1]

For Sandra Nickerson and Bill (Krumske) Mayer.
For Red Sun Press.
For Peter Irons.

CONTENTS

LIST OF MAPS

LIST OF ILLUSTRATIONS

PREFACE

"Pray for the dead and fight like hell for the living!" proclaimed the intriguing, charismatic United Mine Workers organizer, "Mother" Mary Jones (1837–1930). In the early years of our century, the white-haired Mother Jones became a legendary figure in the coalfields of the United States. It was by way of retracing her steps that I first traveled to places like West Virginia and Colorado to examine the lives of the men and women who lived in the coal camps at the turn of the century.

This work originally took shape in my mind as her biography. But I found it difficult to separate Mother Jones from those for whom she lived and worked. Increasingly, her book became a collective biography of her people, an extended analysis of the material and cultural conditions of their lives, a journey of discovery into the forces that swept them along, and which they in turn helped to shape. Many of these forces and events were occurring everywhere in society, not just in the coalfields. They stand at the center of our history as an American people. In its broadest sense, then, this book is a history of the United States before 1920, as magnified through the highly revealing lens of the coal industry.

For the writer it has been a long and circuitous journey that began years ago. Along the way, many have contributed ideas, critical attention, and much appreciated support. But the first debt of any historian is to the articulate voices of the past. I am particularly beholden to two classes of contemporary observers.

In the 1870s and 1880s, many states established Bureaus of Industrial and Agricultural Statistics. The diligent, fussy, overworked statistical clerks employed by these bureaus had among their duties the preparation of annual reports covering the subjects of their respective departments.

These forgotten clerks were writers any writer would be proud to emulate. They were social and political analysts well schooled by their primary task—the gathering of statistics and of statements of fact. Pioneers of empirical observation as it applied to society, they also strove to uncover causes. They used their data to imagine the past and the future, skillfully describing long-term historical trends. The vivid essays found in the pages of their reports, though not entirely free of prejudice, continue to inspire and inform explorers of the American past.

The observers to whom I owe the greatest debt, however, are the men and women of the coalfields. They were often literate, more often articulate, at times even eloquent. They spoke to federal commissions, to congressional committees, to state investigative bodies, and to the press. They wrote prolifically to labor newspapers. Occasionally, a coal miner would sit down at his kitchen table to compose a letter to the president of the United States. Or a miner's wife, moved by tragedy, would open up to a reporter. There were many avenues of expression. In 1899, the Colorado Commissioner of Labor Statistics wrote of the questionnaires distributed among the working people of the state: "Many of the replies received are of a character that in point of excellence in literary composition would do credit to authors of national renown. They give ample evidence of the earnestness, the directness, and the intelligence with which the hard-fisted toilers of Colorado discuss sociological questions."[2] Among the respondents were coal miners who answered the questions with careful compositions on social ills and their possible cures.

If the historian feels reluctant to leave the past because its actors have become like old friends, the present holds its own compensations. To friends, supporters, critics, and readers, to those who have their own work and severe time constraints but who have nevertheless freely contributed to this work, I am grateful beyond words.

I think especially of Milton Cantor, whose consistent support over the long haul has eased the task of bringing this book to light. Neither can I ever repay Ava Baron, a leading scholar on the interaction of gender and class in American working-class history. As her thinking has enriched this work, her friendship has enriched my life. Maury Klein freely shared his own research, and meticulously commented on the manuscript, to its everlasting benefit. His contributions are a tribute to his intellectual generosity. If there really is such a thing as a community of scholars, he is one of its exemplary members.

My thanks go to old friends who believed in this work all along. A partial list includes Sandra Nickerson at the top for her support and encouragement over a long succession of dripping submarine sandwiches. It includes Roz Zinn and Howard Zinn for their inimitable zest for life. Howard's work on the Colorado Fuel and Iron strike and his passion for writing history in a way that matters was the spark that set this work in motion. Meredith Tax has been a dear friend, an important example of a life consciously lived, and a critical intellectual influence.

The list of those I want to thank includes Deborah Lee Gould and Stephen Jay Gould, whose friendship started back in the good old days and hasn't ended yet. Pat Cockrum, erstwhile neighbor and good friend, was a welcome visitor at the end of many a day. Our talks on the process of carrying out self-directed work, hers and mine, contributed spiritual nourishment essential for persevering. The friendship of Farah Ravenbakhsh and Darian Meachem has been indispensable to me and therefore to this book.

Reading and commenting on an unrefined book manuscript is a thankless task, about as gratifying, I suspect, as picking slate from coal. I am therefore particularly indebted to Judy Rosselli for perceptively picking through every dreadful page of an early draft. My sister Pamela O. Long took precious time from her own work to make numerous judicious comments on mine including, occasionally, the rather startling "eek!" Other readers of all or part of the manuscript gave their time freely and their reactions carefully. They include Claudia Mejetic, whose many talents include an uncanny aptitude for spelling. They include Melvyn Dubofsky, Steve Erie, Ann Jefferson, Dan Hallin, Robert Horwitz, Tom Dublin, Mary Person, and Paul Ronsheim. I especially appreciate the reading of Chapter 2 by a mining engineer, Mary Beth Frost, and by a skilled working miner, Sam Witherspoon.

One of the pleasures of research in Denver is the mind-stretching talk with John Graham whose insights into Colorado coal mining history have aided this work immensely.

Mike and Ruth Yarrow's extensive knowledge of coal mining and their convivial sharing of it has deepened my understanding of many issues.

Roseanne Costantino did a fine job of drawing the maps. In so doing, she discovered her own connection with this history: Her great grandfather, an Italian immigrant, was a coal miner who worked in the mines of the Rocky Mountain West.

My agent, Rick Balkin, was an early believer in this book and an able advocate for it. To him heartfelt thanks are due.

I am grateful to the Louis Rabinowitz Foundation for support in the early stages of this project, and to the Bunting Institute of Radcliffe College for support in its later stages.

I am in the debt, as every historian must be, to archivists, company officials, and union officials who were efficient and generous in providing access to papers. The late Arnold Miller, coal miner, activist in the Black Lung Association, and past president of the United Mine Workers of America, introduced me to the coalfields of southern West Virginia and later provided access to the all-important UMWA papers.

The enthusiasm for this project of Mary Walton Livingston of the National Archives has meant more than she could know. I am grateful for the support of the staffs of the Western History Department of the Denver Public Library, the Colorado State Historical Society, and the Nebraska State Historical Society. I must thank Larry Cruse and his fellow staff members in the Documents Department of the library at University of California at San Diego. They remained helpful and cheerful in the face of my constant presence and never-ending requests.

This work benefited enormously from the generosity of Ken Longe of the Union Pacific Railroad Corporation; Don Snoddy of the Union Pacific Museum; and Emogene Coulter and Tony Pregnell of Rocky Mountain Energy. Finally, I want to thank Marion Hanscom of the Special Collections division of the Bartle Library at the State University of New York at Binghamton, as well as the excellent staff of the Rockefeller Archive Center.

Peter Irons deserves more than thanks. He brings to a lifetime companionship a regard for the integrity of the other that is truly gratifying. It pleases him to quote Horace Mann, who said, "Be ashamed to die until you have won some victory for humanity." He has done his own part toward such victories. It is also a victory for humanity, I insist, that his important work was never done at the expense of mine, and that while doing it he also accomplished all of the family shopping and cooking. His faith in this project has sustained it, from beginning to end.

NOTE ON STYLE. Following other historians of working-class life, I have done away with the practice of interjecting the snobbish little word *sic* into the quotations of people who expressed themselves in non-standard English. The utmost care has been taken to reproduce the quotations accurately. Readers should also be aware that the word *employees* used to be spelled *employes; Finns* used to be *Finlanders;* and that *Pittsburgh* has sometimes done without its final *h*.

INTRODUCTION

This book relates the history of coal mining in the United States from early times until 1920, the year John L. Lewis took control of the United Mine Workers of America. In it I offer a vision of American history that encompasses many decades and several regions. I have worked to achieve a picture that reflects some of the complexity of the past as it actually existed. Then, as now, men usually did not live without women. Technology did not develop apart from conflicting human values. Perceptions and loyalties regarding class have always been enmeshed in changing concepts of masculinity and femininity. Working-class communities were built near workplaces; the work itself was central to the experience of the community. The history of firms and of their management did not occur apart from the men and women who worked within them. Labor's struggles did not occur in a business vacuum.

To do justice to the complexity of the real world as it was, I have utilized the approaches of several specialties found within the grab bag of American historiography. In the tragic and triumphant stories, and in the analysis contained in these pages, readers will find more than a glimpse of women's history, labor union history, business history, the history of work and of technology, and the history of working-class communities and cultures. Each approach is a lens that throws its own special light on the past, and each becomes indispensable to a picture of the whole.

Coal mining took place in semidarkness, but its history illuminates the grand sweep of the American past remarkably well. I designed this work with the goal of achieving a new historical synthesis that would give intellectual pleasure through the breadth of its vision. I, for one, am weary of narrow studies. In this history of coal mining, the time span covered is long enough to allow a view of major shifts in American society; the geographical net is cast wide enough to permit comparison among regions and a view of national trends; and the subjects engaged are varied enough to allow a multidimensional view. Let me assure the reader that this was not done at the expense of the research—the bone and sinew of any historical work—but only at the expense of time.

Where the Sun Never Shines begins with the truism that our history as a nation evolved as part and parcel of the history of the world. The story begins where the American coal industry began—in Great Britain. Coal was the central fuel of the industrial revolution, that series of far-reaching changes that emerged in Great Britain two centuries ago, spread to Europe and to the United States, and continues to transform the world. In important ways, the industrial revolution emerged from the early coal industry rather than the other way round. Chapter 1 elaborates on this, describes the world's first coal-mining people, and then shifts to the American coalfields along with a great migratory stream of British miners.

Once in America, we will explore the nineteenth-century miner's workplace, "away down in the bowels of the earth," as a miner put it, "where one ray of God's blessed sunlight never shines."[1] The actual form of the workplace, as well as work methods and procedures, grew out of social relationships and human values. The analysis moves from physical and then social conditions to political responses—by working people in all their diversity and by owners and managers, whose singular pursuit of profit led to the industrial development of the country.

Part II relates the history of the Rocky Mountain coalfields (to 1920), with Colorado at its center. This book, then, contains a regional history set in the context of the national history explored in Part I, and the depth of the regional history complements the breadth of the national. The progress of industrialization in the Rocky Mountain West, with its unique geography and particular people, reflected national and international trends. Yet the region's history was more than a mirror. Some local events stand at the heart of a nation's past. Among them must be placed a great coalfield war: the Colorado Fuel and Iron strike of 1913–1914.

The strike was one of the most grueling, long-lasting, and widely known industrial conflicts in the history of the United States. For more than a year, it crowded other domestic news off the nation's front pages. Finally, it exploded, on 20 April 1914. That Monday morning, two companies of the Colorado National Guard attacked a tent village inhabited by strikers and their families—more than one thousand men, women, and children. The troops fired into the Ludlow Tent Colony throughout the day, and at dusk burned it to the ground. Eleven small children and two women were among those who died in what electrified the country as the Ludlow Massacre.

Because the strike lasted so long, attracted so much attention, and drew so many actors into its storm, its history brings into sharp focus the society it tore apart. The society thus brought into view is our own. The strike was a bitter class conflict coming at the end of a thirty-year period in which the growth of industrial capitalism had radically altered the class structure of society. Before 1880, the predominant economic institutions in the United States were the family farm and small-scale enterprises owned by craftsmen such as blacksmiths. By 1910, giant corporations dominated the economy and their influence permeated the social and political life of America.

Coal catalyzed the change. What the geologist Thomas Huxley said of Britain in 1870 fits America as well. "Wanting coal," he stated, "we could not have smelted the iron needed to make our engines, nor have worked our engines when we had got them. But take away the engines, and the great towns . . . vanish like a dream. Manufactures give place to agriculture and pasture, and not ten men can live where now ten thousand are amply employed."[2] Coal fueled the new industrial capitalism. Wood, a less efficient fuel, could never have supported the same development.

The change from the small-scale enterprise to the large corporation entailed an enormous shift in values—an agonizing disruption of a world-view that prized the legacy of skill handed down from father to son, that strove for a life-style that combined prideful work with a rich and communal leisure. The miner's self-esteem was tied up with his skill and with a notion about his manhood involving the ability to control his work and how he did it. Opposing him, as the new industrial system developed, was a new kind of manager and a new kind of management. The manager's tools were the machine and the influx of unskilled workers into the coal industry from all parts of the world. The perceptions of these new workers did not coincide with those of traditional craft miners. If working people were to organize with the purpose of achieving common goals, they would have to do so across a cultural and ethnic diversity that existed nowhere else in the world.

These days, any labor history worth its salt must consider the cultural and ethnic dimensions of working-class life. But this book also returns to a more traditional study of trade unions to describe not the heroic deeds of the leaders (the old way) but the internal dynamics of coal miners' unions and their interaction with the cultural and political life of the community.

Labor history is conventionally practiced apart from business history, while business history flows in its own separate stream. Yet working people and business people have always been inextricably linked. The developing consciousness and institutions of the one group were nothing less than the very context in which those of the other existed. However the classes of society might be described, they cannot even be conceived except in relation to each other.

This book examines the growth of the large-scale corporate enterprise as it related to the coal industry. This growth involved changes in owner-ship patterns and the rise of management as a new kind of activity. At the same time, the working population shifted away from being essentially an aggregate of craftsmen and began to resemble instead a new industrial working class. The emergence of a new system of classes, and the conflict, accommodation, outright war, and ongoing struggle between working people and their employers is central to the history of coal mining in the United States, and to this book.

Immigration is another cardinal fact of the American past. Race, na-tional origin, culture, and religion divided working people a hundred times over. If America was not precisely a melting pot, because in the cities new arrivals lived and worked within ethnic and language enclaves, this was less true of the coal regions. Isolated coal towns with their own ethnic neighborhoods were perhaps miniatures of urban society, except that people of necessity lived closer together. Coal operators mixed national and racial groups to keep workers divided, but the effect of doing so was to bring them together, and the result could be an exceptional ability to fight back.

The women of the mining communities were more than a colorful sidelight to this history. That the most prominent organizer of coal miners was a woman—"Mother" Mary Jones—was no freak accident but grew out of the strong role women played in the struggles of the coalfields. Their activism was based not on vicarious support for their husbands, fathers, and brothers but on their own direct experience with the com-pany, gained in the company towns. When the miners went on strike, as they did more frequently than most or perhaps all other groups, the action was commonly undertaken by the entire community.[3] The aggressive participation of wives and daughters was a root cause of the particular militance of the mining communities. The women's work in the coal camps, their perceptions, and their consciousness and the activism that grew out of it are vital to the history of coal mining in the United States.

Finally, I have worked to place the history of coal mining within the framework of American history as a whole. To give one example, the chapter on coal miners during the Civil War reminds readers of that traumatic national experience before showing its effects in the coalfields.

By definition, a historical synthesis, like any work of art that succeeds, is greater than the sum of its parts. It is more than a jumble of aspects, a hodgepodge of factors. A new synthesis must please the intellect by its completeness and by its essential accuracy. It must shed the light of truth on the darkness of the past and do so with literary merit. This has been the standard toward which I have guided this work. To what extent I have reached it is of course for the reader to judge.

Let us begin our particular journey into the American past with a look at coal, the substance, and its place in our world. This will give us an immediate connection with the coal miner, for the coal miner knows coal. As his body is saturated with its dust, so his language is saturated with its lingo, and his culture with its lore. He knows the feel of coal, he knows its sound as it shifts and cracks, he knows the gas it constantly emits. To understand the physical facts of coal is to open the door to his world.

I
COAL IN
AMERICA

A HISTORY OF WORK,
VALUES,
AND CONFLICT

PROLOGUE:
COAL IS A ROCK
THAT BURNS

I have examined by the light of my miner's lamp, exquisite designs of fern leaves impressed on pieces of fallen roof. I have seen the exact shape of petrified mice in the sides of the underground roadways, and at one colliery used to pass a place daily where a perfect figure of a stone snake with a lifted head seemed to indicate the way to our new workings.

B. L. Combs, Those Clouded Hills, *1943* [1]

Coal is stored energy from the sun. [2] Composed mainly of the element carbon, it resulted from photosynthesis: the conversion of the sun's energy into organic compounds that takes place in the leaves of green plants. Coal is fossilized plant matter—the leaf and wood remains of any kind of plant, but especially of the ferns and rushes that grew as tall as trees in the warm, moist atmosphere encircling the earth 350 million years ago. Coal beds originated as swamps that evolved into vast, thick peat bogs. Gradually the bogs sank; often they were flooded under shallow seas.

These bogs were buried under sediments, compressed, and subjected to heat and to the pressure of the earth. The process of decay continued long after burial. Pockets of gas released during putrefaction were trapped under the earth with the coal. Coal itself, even the smallest lump of it, continuously emits marsh gas, or methane.

Coal, oil, and wood have in common that each is composed mainly of carbon. Wood was the first to be used as a fuel and indeed, for most of human history people burned firewood for cooking and heating. The change to coal occurred in the growing emergency of a wood famine. The significance of this change was that coal gave off twice as much heat as a comparable amount of wood.

Coal is a rock that burns. Geologists rank it into three major types according to the increasing concentrations of carbon. Lignite coal contains the least amount of carbon; bituminous contains more; and anthracite contains the most.[3] The differences are caused by the variability of earth forces (heat and pressure) upon it. Lignite, a soft brown coal with visible plant components, often occurs under plains or prairie lands. Bituminous, or soft coal, the most common type, burns with a smoky yellow flame. Anthracite, or hard coal, is dense, rocklike, and glossy black. The richest anthracite field in the world lies under the Wilkes-Barre/Scranton/Pottsville region of northeastern Pennsylvania. Anthracite is not necessarily older than other coals, but it formed under the extreme pressure of mountain building. It burns with a blue flame and little soot. It is difficult to ignite, and long after soft coal had become an ordinary fuel, people did not suspect that it would burn. They referred to it as stone coal.

Coal is not only ranked but also graded. The rank pertains to the degree of alteration in the coal (from lignite to anthracite), while the grade describes its quality. The two chief impurities that reduce the quality of any rank of coal are sulphur and ash. Ash is what remains in the fireplace after the fire goes out. It inhibits burning and later has to be removed. Sulphur is a noxious pollutant of both air and water. In the air it combines chemically with water and oxygen to form acid rain, which can be as strong as vinegar and lethal to lakes, forests, and crops.[4]

The flame of industrialization spread from a coal fire. In 1875, a writer could justly say: "Coal is to the world of industry what sun is to the natural world."[5] In the nineteenth century, it fueled the steam engine, the new source of power that drove machinery and turned the wheels of trains. Flickering coal gas lamps lit the streets of Dickens's London, and of cobblestone Baltimore. The rattle of coal pouring into the cellar bin was by the end of the century a familiar sound in Great Britain, Europe, and the United States: Coal had replaced wood as the predominant home heating fuel.[6]

The coal industry's most important manufactured product was and still is the metallurgical fuel coke. Manufactured by baking certain types of bituminous coals, coke is a nearly pure form of carbon. The baking drives off the volatile matter—that which turns into gas—and chemically alters the minerals. Coke is a blast furnace fuel used in the smelting of iron ore.

As such, it is one of the essential components of industrial civilization.[7]

Coal was a latecomer on the stage of human history, and oil arrived even later, to a world already transformed by the industrial revolution. Oil's commercial use began with the gusher of 1859 at Titusville, Pennsylvania. At first, oil did not compete with coal as a fuel, but replaced whale oil in lamps and was used to grease the ever-increasing number of machines in the world. Only at the dawn of the twentieth century, the century of the automobile, did the internal combustion engine supersede the coal-burning steam engine. Not until the 1940s did oil and its consort natural gas succeed in toppling Old King Coal off the energy throne.[8]

Today, three-quarters of all coal mined in the United States is used to fuel electric power plants, and most of the rest goes into making coke. In addition, the volatile by-products of coke manufacture are one important source of raw material for the petrochemical industry. They form the basis of a myriad of products from aspirin to nylon to dye.[9] But today coal is only one of several sources of energy, and it is not even the most important one. As the reigning fuel of industrialization, however, its impact on the world was profound.

1

THE TURN FROM WOOD TO COAL: A REVOLUTION IN BRITAIN AND AMERICA

In some of the lost causes of the people of the Industrial Revolution we may discover insights into social evils which we have yet to cure.

E. P. Thompson, The Making of the English Working Class[1]

The horse work and the mule work were done by the human beast of burden.

Elderly English coal miner, Flint, Michigan, 1910[2]

The turn to coal began in England in the thirteenth century; by the nineteenth century, the British coal industry was by far the world's oldest and most productive. In the 1840s, some two thousand Welsh, Scottish, English, and Irish coal miners voyaged across the sea to the United States.[3] These seasoned industrial workers found the New World a rural place with few smokestacks. Farmers dug coal in the winter where it outcropped on their land. They hauled the stuff to their buyer in carts or wagons, discontinuing the operation with the spring press of plowing and planting. The only bituminous industry worth the name existed in eastern Ohio and in Virginia along the James River, where enslaved blacks dug coal under the supervision of English and Welsh miners.[4] The anthracite industry had been developing since the 1820s, with Britons involved from the beginning. Even so, in 1840 the entire product for the United States amounted to less than two million tons, a small fraction of the thirty-one million tons produced in the British Isles that year.[5]

In the 1850s, some 37,000 British miners migrated to the United States.[6] A traveler in Wales remarked that every collier (coal miner) with whom he spoke had a "father, brother, son, uncle, nephew, cousin, or friend in America and had been cogitating about going himself." Of the American industry, an observer wrote, "[T]he mining population of our Coal Regions is almost exclusively composed of foreigners, principally from England and Wales, with a few Irish and Scotsmen."[7]

In mid–nineteenth-century America, the world of the coal miner was largely a British world. "I am a miner," explained English-born John Hall, speaking in 1871 to an Ohio mining commission. "I was a miner in the old country, from which I emigrated in 1848. I have mined coal in Pennsylvania and also western Virginia. I began mining work when eight years old." His life pattern followed that of most Ohio miners. His language and culture gave American coal communities a distinctly British flavor.[8]

Not only the cultural ambience of the coalfields, but the capital for opening new mines, the tools and technical know-how, and the labor of the early American industry were predominantly British. The industrial revolution itself, with coal at its center, emerged in Great Britain in the late 1700s. One of the ways the revolution came to the United States was in the person of the British coal miner.

COAL IN THE FIRST INDUSTRIAL REVOLUTION

The world's oldest coal industry produced both the steam engine and the railroad. To imagine the industrial revolution without these is to begin to understand the impact of coal mining on this great transformation in the history of humankind.

The crucial question of why the industrial revolution began in Great Britain and not elsewhere cannot be answered fully here. Cultural and political explanations range from attitudes deriving from the Protestant ethic, to the development of towns free from the suffocating authority of church or state.[9] But any satisfactory explanation must include in its probing another question: Why did the coal industry begin in Great Britain and not elsewhere?

Great Britain is a small place. Its tiny land area supported limited forestlands. Yet for centuries wood was the primary construction material, as well as the primary fuel. Consider how long the British navy dominated the high seas in wooden sailing ships. Scottish salt makers burned prodigious quantities of wood to evaporate brine, wasting the forests surrounding their works. (Salt, an essential of the food economy, was used to preserve meat in the days before refrigeration.) The iron industry, too, depended totally on wood. Iron makers smelted the ore in furnaces fired with charcoal, and they then sold the iron to blacksmiths, who pounded the metal into the shape of tools or implements after heating it in a charcoal-fired forge.[10]

Since ancient times, human beings have occasionally built coal fires. There are biblical references to coal, and archeologists have discovered the remains of coal fires built by North American Indians. In 1275, Marco Polo, the Venetian traveler, reported to his disbelieving countryfolk that in China there was "a black stone existing in beds in the mountains which they dig out and burn like firewood."[11] But coal was not consistently mined until the thirteenth century, when serfs began digging it from outcroppings and shallow pits in the north of England and in Scotland.

Coal is heavy, and it is also bulky. Mining it and transporting it were (and are) two parts of one continuous process, for if it can't be transported to where it will be burned there is no point in mining it. The industrial revolution produced the railroad, but Britain's need for coal predated this means of transporting it.

England's unique geography deprived its people of wood but provided them with coal. Equally important, transporting coal did not present difficulties as great as it would elsewhere. The land is flat, the rivers are numerous and navigable, and no town lies farther than sixty miles from the sea. The small land area made it possible to utilize coal before an efficient transportation system was devised to transport it. Geography was not the only factor: After all, Britain's wood famine occurred in the context of a developing industrial economy. But it is difficult to imagine how a mountainous and relatively landlocked country, even one like China with an abundance of coal, could have produced the world's first industrial revolution.[12]

As early as 1300, the scarcity of firewood drove Londoners to heat their houses with coal, despite the "intolerable smell" of the smoke. Blacksmiths began to burn it in their forges. By 1615, four hundred English sailing ships were engaged in transporting coal, half of them back and forth to London.[13]

In England and Scotland, the world's first coal miners dug the fuel on the great estates of the landed aristocracy. In Scotland, they mined the immense land holdings of the monasteries. They were serfs; families bound to lords and accustomed to working together on the land as a team began to dig coal as a team. Here is the origin of the fact that women as well as children toiled as coal bearers in the British mines for the first five hundred years.[14]

At first, they dug the coal out of shallow holes with wooden shovels. But, as the demand grew, quarrying gave way to underground mining. Solving the exasperating difficulties of subterranean mining took centuries but resulted in technological breakthroughs of immeasurable importance.

Shafts were sunk. By the early 1700s, the average shaft penetrated two hundred feet into the earth, while the deepest reached nearly five hundred feet. A century later, one shaft descended 993 feet, nearly a fifth of a mile.[15] Mines extending below the water table continuously filled with water. As time went on, the underground workings extended farther and farther from the bottom of the shaft. The problems of hauling the coal from the face (the part of the seam from which coal was being removed) to the shaft, and the problems of hoisting it, became acute as shafts were dug deeper and the face receded farther from it. The steam engine and the railroad were developed as solutions, worked out over two centuries, to the problems of drainage and haulage in the coal mines of Great Britain.[16]

Water was such a problem that at some mines more water than coal was brought to light. At one colliery, it took five hundred horses drawing up water in buckets to keep it under control.[17] A suction pump could siphon water up only thirty feet; until the mid-1600s no one understood why this was so. The problem was referred to Galileo and solved by his student Torricelli, who announced in 1644 the discovery of atmospheric pressure. In other words, air was a substance with weight. Air weighed on a pool of water enough to push it thirty feet up a pipe, but no more.

The grandfather of the steam engine was the pump, invented on this principle that air was a substance with weight. Thomas Savery patented his invention in 1699 and called it "the miner's friend." An improved version designed by Thomas Newcomen pumped water out of British mines for a century thereafter. It consisted of a large piston working in a cylinder. A vacuum inside the cylinder enabled the atmospheric pressure outside to press the piston down. Steam was then injected into the cylinder to push it back up. (The steam was made by boiling water over a coal fire.) To make a vacuum again, the cylinder was cooled by pouring cold water on it. The steam condensed, and the piston dropped into the newly created vacuum.[18] The up and down motion of this piston, transferred to the moving parts of machines and especially to the wheels of trains, would change the face of British society.

James Watt radically improved Newcomen's pump to invent the steam engine, first put to commercial use in 1776. It provided the world's first source of power greater than muscle or the waterwheel. Although water-power long remained appropriate in many industries, the wheel could only turn in a river or stream and the energy it provided was limited.[19]

Watt's engine was superior to the Newcomen pump, which consumed prodigious quantities of coal. This wastefulness concerned coal owners not at all, but prevented the pump from being installed at any great distance from a mine. The pump tied to its source of fuel was like the waterwheel tied to its source of water, limited. The steam engine burned less coal. Industries operating far from the coalfields (such as textiles) gradually adopted it as a source of power. But its most far-reaching effect came in the early years of the nineteenth century, when it formed the kernel of another invention: the railroad.[20]

If in every essential way the coal industry contributed the steam engine to the industrial revolution, it also contributed the first reliable transportation system in history—the steam locomotive.[21] Both the locomotive and the track were devised in the process of solving the problem of hauling the heavy, bulky fuel from the face to the bottom of the shaft.

The track came first. Most likely, as Thomas Ashton and Joseph Sykes speculate in *The Coal Industry in the Eighteenth Century*, it originated as two parallel planks laid along the uneven floor of a coal mine tunnel. This crude arrangement eased the pulling along of a wagon load of coal. Later, a pin projecting from the wagon into the space between the planks kept the wheels from falling off. Later still, wooden tracks were laid above ground, from the mine opening down to the sea. In 1676, an astonished traveler noted that on these parallel timbers one horse could draw carts holding four or five chaldrons (about eight tons) of coal.[22]

The locomotive came later. A century after the first track had been laid above ground, a coal mine manager applied the steam engine to the problem of underground haulage. The inventors of the locomotive were closely associated with the owners of the largest mines in Britain.[23] The railroad, devised in the early 1800s to haul coal, became the overland transportation system of the new industrial capitalism.

Slowly the railroad replaced the horse-drawn wagon. This is one illustration of how, as David Landes has aptly put it, the industrial revolution in its physical dimension brought the substitution of mineral forms for animal and vegetable forms. The steam engine replaced human and animal muscle power. Iron replaced wood as the chief construction material. Coal replaced wood in the fireplace. Coke replaced charcoal in the blast furnace.[24]

The adoption of a new blast-furnace fuel was at least as important as anything else, because it relieved iron makers of what certainly would have been a fatal dependence on wood.[25] Coke functions exactly as charcoal does in the blast furnace. The fuel is set to burning in the bottom of the furnace, and the iron ore, usually an iron oxide, is dumped on top. The fire melts the iron, but the fuel has a second function as well. During combustion, the carbon in the fuel combines with the oxygen in the ore, freeing the iron to exist in pure form. The fuel must provide not only heat, but carbon; in this function, it is called the reducing agent. To this day, the iron and steel industry has found no good substitute for coke.

In theory, once in the furnace, there was no difference between charcoal and coke. Either provided a good source of carbon. But in practice it took more than a century of experimentation to make coke work as a blast-furnace fuel. Coke burns less easily than charcoal; it requires a hotter fire. Just getting the temperature up required years of effort. This succeeded only after the steam engine replaced the waterwheel in working the bellows that blew the air (the blast) into the fire. But once higher temperatures were achieved, the heat enabled impurities like sulphur to pass to the iron. This made a brittle iron that blacksmiths refused to use. After all, they depended on the stuff not breaking while they hammered it into shape on the anvil.

The use of low-sulphur coal to make coke, and a new process called puddling that removed impurities from brittle, coke-smelted iron, opened the door to the widespread use of coke in the blast furnace. Coke enabled the British iron industry to survive, because it was not made from trees. By 1806, all but 11 of Britain's 173 blast furnaces were fueled with coke. No further impediments stood in the way of rapid industrialization.[26]

The technical side of the industrial capitalism that took off in the late 1700s was the marriage of coal and iron—in the blast furnace, the steam engine, and the railroad. The marriage was consummated when coal replaced wood as the fuel in stationary steam engines working at a distance from the mines, and when it replaced wood fuel in locomotives. But the meaning of the industrial revolution extended far beyond the technical.

It is difficult for us, living as we do with cars, airplanes, telephones, radios, and televisions, to fully appreciate the meaning of the word *local*. For all of the centuries before the railroad, the weather hampered transportation. Mud or snow impeded overland travel. Before the radical improvement of road construction in the early 1800s, the usual speed of a stage coach was five miles per hour.[27] Rivers rose in the spring but dried up in the summer months and froze in the winter. Canals were a great advance over rivers, but they, too, froze. They could be used only during daylight, and even then the barges moved no faster than the horses and mules pulling them along from the side. Wooden sailing ships, powered by unreliable winds and tides, provided the only transportation by sea. Communication, by letter or word of mouth, proceeded at the same speed as the available transportation.[28]

Before the railroad, society carried out production on a small scale, at the local level. One couldn't order supplies from far off and expect them to arrive at a particular time. In the mercantile capitalism that preceded the industrial capitalism of our own time, people made most things at home, by hand. Nevertheless, trading was international. Merchants bought manufactured goods, sugar, tobacco, and human slaves cheap, and transported them across the sea in order to sell them dear. The great trading companies accumulated wealth—a matter of immense significance for the period that followed—but did not centralize production in the factory. Nor did they alter the traditional, local manner in which goods were produced.

Family members, typically everyone past toddler age, carried out traditional craft production in households or in small workshops. To describe this system of production in pure form will help us to gauge the shape it took in the caverns beneath the earth. A craftsperson obtained the raw materials and supplies locally and sold the finished goods locally. This producer, the man or woman who did the work along with a few assistants, made all decisions about procedures and also determined the pace of work. No enterprise required a large capital investment, and no workshop employed more than a few apprentices.[29]

The change from craft production to mass production evolved in stages, transforming different industries at different times. The factory system, which first appeared in Britain's textile industry, was at the center of the change. The first thing to be altered was the physical work space: The capitalist moved many separate work units under one roof so that more people worked closer together. (As we shall see, it was possible to reorganize the underground work space in a coal mine as well.) The whole job of production was broken into its several parts. Instead of each worker carrying out the whole task, he or she would do only one part of it, repeatedly. Many procedures were mechanized, and the machines ran off a single source of power. The manager determined the pace of the machine, and the machine determined the pace of work. The task became less meaningful to the worker, who could no longer take pride in how well the product was fashioned. (Although coal was not a manufactured product, its quality could vary and was also a matter of pride.) The workers, now supervised, suffered a significant loss of freedom. There was widespread resistance to the idea of continuous labor for fifteen to eighteen hours with no opportunity to step outside for a breath of fresh air.[30]

Ownership of the means of production (tools, machinery, raw products, and supplies) passed from artisans and peasants to the newly forming class of capitalists, which in Britain consisted mostly of former merchants and owners of the great estates. The increasing division of labor meant that knowledge of how to do the whole task passed in the same direction, from those who did the work to those who now owned or managed the means of doing it. The means of production were concentrated into fewer and fewer hands: Setting up a factory required a large capital investment.

Mass production increased the number of commodities (goods to be bought or sold) and lowered their cost. It increased the productivity of the individual worker fifty- to a hundredfold. The new industrial order created a new class of capitalists, small in actual numbers, but far wealthier and more powerful than the class of merchants and estate owners that preceded it. It also created a new class of workers, a class which had been minuscule under feudalism and during the long transition to industrial capitalism, a class whose members had nothing to sell but their own labor.[31]

Whatever the ultimate benefits or disasters of industrialization, it came to artisans and peasants as an economic calamity.[32] Cheap English manufactured goods began to flood first England itself, then Ireland and Europe, ruining the artisans in those places. Added to this, a population explosion occurred in the mid-1700s for complex reasons that scholars are still debating. As a result, peasants felt compelled to divide their land into smaller and smaller plots among too many sons. At the same time, the landed gentry, some of them seeking to improve the efficiency of agricultural production, began to enclose the forests and grazing lands that whole communities had previously shared.[33]

The peasants of Ireland and Germany began to plant potatoes because their tiny plots would support no other crop. In the wintertime, the old people, once highly esteemed in their villages, were sent out to beg. Itinerants and paupers became a common sight on the roads of Europe and Britain. A Scottish writer of the early 1700s estimated that one-fifth of the entire population of Scotland lived in a state of beggary. Lack of data feeds the controversy over whether the industrial revolution caused the standard of living to rise or fall, but those who claim it rose find it difficult to argue that this occurred before 1820 or even 1840.[34]

Vast migrations of the destitute were set in motion. They became agricultural laborers on the Great Estates or sweated labor in London or Liverpool. Many left, first their villages and then their countries, in search of a livelihood. In the century following 1820, eight million persons left their British and Irish homelands.[35]

"The modern industrial proletariat," writes Sidney Pollard of the early decades of the industrial revolution, "was introduced to its role not so much by attraction or monetary reward but by compulsion, force, and fear."[36] This is nowhere more certain than in the case of coal miners. As feudalism disintegrated, the Scottish rulers moved to insure that the serfs who dug the coal did not become free, especially not free to leave. One of the first laws to deal with coal miners, enacted by the Scottish Parliament in 1606, virtually enslaved the Scottish mining people by punishing those leaving their masters without permission as "thieves of themselves." In a harbinger of the new factory discipline, this act also abolished Saturday holidays, providing that all hewers [miners] "lying idle" would be fined twenty shillings and sustain "punishment of their bodies."[37]

Later in the century, the Scottish Parliament gave coal owners the right to seize vagabonds of both sexes and their children and force them to work for life in the mines. The family doom became hereditary. Coal owners bought and sold the mining people with the collieries, inventorying them on the bills of sale.[38]

In 1841, an elderly Scottish miner reflected on the hardship of this life: "When I was nine years of age we were all then slaves to the Preston Grange laird. . . . When the miners did not do their masters bidding we were placed by the neck in collars called juggs and fastened to the wall." Another punishment was to be tied facing the horse, whose circular trotting wound the pulley that raised and lowered baskets in the shaft. In this way, the offending miner was made to "run round backwards all day."[39]

English miners experienced more freedom than their Scottish counterparts. Those working on the great estates were bound by yearlong contracts to the landed gentry. They, at least, could change employers once a year. As Michael Flinn points out in his monumental study of the British coal industry, the yearly haggle over terms of the binding contract provided important experience building up to trade unions. Further, the widespread poaching of miners from neighboring mines, and numerous illegal escapes and departures to new employers provided, if not liberty, more room in which to maneuver. But conditions were not desirable. This is shown by one case in which a coal-owning Member of Parliament accepted seventeen condemned criminals in 1690 to work underground for five years as an alternative to execution. The prospect of certain death upon capture was not enough to deter several of them from escaping.[40]

Some English and Scottish miners were truly independent handicraft workers. Consider the miners who dug the coal underlying the Forest of Dean, an area belonging to the crown. These "free miners" inherited their access to the forest and its resources. They built cottages, cut wood, and grazed their animals there. In exchange for one-fifth of the coal they dug, the crown gave them the right to mine it.

With the coming of industrial capitalism, the traditional prerogatives of this group began to erode. In the early nineteenth century, some of these miners formed partnerships with outside capitalists, undermining the system of inherited access. At the same time, the crown began to reclaim the resources of the forest.[41]

The subcontract, or butty system, contributed to the erosion of the free miners' way of life. At first perhaps, a group of miners would go to work for one of their number who owned the rights to the coal. In the beginning, they probably went underground as a group of equals. But as the demand for coal increased, and one owner gradually controlled a greater number of claims, he would contract with another miner to actually mine the coal. The subcontractor, or "butty," would in turn hire miners to work for him. He might be a skilled miner working with a few assistants, or a subcapitalist with a hundred men working under him. His profits depended on reducing wages below the bargain price he had made with the coal owner. In certain parts of England, especially where work methods (the long wall system of mining) called for gangs of men working together, the butty system became the common form of underground management.[42]

Children, chiefly though not exclusively boys, entered the mines at ages six, seven, or eight. Some were the offspring of mine workers; others were paupers—orphans taken from nearby workhouses and apprenticed to a hewer for a period of ten to twelve years. The practice of using boy orphans in the mines was so common that near some English coalfields the workhouses had only girls. These small workers labored six days a week, twelve to eighteen hours a day, opening and shutting doors in the underground passageways and, as they grew stronger, loading and hauling coal, sometimes through roadways less than two feet high.

The hewers who apprenticed the children often treated them badly, frequently forcing them to work in the most dangerous places. The disobedient ones were brought before a magistrate who committed them to prison.[43] One boy who had run away from his master testified in 1842 that the man had "stuck a pin in me. He used to hit me with a belt and a maul, or sledge, and fling coals at me. He served me so bad that I left him. . . . I used to sleep in the cabins upon the pit banks. . . . I ate . . . the candles that I found in the pits that the colliers left overnight. I had nothing else to eat."[44]

One early historian of the industry wrote that the children were sent underground with no more to eat than bread and cheese, "and this they sometimes could not eat, owing to the dust and badness of the air. The heat was at times so great as to melt candles, and many of the roads were covered with water." The children grew up crooked and physically deformed. The girls (writes Boyd in 1895) soon became "as rough and uncouth as the men and boys, fighting and swearing like them, and many bastards were born in the colliery villages."[45]

The children of the coal mines, Frederick Engels observed as late as the 1840s in *The Condition of the Working Class in England*, "upon arriving home throw themselves down on the stone hearth . . . fall asleep at once, without being able to take a bite of food, and have to be washed and put to bed while asleep. It even happens that they lie down on the way home, and are found by their parents at night asleep on the road. It seems a universal practice among these children to spend Sunday in bed to recover in some degree from the over exertion of the week."[46]

Some collieries, particularly in eastern Scotland, employed women and girls extensively. In Lanarkshire, women mined one seam exclusively, "the women's coal," but more usually wives and daughters worked with husbands and fathers as coal bearers. The mother would leave her infant with an old woman in the village who fed it whiskey mixed with water during the long daily absence. A minority of the women, called "fremit bearers," or "frembs," were unrelated to miners. Each day, the mine manager assigned the frembs to work for different unmarried hewers. "Their lot was unspeakably arduous," explain Thomas Ashton and Joseph Sykes, "and their status so low that few would accept the service."[47]

The hewer, the most skilled underground worker, entered the mine about three hours before the women to hack and (after 1719) to blast the coal off the seam.[48] The women and girls arrived later to haul it out. The mother and her older daughters would enter the mine carrying baskets, which they then laid down so the miners could roll in the large pieces of coal. Frequently, a woman's burden was so heavy that it took two men to lift it on her back. She brought up approximately two tons per day. In 1808, an observer wrote:

> The mother sets out first, carrying a lighted candle in her teeth; the girls follow, and in this manner they proceed to the pit bottom, and with weary steps and slow, ascend the stairs, halting occasionally to draw breath, till they arrive at . . . the pit top, where the coals are laid down for sale, and in this manner they go for eight or ten hours, almost without resting. It is no uncommon thing to see them ascending the pit, weeping most bitterly, from the excessive severity of their labor; but the instant they have laid down their burden on the hill, they resume their cheerfulness and return down the pit singing.[49]

Towards the mid-nineteenth century, female coal bearers at another mine were seen carrying coal in wicker baskets fitted to their backs, held by a leather strap which passed around the forehead. Bent far over, steadying themselves with a short stick, they hauled 150 to 200 pounds of coal in each load.[50] At another place, coal bearers, including men, women, and children, were hitched to tubs of coal in a contrivance known as "the painful harness." It consisted of a leather belt that went around the waist. Hooked to the belt in the front was a chain that passed between the legs and attached to a tub of coal. The bearers dragged the tubs through passageways on hands and knees, sometimes through water several inches deep.[51]

As grueling as the woman's work was, she carried it out in the context of a preindustrial, family system of production. When Victorian lawmakers banned women's underground labor in 1842, many families languished to the point of near starvation. The legislation barred the women from their only means of livelihood, as Angela V. John notes, with no compensation to them or their families. Many continued on illegally, and it was another twenty or thirty years (and longer in Europe) before women's underground work was eradicated.[52]

As long as the workers dug coal from shallow pits, accidents were minimal. In the early 1700s, this began to change. Exact figures are unknowable, since few records were kept of the frequent deaths and severe injuries that occurred in the mines: As late as 1842, there was no such thing as a coroner in Scotland.[53] It was to the capitalist's advantage to keep these grisly facts hidden from view. At one English colliery, the working people were forbidden even to talk about the accidents that happened. In 1803, the local coroner in White Haven took it upon himself to investigate the death of a woman killed there. His probing infuriated the manager, who believed such inquiries were "calculated to frighten the ignorant and discourage them from going into the pits. . . ."[54]

The people lit their way in the dark with candles. As the mines grew deeper and more gaseous, the workers ignited explosions with their open flames with increasing frequency. The Davey safety lamp (1815), which enclosed the candle in a wire mesh to separate the flame from the gas, worked in principle but in practice made matters worse. It enabled the diggers to enter places previously rendered totally inaccessible by gas. The introduction of the Davey lamp actually caused an increase in the number of explosions.[55]

Falls of coal crushed some mine workers; others died falling down the shaft. The workers entered the shaft dangling on ropes or in baskets, while the counterweight rising was a basket of coal. On the way down, the basket swayed and smacked against the sides of the shaft. At times, the coal coming up collided with the basket of workers going down, a form of death called a "wedding." Death also came to exhausted workers letting go of the rope. Or rotten ropes could break, resulting in what one early writer called "a terrible fall of the living load."[56] In 1777, owners began installing the Newcomen engine to power the hoist for raising and lowering coal and people. The practice of employing ten-year-old boys to operate the mechanism frequently resulted in the workers being drawn up into the gears and crushed, or suddenly dropped to their deaths.[57]

There is little doubt that the mining people resented their condition. An early hint of their animus was an Act of the Scottish Parliament (1592) condemning to death anyone who willfully set fire to a colliery.[58] Nonminers felt repelled by what they knew of miners' conditions, and as the coal market expanded in the 1700s it became difficult to persuade free laborers to enter the mines, even with guarantees of continued freedom and offers of higher-than-average wages. The Acts of Parliament that emancipated Scottish mine workers still living in serfdom in the late 1700s actually gave the labor shortage as an important rationale for improving conditions.[59]

British coal mine workers pioneered the modern labor movement, beginning in 1765 with a strike in England to protest the owners' attempt to increase the period of the bond. Union organizing began in earnest in the 1820s. Coal miners founded their first national union in the 1840s— the Miners' Association of Great Britain and Ireland. During this decade, unrest seethed on the continent (culminating in the revolutions of 1848) and in famine-scarred Ireland. In Great Britain, Chartism, a working-class movement for political rights, rose only to be crushed by the government. In 1844, a strike of 34,000 miners ended in a disastrous defeat. This, coupled with a three-year-long depression, obliterated the first national miners' union. British coal miners, highly skilled in their work and by now imbued with trade union ideas, began migrating to the agricultural country across the Atlantic.[60]

COAL IN THE INDUSTRIAL REVOLUTION: THE UNITED STATES

Seasoned British coal miners arriving in the United States found a country in which ninety percent of the population (in the 1840s) lived in rural places with fewer than 2,500 inhabitants.[61] Unlike Great Britain, the country encompassed a vast terrain intersected by mountain ranges. Both transportation and communication moved at a leisurely pace: A letter spent three weeks traveling between New York City and Chicago.[62]

Factories, requiring a large amount of capital and employing more than a few workers, numbered slightly more than one hundred. As Alfred Chandler, Jr., has noted, they were largely confined to the New England textile industry and, even then, they drew power not from the steam engine but from the waterwheel. Many items, such as soap, clothing, and candles, were made at home and consumed there, never becoming commodities at all.[63] In 1840, a time when coal output could accurately measure industrial development, 6,811 mine workers in the United States produced less than two million tons of coal.[64] Despite the availability of the requisite technical knowledge, industrialization lagged at least fifty years behind that of Great Britain.

In America, there was no fuel emergency; in many places, the forest still seemed endless. Blacksmiths and salt makers used coal, but to avoid the transportation problem they established their works near outcrops.[65] As late as 1850, wood provided ninety percent of the country's energy.[66] Firewood was the principal home heating fuel. Early steamships (1820s) and most pre–Civil War locomotives burned wood for fuel. In the 1830s, the Baltimore & Ohio Railroad began to experiment with coal-fueled locomotives. But difficulties such as the hotter fire burning through the fire box, voluminous smoke, and sparks flying out the smokestack and setting nearby barns and fields on fire, delayed their widespread adoption. In the 1840s, the Philadelphia & Reading Railroad began transporting anthracite coal from Schuylkill County to Philadelphia for use as a home heating fuel. Wood fueled the locomotives of this first coal-carrying railroad.[67]

By the 1840s, British iron manufacturers had converted to coke. In contrast, American iron makers, centered in Pennsylvania, continued to use charcoal as a metallurgical fuel. According to Peter Temin, iron makers in western Pennsylvania (cut off from the East by the Allegheny Mountains) had access to bituminous coal, but it was full of sulphur; coke manufactured from it made such poor iron that manufacturers were obliged to sell it too cheaply to make a profit. They stuck with charcoal.

East of the Alleghenies, iron manufacturers had access to anthracite coal. With its high carbon content, anthracite has been called a natural coke, but it is difficult to ignite. Its use in the metallurgical furnace requires an extremely hot fire. Adequate temperatures could not be attained until 1828, when James Neilson, a Scottish iron maker, achieved the technological breakthrough of the hot blast, in which the bellows blew preheated rather than cold air into the bottom of the furnace.[68]

Using the hot blast, George Crane, an iron manufacturer in the anthracite region of Wales, and his manager David Thomas, first succeeded in using anthracite as a metallurgical fuel. A Pennsylvania anthracite firm brought Thomas to the United States, and by 1840 he had fired America's first anthracite-fueled hot-blast furnace, near Allentown. By the end of the decade, the new metallurgical fuel had become the vehicle to cheap, high quality iron manufactured in the United States. Anthracite, as Alfred Chandler, Jr., notes, played the same pivotal role in the American industrial revolution that coke had played in the British.[69]

The anthracite industry had been developing since the 1820s; interest in it dated back to the turn of the eighteenth century. In 1802, from the mountainous wilderness of northeastern Pennsylvania, Jacob Cist and his nascent Lehigh Coal Mining Co. had attempted to float six boatloads of hard coal down the Lehigh and Delaware Rivers to Philadelphia. Four of the arks broke up in the rushing waters, but two arrived safely. Unfortunately, an early writer explained, "it seems that purchasers were not numerous." Finally, the company found a buyer in the municipal authorities. However, these gentlemen could not make stone coal burn. In disgust they broke it up and used it to gravel the walks, "and here and thus ingloriously terminated" the operations of this company for another seventeen years. In 1817, another enterprising gentleman, Colonel George Shoemaker, dug some coal near Pottsville and transported it laboriously by wagon to Philadelphia. He was received, Eli Bowen reported in 1857, "with coolness if not with rigid scrutiny." By now, people knew that anthracite would burn, but some of Shoemaker's buyers could not manage to ignite it. They became indignant and the colonel narrowly escaped arrest for swindling. (He quietly slipped out of town.) But times were changing. In 1820, 365 tons were shipped from the anthracite region to Philadelphia.[70]

But it was not until the decade of the 1840s that the industrial revolution based on fuel coal took off. During that decade, transportation and communication were put on a rapid and reliable basis. The ocean-going steamship put the trip across the Atlantic on a timetable. The railroad was extended and became an important market for iron: Both rails and rolling stock were made of it. Now the vast distances in America stimulated industrialization. Communications were also revolutionized. Samuel Morse made the first practical use of electricity in 1844 by building a telegraph line between Washington and Baltimore and inventing the Morse code for sending messages over it.[71]

Coal was the key to the industrial revolution that finally emerged in the United States. Within three decades, the black rock and those who mined it were central to the economy. A coal miner speaking in 1874 saw the importance of this. "I hope to see the miners of America realize," he said to a group of the same, "that . . . they are foremost in producing wealth, that . . . millions of firesides of rich and poor must be supplied by our labor, that the magnificent steamer that ploughs the ocean, rivers and lakes, the locomotive, whose shrill whistle echoes and re-echoes from Maine to California, the rolling mills, the cotton mills, the flour mills, the world's entire machinery, is moved, propelled by our labor."[72]

Pennsylvania, America's leading coal state for the entire century before 1930, cradled the country's industrial revolution. The bituminous coal surrounding Pittsburgh made that far-western town an early center of the

Colliery scene in anthracite region, 1880's. Photo by George M. Bretz, Photography Collections, Albin O. Kuhn Library and Gallery, University of Maryland Baltimore County.

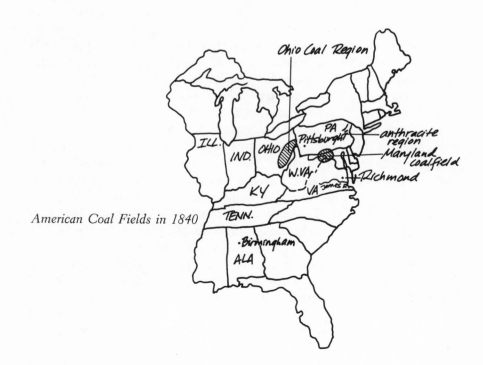

American Coal Fields in 1840

steam engine, and, by 1833, ninety steam engines choked the place in a haze of smoke.[73] But it was anthracite—used to smelt cheap iron and to heat houses on the eastern seaboard—that catalyzed the revolution. Until the 1860s, Pennsylvania's anthracite region produced more coal than the rest of the country combined.[74]

Except for the rich natural resources found in the United States, the essential components of industrialization came from Great Britain. In the 1840s, British capitalists, flush from the previous decades of rapid economic development, reacted against the social unrest in Britain and on the continent by escalating their investments in the United States.[75] Britain also contributed much of the technology and a significant portion of the labor to the American industrial revolution. British miners arrived with the most complete technical knowledge of coal mining then in existence, and it was they who introduced the skill of craft mining to the United States. The early methods of mining in this country were literally those which had been developing for centuries in Great Britain.

2

MY LAMP IS MY SUN:
THE WORKPLACE AND
THE WORK

Human work is conscious and purposive, while the work of other animals is instinctual. . . . In human work . . . the directing mechanism is the power of conceptual thought. . . .

Harry Braverman, Labor and Monopoly Capital[1]

. . . the miner is a free, intelligent, thinking being, and entitled to be treated as such, and not as a beast of burden to be knocked and kicked around. . . .

John Siney, coal miner and union leader, 1874[2]

A coal mine is warm, dark, and often wet.[3] It is devoid of common outdoor sounds—the chirping of birds, the rustling of wind through leaves. In a strange kind of silence, the nineteenth-century miner listened for the cracking of the roof, a sound that rolled like thunder for days before a fall. He listened to the significant scuttling of mine rats. He watched the flame of his lamp for a blue cap indicating the presence of gas.

The miner reached his workplace in one of several ways, depending on the mine. The four types of underground mines are named for their method of entry. Consider the drift mine, one of the earliest and simplest types. Where coal outcrops at the base of a mountain, an opening can be driven into the coal itself. The miner enters the mountain on an under-

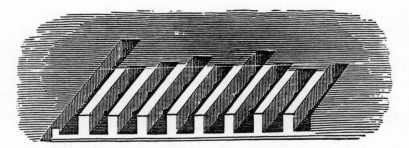

Breasts, chambers, or rooms.

ground road that inclines slightly upward, following the angle of the seam. The main tunnel of a drift mine can go straight through a mountain. In one Maryland coal town, the women took a shortcut through the mine to get to the store on the other side of the mountain. The drift mine is the nearest thing to a quarry. Virtually anyone can begin digging into a mountain where the coal outcrops, without incurring the heavy expenses of hoisting coal or pumping water. Because the mine slants upward, drainage is natural.

The second type of mine, the slope mine, slopes downhill, following a coal seam that lies upended in the earth. Like the drift mine, the slope is driven into the coal itself at the outcrop. The tunnel mine, the third type of mine, was not driven into coal but into the dirt or rock of a hillside toward the coal seam at an angle.[4]

In the fourth type of underground mine, the shaft mine, the miner descended a shaft in an elevator, called a cage. A deep shaft can provide access to more than one coal seam as it passes through coal interlayered

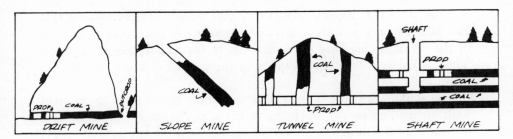

Four types of mine entrances.

with sandstone, shale, or clay. Digging a shaft requires considerable work and expense before the first shovelful of coal can be extracted, and then elaborate arrangements must be made for hoisting coal and for pumping out water. The shaft mine is capital intensive, but, once built, it is the most efficient way of gaining access to more than one seam. Today, mine shafts should be counted among the wonders of the world. The deepest shaft (in the 1980s) goes down 2 miles, while the deepest borehole (as of 1985) penetrates 7.5 miles into the earth.[5]

Traditionally, miners in the United States dug coal by the room-and-pillar method. To open the mine, they drove a "main artery" or "main entry" through the seam, and this became like the main street in a town. (In the anthracite region, the main entry was called a "gangway.") Miners then drove cross streets, called "entries" or "headings," off the main entry at right angles to it. Opening off the entries were "rooms," sometimes called "chambers" or "breasts." These were the underground workplaces at the face. A room might be eight yards wide. Between rooms, solid pillars of coal—twenty to one hundred feet wide—were left to hold up the roof. Props made of timber supplemented the pillars of coal.[6]

To reach his room, a miner might walk underground a mile or more. The workings could be extensive. In 1870, one anthracite mine, in operation more than thirty years, had nine miles of underground passages. Another extended underground for fifteen miles. Although such mines had more than one entrance, the time required for traveling from the entrance to the individual workplace could be considerable.[7]

In some cases, crawling and climbing would more accurately describe how a miner got to his underground room. This was especially true of the steeply inclined seams of the anthracite region. Picture a mountain with several coal seams lying upended in it. The coal seams lie vertically like chimneys, and the tunnel dug into the side of the mountain is a flat road that crosses several upended seams. Miners walk or ride into the gangway (which is driven through rock) until they reach one of the seams, where they then climb up to their workplace at the face. The blasted coal falls down chutes to the gangway, where it is loaded into cars.[8]

The workplace at the face could be anywhere from two to twenty feet high, depending on the thickness of the seam.[9] Usually, the seam was shorter than the miner. "One of the most exhausting things about mine

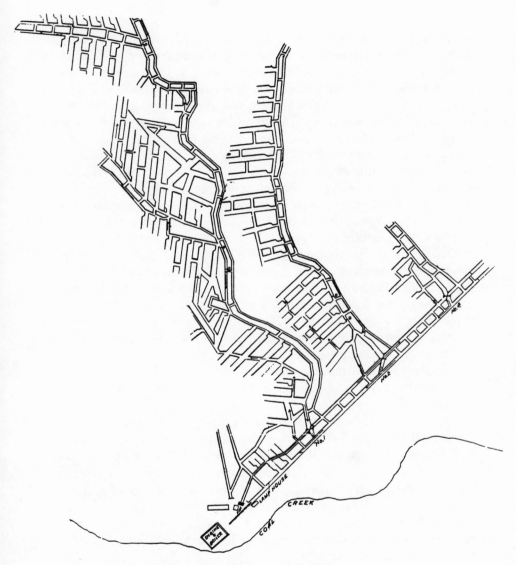

Map of a coal mine at Crested Butte, Colorado.

work," John Brophy explains in *A Miner's Life,* "was the necessity of working a ten or eleven hour day without a single chance to stand erect and stretch. It was a great advantage to be short."[10] Continuous hunching permanently altered a miner's posture. In *King Coal,* novelist Upton Sinclair graphically described a group of miners emerging at the end of the day: "[T]hey walked with head and shoulders bent over and arms hanging down, so that, seeing them coming out of the shaft into the gloaming, one thought of a file of baboons."[11]

Descending into the gloom, the miner encountered warmer and warmer temperatures. At ninety feet, surface extremes of summer heat or winter cold don't affect the air underground, which stands at about sixty degrees. Going lower, the temperature rises one degree for every seventy-five feet descended.[12]

Darkness and silence prevailed. "The lamp on a miner's hat peak," wrote a superintendent in 1896, gives "enough flickering light to show the low, overhanging roof beams, bent here and there with the significant emphasis of the superincumbent 3 or 400 feet of surface."[13] Another writer gave this eerie picture: "The lamps only shine at certain points, lighting up the faces of the men, the shape of the waggons, and the coal which glistens here and there. The rest is cast in shadow, and yet the whole effect is animated and startling."[14]

"My lamp is my sun," a Belgian miners' song goes, "and all my days are nights."[15]

The chief sounds were the clinking of picks and shovels, the rattle of the mule-drawn train (called a trip), and most importantly, the occasional cracking, pounding, and rumbling of the roof. "When [the miners were] at work the other night, the roof started pounding," reported a Wyoming superintendent to his superior, "and sounded as if it was 40 or 50 feet above them. They were half scared and thought it was going to cave in."[16] Indeed, the wise miner paid strict attention to the sounds emanating from above his head. Brophy writes: "After a while the timbers begin to splinter and you can hear the roof 'working.' This means that the strata of slate in the roof are beginning to break. It makes a sound like thunder, which can go on for as long as two or three days. An experienced miner can tell from the way the roof is 'working,' and from the splintering of timbers just about when the roof is ready to fall. Then he gets his mine car and himself out of there fast."[17]

Fungus thrived in the hot, damp atmosphere of nineteenth-century coal mines. It grew on timbers in the form of beautiful, light, cottony filaments. In the words of Louis Simonin, writing in 1869, it consisted "mostly of a . . . snow-white; and occasionally of substances like tawed leather, and of a yellowish color. The rotting of the timber . . . gives out a particular odour like that of creosote . . . by no means a disagreeable smell. Particular insects—moths, flies, and gnats—collect near the parasitic vegetation."[18]

Fine coal dust permeated the atmosphere, which could also become contaminated by what one miner called "the arch-fiend, explosive air." In the early period, it was thought that some mines were gaseous while others were not. But a modern (1973) manual on coal mining warns, "[T]he most common explosive gas in mines is formed to some degree in almost all mines at some time or another during their life. This is important to remember, as there have been several mines which have been considered nongaseous but in which gas had later been ignited with serious results."[19]

The miners called the gases "damps," a word probably derived from the German word *dampf,* meaning vapors or gases.[20] Methane, or marsh gas (CH_4) is the most common gas encountered underground. (It is the main ingredient of natural gas.) Intermixed with a sufficient quantity of oxygen (we now know five percent), methane was known as firedamp and is highly explosive. The gas, trapped in pockets and crevices hundreds of feet underground, constantly bubbles and seeps into a coal mine, sometimes under great pressure.[21]

A less common gas, usually present only after an explosion of firedamp (or sometimes after routine blasting), was known as carbonic oxide gas, or white damp. We now know that "white damp" is a mixture of gases including carbon monoxide. White damp will burn with a beautiful blue flame. Deadly in minute quantities, it comes to the victim, an early writer explained, with the faint smell of violets.[22]

Carbonic acid gas—called black damp—occurred more commonly than white damp. Black damp is understood today as simply an atmosphere deficient in oxygen. "Its effect on the miner," a Colorado coal mine inspector explained in 1884, "is such as to produce a feeling of numbness, or dull pains in the joints of the legs and arms, followed by a sometimes violent headache and a . . . drumming sound in the ears, accompanied by deafness. . . ."[23] It could also cause death by choking.

Slope entrance to an anthracite mine. Photo by George M. Bretz, Photography Collections, Albin O. Kuhn Library and Gallery, University of Maryland Baltimore County.

Mine interior, Shenandoah, Pennsylvania, 1884. Ibid.

The air underground, all of it, has to come from the outside. In mines that were not very deep or extensive, natural ventilation worked by the warm air of the mine rising, while the colder air of the outdoors passed down an air intake shaft to replace it. This worked well only in the wintertime, and, even then, the more extensive the workings, the more inadequate any natural ventilation became. On a daily basis, poor ventilation affected no one but the miner. "I have known men become so weak, working among the bad air," complained an Ohio miner in 1871, "that they could not walk up out of the slope." "The bad air affects me so that I cannot eat or sleep," lamented another miner. "It makes me dizzy. I pray God for the passage of a law compelling owners of mines to provide fresh air for the working places."[24]

Lack of ventilation also affected the operator, who could lose an entire mine in an explosion. But this was a situation calling forth hindsight, a worthless commodity under the circumstances. The operator's insensibility grew out of his anxiety to profit quickly, often in combination with insufficient capital to sink the mine properly in the first place. Many operators worked on, Anthony Wallace notes, relatively untouched by problems of mine ventilation "until gasses, great depths, and heavy loss of life called the law down upon them."[25] Even the law, we must add, failed to adequately protect the miner. In 1884, the Colorado coal mine inspector wrote:

> I have found miners at work 200 feet ahead of the ventilating current. When I looked in [one] close and dirty hole, I could hear the miner's pick striking the coal, but at that distance I failed to see the miner, but [knew] from experience if I went in there I would find him. I ventured in a little over 100 feet . . . and I could just see a glimmering light. . . . The air was almost stagnant to the feeling and densely charged with dust, and in a few minutes I could feel the effects of . . . black damp.[26]

An early ventilating technique involved installing a furnace at the bottom of the shaft. Some distance away, an air intake shaft was sunk. The furnace shaft acted as a chimney, drawing the warm air up and out. Fresh air entered the intake shaft to replace the air drawn out of the chimney.[27]

Whatever the savings to the operator, the disadvantage to the miners of utilizing the only escape shaft as a chimney was demonstrated at the Avondale (Pennsylvania) Mine disaster of 1869, the first great anthracite coal mine disaster. The company had built a rickety wooden coal processing plant—a high building known as a breaker—directly above the shaft. In the early morning of 6 September 1869, the flue partition in the shaft caught fire from the furnace. The work force, 179 men and boys, had just descended. The fire roared up the shaft and ignited the breaker. "The whole of the immense wooden structure was wrapped in flame which arose to a height of 100 feet, swaying to and fro in the wind. . . . The hoisting ropes and all the non-combustible material fell crashing down the shaft, followed by pieces of burning timber." Ten thousand people rushed to the scene from nearby towns to assist in the rescue, but they could not extinguish the fire in time. Rescue workers discovered the entire work force dead behind a barricade they had built to hold back the noxious fumes and smoke. "Fathers and sons were found locked in each others arms, some of the dead were kneeling in prayer . . . some appeared to have fallen while walking."[28]

By the 1870s, operators had begun replacing furnaces with fans. (In the winter, these blew frigid air into the depths.) Forcing air into workings that went for miles underground required a complex system of barriers and trapdoors to direct the "air course." Partitions, called brattices, closed off abandoned workings, which were no longer ventilated. Made of wood and canvas (and later of concrete), the brattices prevented the stream of fresh air from dissipating, and prevented gases and sometimes even smoke from smoldering fires from entering the functioning parts of the mine.[29]

Doors connecting the various underground passages were critical to the controlled flow of air. They had to be opened frequently to let through "trips"—mule-drawn trains hauling coal or miners. Yet they could not be left standing open, for that would permit the air to escape or dissipate. Boys as young as seven or eight, called trapper boys, did the work of opening and closing the doors.[30]

In some coal mines, underground fires were not unusual. A mine fire can be an extreme emergency; in some cases, it makes an explosion within minutes or hours inevitable. And mine fires are difficult if not impossible to put out. In anthracite Pennsylvania, the Lehigh Navigation Coal Company fought a fire ignited in 1859 for eighty years. A mine under Centralia, Pennsylvania, caught fire in 1962; more than twenty years later, the fire has destroyed the town, but it continues to burn. A mine fire feeds on a virtually infinite supply of fuel. Oxygen, difficult to eliminate from miles of underground workings, fans the flames. A fire can start through

Trapper boy. Photo by Lewis Hine, courtesy National Archives.

human error, of course, but it is also possible for the gob, the waste thrown to the edge of the tunnels, to ignite spontaneously like a wet haystack. Around the turn of the century, most of the lignite-subbituminous mines in the high plains north of Denver smoldered continuously. "The coal ignites itself," a labor journalist wrote, "and when there is a draft it smokes like a chimney."[31]

Mines extending below the water table were wet. Anthony Wallace estimates that pumps lifted twenty tons of water from the Pennsylvania anthracite mines for every ton of coal mined. Water could make working conditions unpleasant, for sometimes miners had to kneel in it all day, but only rarely did wet conditions lead to a debacle such as that which occurred in the Diamond Mine near Braidwood, Illinois. The land there is low and flat, and the coal is close to the surface. In the wet spring of 1883, a sudden rush of water into the mine drowned sixty-nine miners.[32]

The miner began his day by walking to the room assigned to him by the mine boss (underground foreman). He took along his tools, supplies, and dinner pail, because it was usually too far to return to daylight for lunch. John Brophy first went underground in the 1890s, at age twelve:

> My father led the way, each of us carrying flickering oil lamps that empha-
> sized the shadows. He explained that the scurrying sounds we heard were
> made by mice in the gob, the edge on either side of the track where waste
> was thrown. Mice and rats came into the mine to live on oats spilled by
> the mine mules and on other refuse. . . . We took off our coats and put
> down our dinner pails, tools, and supplies. The drift was near the surface
> so the temperature was low—about sixty degrees—unlike the heat of
> deeper mines. . . .[33]

Brophy's father gave him a short lecture on the care and use of supplies—oil for the lamps, blasting powder, and squibs (fuses)—and on the
care of tools. He then instructed him on proper behavior: "A man was
expected to conduct himself so that he would not make conditions worse
for his neighbor. For instance, if he had need for a bowel movement, he
took care of that need over in the gob, and covered the waste with dirt
and slack, to minimize the contamination of the air."[34]

This need for a bowel movement in an underground workplace with no
sanitary arrangements, shared by up to several hundred miners, dimin-
ished the quality of the working environment. "Complaints are made,
even" reported a Colorado mine inspector in 1884, "that in some of the
mines, the air is rendered horribly foul, not only by its intermixture with
the ordinary gases encountered, but also by the fact that the commonest
sanitary laws are neglected, and the mines are allowed to become reposito-
ries for ordure, emitting the most stifling stenches."[35] Aesthetics were not
the only reason for concern, as diseases such as miner's hookworm resulted.[36]

A more pleasant feature of the cavernous workplace was the company
of a certain creature who shared the space with the miner, namely the
mine rat. According to an 1891 essay in the *United Mine Workers'
Journal*, the miners viewed them with feelings of satisfaction. The fact
was that "among the tenets of a miner's creed is belief in the sagacity of

rodents." As long as they were running over his feet and hanging around his dinner pail, the miner believed himself secure. But as soon as they began leaving, many workers would cease their tasks immediately and follow them out. Probably the rats could detect the settling and working of the roof that preceded a cave-in before the men could, due to the constriction of their holes and hiding places.

The miners' kindly sentiments toward rats led to rather extensive relationships. "It is a common sight to see a miner feeding half a dozen or more from his dinner pail. Frequently they become so tame that they will climb on a miner's lap as he sits at his lunch and crowd around him to receive such portions of his meal as he has taught them to expect." This observer swore he had seen one as large as a powder keg.

The complete underground stable, maintained for the mules, also provided rodents with oats. During strikes, when mules could be hoisted up and put out to pasture, and when miners failed to appear daily with their dinner buckets, the rats would sometimes emerge and overrun the mining village, devouring the food supply and endangering the children. One writer told the story of a strike lost because of the animals swarming around the dwellings and becoming a terror to the families.

Even underground, they could become intolerable. In one mine near Hazleton, Pennsylvania, the huge rat population ate the soap, drank the lamp oil, and dined on the miners' dinners, and not only when offered. To keep them away, the workers buried their dinner buckets under coal and slate, often in vain. The meal itself was completely taken up with the problem: "Many a time a miner would be compelled to fight with a horde of hungry rats that disputed with him for the possession of his lunch." In the stable, the rodents fed not only on the oats but on the mules themselves:

> It was a common thing for miners, going to work in the morning, to find the stable floors covered with hundreds of rats that had been trampled to death by the mules, as it seemed to be a favorite act with the rats to gnaw the fetlocks of the mules, frequently eating them away, notwithstanding the frantic tramping of the mules and the scores of their own numbers that were crushed beneath the mules' feet. I have myself seen mine rats covering a stable floor a foot deep, having thus fallen victims to their greed for live mule flesh.

This was too much! The operators finally called a halt to mining, brought the mules up, and had poison put around the workings. Three days later, the miners returned to load dead rats into coal cars. The carcasses filled three mine cars holding a ton and a half each.[37]

THE WORK OF MINING

In room-and-pillar mining, the men worked exclusively with hand tools. To describe the work is also to illuminate who did what in the workplace. It is important to realize that at no time before 1920 was the term "miner" used to signify all underground workers. Only the most skilled workers were called "practical miners" or simply "miners." The miner was generally paid by the ton (or by some other unit, such as the yard or the wagon). He, in turn, employed his own laborer or "helper," whose main task was to load coal. The miner worked in relative isolation with one or two helpers in the room. Here was the craftsman and his apprentice in the underground workshop.

The operator employed the other workers. They included mule drivers, trackmen (who laid the track to the working face for the coal cars to run over), trapper boys, and engineers. Most engineers were not professionals as we use the term today, but were tenders of engines. A shaft engineer tended the engine that powered the hoist, a fan engineer tended the engine that powered the ventilating fan, and so on. A stable boss and his assistants cared for the livestock. A blacksmith sharpened the miners' tools, shod the mules, and repaired coal cars.[38]

Typically, the skilled or practical miner had begun his learning as a boy, either at his father's side as a helper or as a trapper boy. From trapping, he advanced to mule driving, a job usually done by adolescents. Next, he would become a laborer, working for a miner, and finally he would be assigned to his own room, paid by the ton, and in a position to hire his own laborer.

The miner's first job of the day was to undercut (undermine) the coal.[39] Lying on his side, holding up his head, he used his pick to make a three- or four-foot cut into the base of the seam. This might take as long as two or three hours. As he worked, of necessity he would slip underneath the now hanging block of coal, which might weigh a ton or more. To prevent it from dropping prematurely, he propped it with short wooden blocks known as sprags.

Next, in an early variation of the room-and-pillar method, the miner knelt to make two vertical cuts from the top of the seam down to the undercut. He then hammered wedges into the top to drop the coal. He attempted to take the coal without shattering it, substituting explosive powder for wedges only if the coal was quite hard. "The use of [blasting powder] is objectionable," we read in an 1875 treatise on coal mining, "on account of the coal being thus broken too fine and causing too much waste. . . ." Anthracite, being too hard to undercut, was the first to be routinely blasted down. "Sometimes a miner with a pick can make practically no impression on the coal, it is so hard," an erstwhile miner explained. "It takes powder to move it from its original bed."[40]

But even in soft coal mines, blasting increasingly replaced wedging. After undercutting, the miner would drill a long hole into the top of the seam, angled up. The hole was nearly as long as the undercut was deep. He then made a cartridge by wrapping a six-inch piece of newspaper around a stick. He filled the cartridge with blasting powder, stuck it on the end of an iron rod five feet long (the miner's needle), and inserted it deep into the hole. Next, he withdrew the needle and made the squib—a short roll of waxed paper with a sprinkle of powder at its point that acted as a fuse. He inserted the squib into the mouth of the hole, lit its exposed end, and promptly departed the room. The flame traveled up the hole and exploded the cartridge. A single blast brought down a ton or more of coal.

Blasting, done properly, required considerable skill but less, in the view of the traditional miner, than wedging. The increased use of blasting constituted an attack on the value of the miner's skill, according to a miner signing himself Jack in 1875:

> A practical miner works all such coal without the use of [blasting] powder, unless there is some trouble in his place where he cannot take it down by the use of the wedge. On the other hand some Tom, Dick, or Harry that perhaps knows very little about coal mining brings a drilling machine and a keg of powder and by mere force of blasting puts out perhaps as much coal as another man, not caring who is suffocated by his powder smoke. The skilled miner has no advantage over the greenhorn.[41]

After the coal was mined, that is, removed from the seam, the miner and/or his laborer loaded it into a coal car. The miner hung his identifying number tag on the car and pushed it out to the main passageway, where the mule driver collected it into his train. Mules pulled the trip along the track out to daylight (in a drift mine) or to the shaft bottom, where it was lifted to the top by a steam-powered hoist. From slope mines, the coal cars were sometimes hauled to the surface by a system of pulleys and ropes.

Above ground, the coal was inspected for impurities such as slate, a general term referring to one of several types of shale. Then the weigh boss weighed the coal and credited it to the appropriate miner according to the tag on the car. At this point, the procedures for treating bituminous and anthracite coal diverged. Anthracite, used extensively as a home heating fuel, was broken, cleaned, and sorted above ground before being sent to market. At a bituminous mine, the coal car was drawn up the tipple, a kind of rocking platform built over a track. The car was locked onto the platform, which was then tipped over to dump the coal into a waiting railroad car. Sometimes, bituminous coal was also separated into different sizes: lump (the largest), nut (one-inch pieces), and slack, or small coal. Unseparated bituminous coal was known as "run of mine."[42]

Because the miner was paid by the ton, he called any work he had to do but which did not directly result in coal mined, "dead work." As work progressed, the face of the coal steadily receded from the entrance of the mine. Track had to be laid up to the face so the cars could be moved close to the fallen coal for loading. The room had to be timbered. Props, provided by management, had to be set in place to supplement the pillars of coal. Brushing the roof—blasting down overhead rock to gain headroom in the work place—was sometimes necessary. Then, of course, the rock had to be removed before mining could begin. (The term "brushing the roof" also referred to the ineffectual practice of waving one's shirt above one's head to disperse an accumulation of gas.) Sometimes, rock fell of its own accord, and it, too, had to be cleaned up before mining could begin. In some mines, water had to be bailed out of the room each morning. Sending out "clean" coal was more difficult in some mines than in others. Miners had to handle any clay, dirt, or rock that came down with the coal.

In this way, dead work could add up to large segments of work time devoted to unpaid labor. In the period before 1920, the demand "payment for deadwork" echoed through hundreds of strikes. Dead work also increased the dangers of mining, because in order to make wages miners felt compelled to hurry through or sometimes skip unpaid tasks such as propping the roof. Under union conditions, the operator paid for much of this type of work by hiring men by the day to do it.

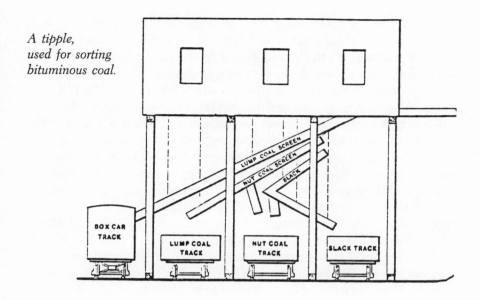

*A tipple,
used for sorting
bituminous coal.*

Miner using a jacket to brush out gas, circa 1916.

The problem of transporting the coal began at the ever-receding face. A vast railroad system operated underground, consisting of wooden or iron coal cars running on tracks and pulled by mules. The system required tracklayers, mule drivers, and a complete underground stable. At the end of the nineteenth century, the coal industry utilized some 25,000 horses and mules—mostly for underground haulage. This contrasted with some 1,000 vehicles powered by steam engine, electricity (a new factor at the end of the century), or compressed air.[43] If the industrial revolution substituted engine power for muscle power, it had barely touched the coal industry's haulage system (except for hoisting) by the twentieth century.

For the operator, mules constituted an expensive item of mining. A good mine mule cost $175 dollars (in 1876), and its working life lasted about seven years. The average mine around Pittsburgh worked fifteen mules. Mules, like men, could be killed or injured. "The slaughter of mules is fearful," scolded the manager of a large anthracite coal firm (working more than twenty mines) in 1875. "We must kill 20 mules a month."[44] Mules were killed, or injured beyond recovery, by falls of rock or coal, and especially by haulage accidents such as a loose coal car hurling backward down a slope into the upcoming trip.

Mining the coal, loading it, and hauling it out made up the essential tasks of room-and-pillar mining. After the rooms in one section were worked out, one job remained. The pillars—columns of coal supporting the roof—had to be mined. "Robbing" the pillars was a high-paying job given to experienced miners only. A highly skilled miner, most operators and miners believed, could take down pillars under the protection of his knowledge about roof noises and movements. But the fact was that when a pillar was removed the roof it supported tended to collapse. Keith Dix has demonstrated that in West Virginia one-fifth of all miners killed by roof falls in the decade before 1907 had ten or more years of experience. More than half the men crushed this way had two or more years of experience. No wonder: Only the most experienced miners did the treacherous work of pulling the pillars.[45]

The room-and-pillar method wasted coal as flagrantly as it did lives. There was no possibility of extracting pillars entirely. In the 1870s, a mining engineer estimated that one-third to one-half of the coal was abandoned in the mines, mostly in the form of pillars left standing.[46]

More efficient and safer procedures existed, but the room-and-pillar system predominated so completely in the United States before 1920 that other methods fade into insignificance. Eventually, two alternatives—longwall mining and stripping—supplanted the traditional method. As early as 1886, two industry specialists advocated longwall mining for use in America; it had been employed in England since the early 1700s. Nevertheless, by 1916 only a tiny percentage of American mines utilized the method, despite the fact that the procedure made it possible to extract more coal from the same mine.[47] Why did the more wasteful method dominate for so long? As we shall see in the next chapter, room-and-pillar mining fit nicely with certain attitudes and beliefs about work that went along with a prefactory or craft-oriented worldview, whereas both longwall mining and strip mining grew out of a different thought process and a different economic system.

Longwall mining can easily be understood if one thinks of a bed of coal as an underground field extending for hundreds or thousands of acres. When coal is reached by shaft or slope, a main entry is driven as before. Then two parallel tunnels are driven off to one side, four hundred to six hundred feet apart. (Today, this job can take a year or more.) Between the tunnels, made permanent with timber and other construction material, the longwall mining is done. The miners (a whole gang rather than just two or three in a work space) work at the face under a set of movable pillars, which they move forward (one at a time) as the face recedes. (Today, the supports are gigantic ten-ton jacks.) As miners dig the coal and move their supports forward, the roof behind the supports collapses into the space left by the mined-out coal. Miners gain access to the exit (the main entry) through the parallel tunnels to which their forward-moving work space is always connected. Longwall mining cannot be done in every geological situation. It requires a coal seam with a uniform thickness and a regular top and bottom (roof and floor). It also requires a greater capital investment because of the equipment involved, and because of the elaborate work driving the two tunnels before mining can begin. Because it brings more workers closer together, enabling greater supervision, longwall mining brings the underground workplace a step closer to the factory system.[48]

The second alternative to room-and-pillar mining available in the early days—strip mining—is today used to extract more than half of all the coal mined in the United States. Strip mining is done by removing layers of earth and rock, known as the overburden, from the coal with earth-moving equipment. It is akin to quarrying, the oldest method. In 1877, steam-powered shovels were being used near Pittsburg, Kansas for strip mining. A few years later, in Illinois, miners removed the overburden with scrapers pulled by horses. Yet, as late as the 1930s, strip mining accounted for only three percent of United States output.[49]

WOMEN'S WORK

The women's work in the coal communities supported the miners' work underground. The lot of the miner's wife was hard. Physically taxed by frequent childbearing, worn down by unremitting labor from dawn to dusk, these women produced, for the time, a significant number of daughters who chose to remain single rather than follow in their mothers' footsteps. Around 1900, Peter Roberts, a scholar of the anthracite region, remarked upon the fact that every coal town had "a class of females . . . who lead a single life." These daughters of miners "prefer to remain spinsters rather than join their lot with mine workers." Spinsterhood was not an admired state, but even the bundle of prejudices that was Peter Roberts had to admit that "[n]o one familiar with the drudgery and toil of the miner's wife can say anything to these women who prefer single life."[50]

In the early 1920s, the U. S. Women's Bureau issued a remarkable report on the condition of women living in the coal communities.[51] By then, they comprised some 500,000 women, of whom some 370,000 were wives. The rest were unmarried daughters over the age of 15.[52] Collectively, the wives "cooked and cared for" more than half a million mine workers, more than 100,000 of whom were boarders and lodgers. This is not to mention the work of caring for more than a million children.

To bring this down to the individual level, for more than half of the wives, "the daily tasks were measured by the demands of households ranging from 5 persons to 11 or more." About twenty percent of the wives contributed money to their households, mostly by taking in boarders— single mine workers or married men living without their families. Keeping boarders involved cooking and cleaning for these men as well as doing their laundry. In many camps, it was customary for each boarder to bring his own food for the woman to cook, thereby increasing the task of cooking far beyond what it would have been if she were preparing the same meal for everyone. Married women also contributed money to the household by cleaning and washing for the households of superintendents and mine bosses. Finally, the textile industry moved silk mills and shirt factories to the anthracite region to take advantage of the female labor— mostly that of the unmarried daughters.

Cleaning house in a coal town was a Herculean task, especially given the scarcity of water. "Coal mining is dirty work," we read in the Women's Bureau report, "not filthy, but oily, soiling, and smudging. A veil of coal dust envelopes the region, covering the homes and home premises with a black deposit." But, as late as 1920, only fourteen percent of a sample of 71,000 coal camp dwellings had running water. Where it *was* provided, running water often consisted of one spigot in the kitchen. The hauling of water, often from a fair distance, became an onerous task, especially considering that every working male in the household had to bathe (usually in a tub in the kitchen) every night.[53]

The miner's wife carried out traditional women's work, but this work had particular qualities deriving from the grime of a coal camp, the lack of a convenient water supply, and the prevalence of boarders in the coal towns. Besides their typical tasks, as Dorothy Schweider explains of the Iowa industry, many coal-mining women cultivated large gardens, kept chickens and occasionally a cow, and sold the surplus vegetables, eggs, and milk. The Iowa miners' wives in her study contributed from one-fourth to one-half of the family income.[54]

Married or not, a man in the coal industry went to work each morning carrying a dinner bucket prepared by a woman. He came home to a bath she had drawn and to a meal she had cooked. His sphere was separate from hers, but intricately related. As for her, she did her work in the context of his job. Her sphere was in the home, but as we shall see, traditional family life was not her only context. She, no less than her husband, worked in the all-pervasive atmosphere of the coal company and its superintendent.

THE DANGER IS DOUBLE

The possibility of death or severe injury clouded the days of the miner and those of his wife. Mining people sometimes dreamed of disasters and, indeed, such dreams were considered a sufficient reason, even by some bosses, for miners to stay home from work.[55] List upon list of the injured and dead compiled by the states before 1910 (when the newly established Bureau of Mines began keeping national statistics) attest to the frequency of what a ballad popular with miners called "the miner's doom." "The fearful mortality and casualties which closes the year," reported an early anthracite mine inspector, "has cast a silent gloom over many poor families in the district."[56]

The vast, grim literature of death in the mines includes such items as a remarkable list compiled by the Union Pacific Coal Company in 1908. On 28 March the company's No. 1 Mine at Hanna, Wyoming, exploded at 2:59 in the afternoon. The explosion killed eighteen miners, experienced fire fighters who had gone underground to extinguish a fire that had interrupted production earlier that week. After the explosion, a rescue party consisting of forty-two men led by the state mine inspector entered the mine. At 10:30 that night, a second explosion ripped through the workings, entombing the rescue party. Only two were saved. With an underground fire raging out of control, the company was eventually able to remove only thirty-two bodies out of the fifty-nine dead. The mine had to be sealed with twenty-seven bodies inside.[57]

The complications of removing bodies from a gaseous mine raging with fire extended the process for months. As each body was recovered, the company listed every item found on the remains—a kind of archeology of the person. Several miners wore fleece-lined underwear, and some carried white silk handkerchiefs. Robert W. Armstrong had an American flag tattooed on his right arm and wore a gold band inscribed "Ida from Harry." He had letters in his pockets and a notebook, and wore an Odd Fellows pin. Mat Huhtala wore a union button and a "society" button. The men carried pocketbooks, most of which were empty or contained amounts such as forty-nine cents. But one miner had gone underground carrying six hundred dollars in cash—more than a year's wages for a miner.[58] Could this have been a relic of his distrust of banks?

From 1839 through 1914, more than 61,000 men died in the coal mines of the United States. Most—nearly 50,000—died between 1870 and 1914. The most fearsome killer picked off its victims in ones and twos, with little publicity attending their doom. A fall of the roof, the rib (sidewall), or of coal, accounted for about half of all underground deaths. Haulage accidents—being crushed between a coal car and the rib, or between two cars, or being run over—was killer number two. Thirdly, mine explosions and accidents involving explosives escorted an equal number of workmen to oblivion.[59]

If explosions were not the most frequent cause of death, they were certainly the most dramatic. An explosion could obliterate an entire union local. Such was the fate of the "Harwick Local" of the United Mine Workers, reported in 1904 under the headline "Every Member Dead."[60] Explosions shocked the sensibilities, the more so given the knowledge that only the fortunate would be killed instantly. Others survived the blast only to suffocate in the carbon monoxide gas spreading inexorably through the workings. Men could live for weeks or even months behind blocked exits, some to be rescued eventually and others to die of starvation.

Survivors never forget the experience. A loud thunderclap signals the explosion. The entire mine is instantly illuminated with the most brilliant lightning. A "roaring whirlwind of flaming air" destroys everything in its path, "scorching some of the miners to a cinder, burying others under enormous heaps of ruins shaken from the roof. . . ." This whirlwind of flame, "thundering to the shaft, wastes its volcanic fury in a discharge of thick clouds of coal dust, stones, and timber." One survivor, blown to a distance and thrown to the ground, saw *a river of fire pass over him.*"[61]

Bituminous coal dust is explosive by itself, without the presence of gas. In a gas explosion, coal dust fuels the fire. Once an explosion has been ignited, whether from gas or dust, a large amount of additional gas is instantly distilled from the dust. A local explosion quickly becomes a catastrophic inferno.[62]

Another type of major disaster occurs in a situation known as a squeeze—the collapse of the roof (or upheaval of the floor) over a large area of the mine. In Carbondale, Pennsylvania, a squeeze occurred in 1846 affecting nearly fifty underground acres. A survivor recorded the horror of it. "We heard the fall; it came like a thunderclap. We felt the concussion distinctly, and the rush of air occasioned by it put out our lights." He and his fellow miners knew the importance of trying to escape immediately, though "we had no idea that the fall had been so extensive or the calamity so great." Hugging the solid wall of coal, they headed toward a particular exit. They soon witnessed the effects of a great rush of air. "Loaded cars had been lifted and thrown from the track . . . debris [was] scattered throughout the chambers." The seriousness of the situation dawned upon them. Before long, they encountered a group of twenty-five or thirty men who were "very much frightened and were running toward . . . the point from which we had just come." They learned from these men that the only possible escape was the place they had just left. "We were greatly discouraged by this news and we turned back with them. . . . We had little hope of being able to get out through the body of the fall . . . for we knew that the mine had been working and that the roof had been breaking down that morning in the lower level. Indeed, we could hear it at that moment, cracking, crashing and falling with a great noise." In fact, the noise so alarmed them that they ventured no farther, but stood paralyzed with fear in the booming dark. Finally, the writer proposed that they break up into groups of three or four and separately try to make their way through the fallen debris so they would not all be exposed to the same particular danger. But the others were too frightened to go along with this idea, and whenever "some of us started out, the whole body rushed out after us. . . ." Finally, miraculously, the mine foreman appeared. He had descended the mine to attempt a rescue when the thunder of the fall had scarcely ceased. "He made his way over hills of fallen rock, crawled under leaning slabs of slate and forced his body through apertures scarcely large enough to admit it, and under hanging pieces of roof that crashed down in his path the moment he had passed. Finally he came to us. . . . [H]e led us back by the terrible path by which he had come. . . ."[63]

No wonder mine workers commonly believed that ghosts of slain miners would haunt a mine, particularly workings from which bodies were never recovered. Some miners insisted they could hear the ghosts of dead miners working in sealed-off sections. Eight decades after an explosion on 1 May 1900 ripped through the Winter Quarters mine near Scofield, Utah, elderly miners interviewed by Marianne Fraser believed that the ghosts of the two hundred dead miners continued to haunt the mine. George Korson, the folklorist who preserved a wealth of miners' songs and lore in the 1930s and 1940s, found a belief among miners that a few of their number possessed "second sight," which enabled them to see phantom funeral processions. When any such sighting occurred, the news would spread rapidly throughout the mine and the miners would quit work immediately.[64]

In 1907, the U. S. Geological Survey issued a retrospective report on deaths in the coal mines, the first national study of its kind. The most startling revelation in this document was that the death rate in the United States (which averaged 3.39 per thousand) was three times that of Belgium, three times that of Great Britain, twice that of Prussia, and nearly four times that of France. Neither were United States mines inherently more hazardous. The report noted that in Belgium, where the mines were deeper and more dangerous, the death rate had formerly equaled that of the United States. Rigorous safety legislation strictly enforced had decreased the number of deaths in Belgium and elsewhere. Coal mining was dangerous under any circumstances, but the knowledge and technology existed to drastically reduce the danger.

In Europe, for instance, the law strictly regulated the amount and type of blasting powder used. One skilled miner, the shotfirer, did all blasting after the others had prepared their shots and gone home. This practice reduced not only the danger of explosions, but also the number of roof falls, the highest cause of death in the mines. Apparently, using too much blasting powder weakened the roof, making it more likely to cave.[65]

Operators understood but often ignored the practice of inspecting the mine for accumulations of gas, and then forcing enough air through the mine to dispel it. Careful operators employed a fireboss, whose solitary job it was to descend the vacant mine at night and inspect each room for gas. He would lift his safety lamp toward the roof (since methane is lighter than air) to see if a blue cap formed around the flame.[66]

Sprinkling coal dust with water makes it nonexplosive. So does rock dusting—mixing the flammable coal dust with incombustible shale or limestone dust. In 1908, the U. S. Geological Survey recommended both sprinkling and rock dusting.[67] Both practices were well known before then and frequently ignored for many years thereafter.

These and other safety measures were recommended, suggested, and urged upon the operators and upon the miners. Some of the recommendations were incorporated into state mining laws, but, even then, they were not enforced. In 1870, the Pennsylvania Legislature, in establishing the state's first mining law, found that ventilation was often supplied by nothing more than the movement of coal cars. Gas was dispelled by "burning it out," an ancient and lethal practice of lighting a long stick and sticking the flame into a pocket of gas, causing a local explosion, which, the theory went, would dispel the gas. The man saddled with this job hoped for the best but seldom lived for very long.[68]

From the operator's point of view, safety was the responsibility of the miner, and accidents resulted from his carelessness. "The miner is free and can protect himself for he can engage in mining or not," an operator said in 1871 to an Ohio Mining Commission. Another operator concurred. "As to the proposed law, I regard it as unnecessary and unwarranted interference with the business of coal mining." In this operator's view, the safety law being considered by the commission meddled with matters "better secured by natural social laws." Fifty years later, miners were still taking the rap for accidents. "Behind the mounting list of fatalities," wrote the president of the Union Pacific Coal Company in the 1920s, "were ignorance, primitive mining methods, absence of competent mining surveys, mapping, and, as potent as any other factor, the stocky figure of the miner himself, the miner with his traditional unconcern for his own safety."[69]

Carelessness did cause accidents. "It is a well established fact," a Pennsylvania coal mine inspector reported in 1882, "that too much haste often breeds serious blunders the effect of which frequently proves fatal." Miners failed to use safety lamps because they gave very poor light. They failed to timber because they were in a hurry. They ignored the pounding in the roof because they rationalized that the roof would stay up while they did one more thing. A most precarious moment in underground mining occurred after a miner had fired his shot and retreated to wait for the blast. Sometimes, nothing happened. Faced with a "missed shot," some miners would quit for the day. Others, such as Evan Williams in 1881, took the risk of going in to inspect their work:

> Accident No. 66—Evan Williams, a miner, age forty-eight years, was fatally injured November 7, and died the same evening. . . . He was firing a hole . . . and when he lighted the match he retreated to a safe place. After waiting a few seconds in vain for it to explode he concluded it was "missed," and returned to renew the squib but, when he was near the hole, it exploded with the result stated. He was married and had five children in very needy circumstances.[70]

Many questions surround the carelessness of individuals like Evan Williams. Did working in constant danger desensitize a man, lulling him into feeling safe in a situation calling for fear? How much training did he have, and who had trained him? Did the miners' conceptions of manliness involve the idea of bravery or bravado and, if so, to what extent did this lure them into taking foolish risks? Was there peer-group pressure to this end?

These questions deserve study, but what is clear is that Williams's carelessness occurred in the context of a system that did not repay safe practices. Miners rushed through work because they were paid by the ton, and they tried to get away with taking shortcuts, especially during hard times. Even the most careful miner worked within an overall system controlled by the management. The management was obliged to supply timbers to the miners' rooms, but sometimes didn't. The management was responsible for ventilation, and for the general plan of mining. In 1907, the U.S. Geological Survey pointed out how much safer it was for a mine to employ a single shotfirer to light shots at night for all the miners. (The shotfirer had no incentive to rush in to check missed shots; he was also experienced in his particular task. Although his was an extremely hazardous occupation, the fact that he worked in a vacant mine eliminated the danger of blasting for the rest.) In any case, the recommendation of the Geological Survey had no effect on many operators who continued on with the old system of every miner lighting his own fuses.[71]

Then again, operators tended to blame miners for explosions, but often the blame was misplaced. True, an explosion is impossible without a source of ignition, and it was usually a miner who ignited the explosion by lighting his pipe or carrying an open flame where a safety lamp should have been carried. The operator tended to point to the source of ignition—the one factor in the miner's control—as the cause. But in a properly ventilated mine in which dust is controlled by rock dusting or sprinkling, no amount of flame can ignite an explosion. As early as 1907, Andrew Roy, a Scottish-born miner who became Ohio's first coal mine inspector, argued: "As a matter of fact no explosion is ever caused by the fault of the miners, for if enough fresh air is mixed with the gas to dilute it, it becomes harmless. The miners suffer death and the operators relate the history of the catastrophe."[72]

The idea of company responsibility for safety lay dormant for many decades. It did not enter mine safety laws because, as Keith Dix observes of West Virginia, in the major coal states the operators wielded a large influence in the state legislatures.[73] But as early as 1886, the thought had occurred to mine workers. That year, a group of Illinois miners recommended "the adoption of a law, with heavy penalties attached, holding employers of labor liable for any and all accidents that may happen to their employees while in the line of service." The first annual convention of the United Mine Workers of America, held in 1891, passed a similar resolution.[74]

In 1907, a coroner's jury in Pittston, Pennsylvania, took the almost unheard-of step of ruling that the deaths of four miners crushed in a roof fall were the fault of the company, a subsidiary of the Pennsylvania Railroad. The miners had been robbing pillars. The company, not the miners, had laid out the overall plan for doing the work. The jury ruled that "so many pillars had been robbed in the mine that sufficient and safe support had not been left and the roof caved as a consequence." The union miner reporting on the ruling went further and argued against robbing pillars altogether, unless something was put in their place:

> The danger of mining at the best is bad enough, but the robbing of pillars makes the danger all the greater. The robbing of pillars is going on at an extensive scale at present, and men must run the risk to their lives or else get a discharge. It is altogether wrong . . . to rob the surface of support, of pillars of coal, until concrete or other imperishable supports are substituted. Where deaths result from pillar robbing, the coal corporations should be mulcted to the maximum obtainable sum.[75]

If death in the mines occurred commonly, ill health may have been virtually synonymous with the occupation. "[L]ook at the man forty years of age, that has dug coal all his life," we read in an 1886 issue of the *Union Pacific Employees Magazine.* "A deformed wreck, physically, if not mentally. . . . Look at the number of miners with broken bones; the number with burns; with stooped shoulders; with weak and impoverished blood; with rheumatic pains from working in water; with affected lungs from working in bad air. A physically sound man fifty years of age, who has dug coal all his life, is almost impossible to find." The most common of these slow calamities was black lung (pneumoconiosis), the severe damage to the respiratory system caused by coal dust. In the nineteenth century, it was seldom mentioned. In a rare allusion to the disease, an Ohio physician testified in 1871 about two of its victims: "I have made two post-mortem examinations, in which there was carbonaceous solidification in the air cells. I have no doubt the carbonaceous particles caused their death. I examined them after death because before their decease they spit a black substance."[76]

The death rate for coal miners was high, higher than for workers in other occupations.[77] Like working in the dark, or working bent over, danger formed part of the miner's everyday experience. It colored his view of the world. One result may have been an oblique perception that taking militant (at times military) action to improve his condition presented less risk to his life and well-being than did going to work. John L. Lewis exaggerated when he rhetorically described the coal miner's world view in these terms: "The public does not know that a man who works in a coal mine is not afraid of anything except his God; that he is not afraid of injunctions, or politicians, or threats, or denunciations, or verbal castigations, or slander—that he does not fear death."[78]

Miners did fear death. They could feel terror at the potential horrors

of underground work. But the kernel of truth made Lewis's pronounce-
ment effective. The bloodstained history of conflict in the coalfields shows
that before 1920, at no time and in no place did strikes or even bloody
mine wars consume more lives than did underground accidents in the
same region during comparable time periods. The public might become
aroused when gunshot wounds caused death in the coalfields. But for the
miners, nothing that happened above ground, not even a gun battle, was
as dangerous as going to work in the morning.

3

MY FATHER'S UNFAILING KINDNESS: MINERS AND MANAGERS IN COAL'S PREINDUSTRIAL ERA, 1860-1880

I have come up through the ranks of digging coal and I am glad I don't have to do it any more. I have worked my way up and I am now a little one-horse operator and I am on the top of my own manure pile tonight and can crow in Pennsylvania.

Ed Wilton, Pittsburgh, 1888 [1]

To our farmer friends we ask—how would you like to have to ask some boss for permission every time you took a day to go to town? We as miners have to sign a contract to do this very thing. . . . How would you like to ask a boss for permission to take two hours at noon instead of one, if you so desired? This is what we have to bind ourselves to do before we go to work in the mines.

Coal Miner, Mistic, Iowa, 1891 [2]

The traditional procedures of coal mining were intimately connected to social conditions and relationships. In the nineteenth century, the worldview of both miners and managers reflected experience in a preindustrial world. Only haltingly did the industry edge away from small-scale enterprise and craft methods. In many places, the preindustrial outlook described in this chapter lasted well into our own century. Yet by the 1870s the key institution of industrial capitalism—the large-scale enterprise—loomed on the horizon, in the nation and in the coalfields, in the public mind and in the minds of coal miners and their employers.

Pennsylvania's first Commissioner of Industrial Statistics, writing in 1872, sketched the coming of the new day. "Now . . . the place of the little shop at the corner," he reflected, ". . . is occupied by the grand manufactory, glowing with the flame of its hundred forges and roaring with the clash and turmoil of its thousand workmen. Now does he, who today . . . [works] for wages, represent the small operator of yesterday in the two-benched or one-forged shop. . . ."3

By 1860, America was already midway between "the two-benched shop" and "the grand manufactory." The number of independent craftspeople (the commissioner's "small operators") versus the number of wage workers was one instrument for measuring the progress of the journey. The "wage system," as it was called, was still new enough to excite much comment. In 1860, thirty-seven percent of the U.S. labor force was self-employed, while forty percent worked as waged or salaried employees. (Enslaved blacks constituted twenty-three percent of all workers). By 1910, the proportion of waged and salaried employees had increased to seventy-eight percent of the labor force.4 At the same time, wealth was being concentrated into fewer and fewer hands.

Regardless of political outlook, nineteenth-century Americans commonly described their society explicitly in terms of class. An observation like the following would never appear in a present-day government publication: "The absurdity of the old-time assininity that the interests of the laborer and the capitalist are identical is apparent to all intelligent people who understand the real cause of the conflict between the classes."5 But in 1901 this observation by Colorado's Commissioner of Labor raised few eyebrows. Some disagreed with him, believing instead that the interests of capital and labor *were* identical. But speakers on either side of the question saw their society in terms of its classes. Neither liberal ideologies nor public relations firms existed to smooth over widening class divisions brought by industrial capitalism. Class, capital, labor—people talked about them all the time.

Industrial capitalism brought a phenomenal increase in the output of manufactured goods, agricultural products, and mineral wealth. Coal fueled the growth of large-scale industry, which in turn created the need for more coal. In 1860, the year of Abraham Lincoln's election, American coal miners dug fourteen million tons of coal, still only twenty percent of that mined in Great Britain. By the turn of the century, however, U.S.

production had risen to more than three hundred million tons, and the United States had surpassed Great Britain to become the foremost coal-producing nation in the world.[6]

Nationwide, class realities were changing profoundly. Yet new attitudes came only slowly to the coal industry. Class consciousness occurs, to paraphrase E. P. Thompson, when working people experience an identity of interest among themselves, and against their employers.[7] The obverse might just as well describe the class consciousness of employers. Among coal miners, that mix of awareness and loyalty evolved in fits and starts. Class consciousness could not precede the formation of an industrial working class. Neither did the operators of small coal companies feel an identity of interest with the owners and managers of the new large corporations. In 1870, the small-colliery owner and the skilled miner had much in common: In different ways, both felt the large corporation as a threat. At the same time, the mine laborer and the skilled miner, both underground workers, did not see their interests in the same light. The laborer, employed not by the company but by the miner, could justifiably blame his poor conditions directly upon the man for whom he worked. He might even turn to the mine owner for protection.

Yet everywhere capitalism (called just that) was under discussion. In relating the history of an early anthracite union, a Pennsylvania official wrote, "For a long time [before 1868] the subject of the relations between capital and labor had been earnestly discussed through the press, on the lecture platform, and in the halls of legislation. . . ."[8]

One participant in the discussion, a coal miner named George Kinghorn, revealed an acute awareness of class in an 1873 letter to *The Workingman's Advocate:*

> A coal miner, though half his life may be spent in darkness, away down in the bowels of the earth, where one ray of God's blessed sunlight never shines, is of ten fold more benefit to the world than all of the bankers, stock brokers, railroad kings, and other public thieves who are eating the very vitals of the nation . . . who are, through their own self-styled aristocratic, bigoted, selfish, unprincipled actions, hurrying themselves . . . to the very abyss of destruction, who by their greedy, avaricious, insatiable desire to grasp the accumulations of labor . . . are driving cold steel to the heart of every workingman . . . the very creators of all their wealth and power.[9]

The same year, an Illinois miner put this more succinctly. "The large corporations," he wrote, "are, slowly but surely, riveting the chains of slavery around us."[10]

THE MINERS

What fueled such outrage? What sort of men believed themselves to be the "creators of all wealth and power"? What freedoms were the corporations strangling? In short, who were the men who dug the coal? How did the world look through their eyes, and what were the forces shaping their worldview?

Statistics on the coal mine work force bear an extremely tenuous relationship to reality. In 1940 the Census Bureau discovered gross errors in labor force statistics dating back to 1870, errors which had produced a vast undercount.[11] Nevertheless, the figures suggest that by 1860 coal mine workers comprised some 36,500 men and boys, of whom 30,000 worked in Pennsylvania and 25,000 were concentrated in the anthracite region. Bituminous miners were scattered throughout western Pennsylvania and in Illinois and Ohio, the second and third most important coal states.[12] During the Civil War decade, the number of mine workers more than doubled, while coal production rose from fourteen to thirty-seven million tons.[13]

Half of all American mine workers were born outside the United States; many were English, Welsh, Scottish, or Irish. Miners from the anthracite region of Wales emigrated to anthracite Pennsylvania until, by 1890, the Welsh community around Wilkes-Barre and Scranton had swelled to seventy thousand inhabitants. As late as 1880, virtually all of Ohio's 5,575 coal miners were Britons. In Illinois, one-third of the miners were British immigrants; fewer than half claimed the United States as their birthplace.[14]

In addition, about ten thousand experienced German coal miners had emigrated to America from the coalfields of Prussia. A minority of foreign-born mine workers, some three thousand, came from Italy or southern Europe.[15] In 1870, black mine workers also constituted a small but growing minority. Of course, native-born whites also mined coal, especially in regions like southern Illinois, where coal outcropped in farmers' fields, and the mountains of West Virginia.[16] The working people of the coal industry were thus divided by cultural identity, race, national experience and loyalty, and language.

Even the British did not see themselves as one people—the Welsh, Scottish, and English felt strong separate national identities. The Welsh spoke their own language; at one time, miners in the Mahoning Valley of Ohio conducted their union meetings entirely in Welsh.[17] The different groups participated in their own nationality-based organizations. An Illinois miner reported in 1874 from Streator, "Each nationality here has its society. The English have the St. Georges, the Scotch the St. Andrews, the Welch the St. David's, and the Irish the St. Patrick's."[18] Even when thrown together in the same coal town or in the same mine, peoples of different nationalities did not automatically mix together.

The Irish, the first Catholics to enter the coalfields in large numbers, stood out by virtue of religion as well as nationality. But even they cannot be considered a homogeneous group. Two types of Irish mine workers came to the U. S. coalfields.

One was the man who had previously emigrated to England or Scotland, forced there by famine, eviction, or political unrest. Irish immigrants began working in English mines as early as 1798. By the time these miners or their sons arrived in the New World, they were as skilled as their English, Scottish, and Welsh counterparts. The life of John Welsh, president in the 1870s of a Pennsylvania miners' union, reflected this pattern. As a child, Welsh moved with his parents from the famine-ridden Ireland of the 1840s to England. There, he began working underground near Durham. At age sixteen, already well schooled in both coal mining and union ideas, he made his way to the United States. After going West to seek gold—he found none—he returned to Schuylkill County, Pennsylvania, to mine coal. He married a literate Irish woman who taught her husband, then in his early thirties, to read and write. Welsh became prominent in the union movement of the 1870s, returning to the mines after the union's defeat. Well read and a capable speaker, he was elected in the late 1870s to the Pennsylvania Legislature on the Greenback Labor Party ticket. He ended his career, like many miners of similar background, as a coal mine superintendent.[19]

A second type of Irish mine worker came directly from Ireland to Pennsylvania with no mining experience whatever. Beginning in the 1840s, Irish peasants entered the anthracite region en masse. Between 1840 and 1860, the population of Schuylkill County, the southernmost portion of the anthracite region, mushroomed from about 29,000 persons to more than 89,000. A good many of the newcomers were Irish Catholics.[20]

Along with a rich musical and storytelling culture, they brought to the coalfields a long-standing resentment of the English. "Our national holiday, July 4, is kept with great zeal by the Irish," a journalist reported in 1877 of the mine workers around Scranton. "It is an outlet for the expression of their animosity to the English."[21]

National antagonisms fed the class tensions that developed between English and Irish when they met in anthracite Pennsylvania. Skilled English or Scottish miners hired unskilled Irish immigrants as laborers; toward them they often nursed an old hatred of Irish Catholics. Even skilled Irish miners found themselves working under British mine bosses. In pre–Civil War days, every mine boss in the anthracite region was English, Welsh, or Scottish.[22] Furthermore, it was well known that English investors provided much of the capital to develop the anthracite coalfield. Class hatred aggravated the national antagonism that flowed among bosses, miners, and laborers. On the other hand, national hatred could divide the Irish from others who shared similar life experiences due to class.

THE FIRM

The mix of nationalities and cultures in the work force is one lens through which one may view the emerging coal industry. In the early 1870s, Pennsylvania's newly organized Bureau of Industrial Statistics provided a quite different lens: individual portraits of the Commonwealth's coal firms. The Bureau's statistical clerks began their duties by recording detailed information on each and every Pennsylvania colliery, no matter how small, along with the unedited opinions of proprietors, superintendents, and coal miners.

The size of the firms showed an industry in transition. In the 1870s, Pennsylvania coal miners worked in firms ranging from tiny banks worked by farmers in their spare time to immense collieries operated by the anthracite railroads. The largest, the Lehigh and Wilkes-Barre Coal Company, employed (in 1875) nearly eleven thousand men and boys working in twenty-seven different mines. The Philadelphia and Reading Coal and Iron Company employed more than eight thousand men and boys in thirty-six different mines. Both firms were adjuncts to railroad corporations. Although the small enterprise was more common in the bituminous industry, companies such as the Penn Gas Coal Company in Westmoreland County (near Pittsburgh) employed more than one thousand men and boys.[23]

Tiny coal banks exemplified the artisan-operated enterprise characteristic of an earlier time. The archetypical independent craftsman owned his own means of production—the tools as well as the natural resources and necessary supplies. He often controlled the transportation of his product. He required little capital and employed at most one or two assistants. In 1880, the U. S. Census estimated the existence of more than five thousand such mines, mainly in the bituminous industry. All of them together contributed less than one million tons toward a national production total of some forty million tons. Relegating to them space commensurate with their size, the Census Bureau printed little information on them. Yet these rustic coal banks exemplified an economic form that was passing out of existence: They illustrated handicraft production in pure form. In the Census Report, we read:

> There is a class of small producers . . . [which] consists of farmers who having an outcrop of coal or a bed of soft [iron] ore on their land, do a little irregular surface work during the fine weather of the winter or in the interval between seed-time and harvest. The coal produced in this manner is used for a local domestic fuel and does not find its way into the general market. . . . [A]s no capital is employed and but very little labor hired, this mining cannot be said to have any influence on the industry at large.[24]

Lacking the sophistication of its federal counterpart, Pennsylvania's Bureau of Statistics spent as many words on the smallest mine as on the largest. The Bureau's initial reports lucidly describe the early form of the industry in the United States.

Due to lack of transportation, the markets for these mines were strictly local. "The coal is delivered from the mine by teams," remarked one owner on the questionnaire he returned to the Bureau, "very irregular on account of the roads and the weather."[25] Another operated his little mine only during the fall and winter months. He employed two miners who dug the coal out of a clean four-foot seam and loaded it into a small car. It, too, was hauled away by teams.[26] A coal owner from Mt. Morris, Pennsylvania, admitted to having coal but no market, nor any means of transportation. His mine supplied fuel for a nearby steam-powered sawmill.[27]

The proprietor of Yorty's Pike Run Coal Works revealed the close relationship in some regions between farming and (bituminous) mining. Proud of his country mine, Henry Yorty emphasized the growing importance of coal to the rural districts:

The Pike Run Banks are what might be called 'country banks.' We have no railroads or steamboats and of course our customers are farmers and others, who depend on hauling their coal with teams. Some come from 9 to 12 miles distant. . . . [W]e have the man with his six-horse team, who takes 80 bushels down, to the man with one horse who takes but 15 bushels. I will also add that the consumption of coal in the rural districts is increasing every year. So your department will see that although Pike Run Coal Works are country banks to accommodate country people, yet they are not an object to be overlooked when making up the mining and industrial statistics of this great Keystone state.[28]

By the 1870s, these mines were the exception. The average Pennsylvania mine worker carried out his tasks in a larger mine, often in one of several owned by a single corporation. In 1875, all but 17 of the 153 anthracite mines employed seventy-five or more persons. Even in the less-concentrated bituminous industry, more than one-third of the 179 Pennsylvania bituminous collieries employed seventy-five or more persons.[29]

THE CRAFT OUTLOOK: MINERS

"Industrial capitalism begins," Harry Braverman writes, "when a significant number of workers is employed by a single capitalist. At first the capitalist utilizes labor as it comes to him from prior forms of production, carrying on labor processes as they had been carried on before."[30] In the 1870s, the skilled miner resembled a transplanted independent craftsman; certainly he saw the world from that perspective. Like the independent craftsman, he owned his own set of tools and hired his own assistants. Skilled, or "practical," miners were often called contract miners, because most of them contracted with the owner to bring out the coal, for which they were paid by the ton, by the yard, or by the wagon. The contract miner paid his laborer or "helper" out of the tonnage rate that he received from the coal operator. Operators kept no records of wages miners paid to laborers. As late as 1902, the Anthracite Strike Commission disclosed that the accounts of a typical anthracite firm showed "that a certain contractor [miner] had received, say $200 for his pay, but it did not show that out of the $200 he had to pay one or more laborers. . . ."[31]

The laborer's position was a far cry from that of the miner. In 1869, an anthracite mine worker described the work in its social dimension to a friend left behind in Wales. "The method of working is as follows," he wrote. "There is a miner, or as you in Wales call him a collier, and a laborer in each stall [room]. . . . The miner cuts the coal and the laborer fills. Each one works by himself. The laborer has six cars to fill each day, each of which holds about two tons. . . . The laborer's wage is one third of that earned by the miner."[32]

A third category, "company men" or "day men" such as mule drivers, tracklayers, and bosses, were the only group to be directly employed by the company for a daily wage. They alone worked as wageworkers or employees in the sense that most people do today.

In 1870, fewer than half of Pennsylvania's 79,000 coal mine workers worked as contract miners. The skilled miners in turn employed nearly that many laborers. There were only 4,600 "company men" or "full-time hands" working in the state, a mere seven percent of the work force.[33]

The contract miner enjoyed a significant amount of independence in the way he carried out his job. In his excellent study of the coal miner's job, Keith Dix explains: "These independent craftsmen learned their job during an apprenticeship period and, once having mastered the requisite skills, worked largely without supervision and at their own pace. The early pick miner was not only in control of his job but also in control of his own time on the job."[34]

The extent to which miners worked without supervision is indicated by how few mine bosses (underground foremen) were employed. In Schuyl-kill County in 1870, a mere 273 bosses supervised nearly 16,000 coal mine workers. This amounted to one boss for every fifty-eight employees. A former bituminous miner recalled that in a Pennsylvania mine of the 1890s the mine boss visited each miner in his underground workplace only once or twice a week.[35] As late as 1932, the miner employed by the Hudson Coal Company, a large anthracite concern, received a visit from his foreman only once a day. In a book published by the Hudson Coal Company, we read:

> The anthracite miner probably enjoys a unique place in industry due to the degree of independence which he enjoys while at work. In factories . . . workmen are under the more or less constant surveillance of the foreman. Constant supervision is physically impossible in underground mining operations. . . . They [the foremen] have many miners working under their jurisdiction, each working in an individual place, so that it is impossible for the foreman to spend but a small portion of each day in each miner's workplace.[36]

In fact, it was the skilled miner, not the underground foreman, who controlled the work process. He conceptualized the work, making numerous decisions about how to do it (how far to undercut, where to place the drill hole, how much powder to use, and so forth). He supervised his laborer, directing all the work carried out in his own underground room.

Pride in careful workmanship infused the early hand miner. He preferred the particular method he had learned from his father, usually a local variation of the standard procedure. Three mining engineers found evidence of this pride and attachment in the 1920s when they descended a Kentucky mine abandoned in 1844. They were awestruck at the level of skill evidenced in the workings: "[G]eometric exactness marked everything they saw. The ribs [sidewalls] of that main entry running back from the shaft were as true and clean as a carefully concreted entry of today."[37]

Two teams of miners, one from England and the other from Wales, had worked this mine. They had "scoffed at each others' skill as miners," with the rivalry between them periodically breaking out into free-for-all fights. The engineers could discern two distinct styles of mining. Standing at the intersection where two side entries crossed the main entry, they could see that "the Durham entry had been sheared on the right and shot from the left while the Welshmen—whether they were all left handed or not—had sheared on the left and shot from the right."

Many contract miners determined not only their own work procedure but also their own hours of labor. In the late 1870s, a coal owner employing twenty-nine miners found it "impossible to tell the hours they work, for they go when they please and quit when they please."[38] Another employer of forty-two miners explained, "No two days run the same. About one-half of the miners lose nearly one-half time by only working part of days, and not going in at all for about a week or more after pay day."[39] (But the evidence on this is conflicting. Complaints about long hours show that this freedom did not universally prevail, although the low tonnage rate was probably a harsher taskmaster than the boss.)

The miners, not the employers, made collective decisions regarding individual levels of productivity. During the 1860s, the first national coal miners' union (the American Miners' Association) regularly updated its internal rules specifying how much coal a single miner (or his laborer) would be permitted to load out of his room on any given day. The miners voluntarily restricted production partly in order to share the available work. In 1891, an operator near Seattle locked out his employees "for asking for the privilege of dividing our work with 75–80 of our less fortunate brothers who were thrown out of work by a certain part of that mine."[40] This craft solidarity, operating as it did at the expense of men who already had work, reflected an ethos of sharing, a sense of collectivity that ran counter to notions of individualistic self-interest.

Miners also favored restricting production for the control it gave them over the length of the workday. In 1864, one coal miners' union announced that "eight hours a day is sufficient for every collier to earn his daily bread and the second eight hours to enjoy reading and writing and the sunshine in the open air and the third eight hours to rest from his labors. . . ."[41] Miners abhorred long hours because they ruined the possibilities for a good life. Long hours underground exhausted the system, one miner explained, and produced a condition of depression and lassitude. "If hours were not so long . . . the men would have more time and energy, and a greater disposition to read and seek cultivation for their minds."[42]

An Illinois miner lamented the life of the man who left for work early and came home late:

> What for? to instruct his children, to pass a few hours pleasantly with his wife and children? No. But to eat his supper . . . and then to lie down more like a beast of burden, than a being that was made in the image and likeness of his Creator. Thus the weary round of life revolves until . . . his constitution is completely broken down. . . . [H]e is a misery to himself, and a misery to his wife and family, and a burden to . . . society . . . The money he saved has gone for doctors. . . . [H]e is probably thirty-five or forty years old, but he is past work, an old man. . . .[43]

Added to the unhappiness of endless toil was the increased danger. "Long hours," a miner testified in the early 1870s, "make the miners very drowsy and careless. To a great many it is a *terror* when they think of the long hours they have to work the next day."[44]

Miners harbored the "strangely felt desire for additional opportunities for recreation and the enjoyment of life," as an Ohio miners' convention expressed it.[45] In 1870, Scranton miners attended a reading and debating society that met at the Welsh Baptist church. The participants discussed matters such as whether or not the world was created, as stated in the Bible, in six days, and whether the days in question were periods of twenty-four hours. Some members of this society owned many books.[46] A mining village nearby organized a library in its Sunday School. "Our library was neither classical nor extensive," the miner in charge of it recalled. "It might perhaps be said that some parts of it were not even elevating."[47]

Be that as it may, miners did spend time reading. During the 1890s, the leading coal miners' newspaper, the *United Mine Workers Journal*, devoted an entire page (out of four) in each issue to "Reading for Leisure Hours"—a short story. "My father was a slow but very careful reader," John Brophy recalled in *A Miner's Life*. "He never took anything on a newspaper's say-so. Because I could read faster than he, he often had me read the longer news stories or feature articles aloud, and we would discuss them." Indeed, for the illiterate, and for those who couldn't afford a subscription, reading aloud provided a sociable means of obtaining information and a forum for discussion. In one mining camp, Brophy and his father spent Sunday evenings with "some special cronies" from the same village in Ireland. The boy, the most competent reader, would read to the miners from *The Irish World:* "It might be a whole issue devoted to the story of Robert Emmet's effort to seize Dublin Castle in 1803, and of his capture, trial, and execution, with long quotations from his stirring speech on the scaffold. Or I might read the story of the United Irishmen of 1798, the failure of the French troops to arrive in support of the rebellion, the capture of Wolfe Tone, and his suicide." On other nights, these two Irish miners would "regale us with tales from the old country."[48]

The rich cultural life of the mining communities included fiddling, singing and dancing, and the telling of tales. Minstrels would wander from coal patch to coal patch, singing ballads (many composed by miners as they hacked away at the coal), telling "folk tales, legends, and anecdotes," and playing fiddle tunes. The folklorist George Korson preserved many songs and stories of the early coal miners, and Korson relates that in Shenandoah during the winter of 1880 a nineteen-year-old mine worker named Martin Mulhall invented an oral thriller called *The Adventures of the Flynn Gang*. "For four months, Shenandoah's boy workers would gather every evening at the Mulhall home to listen to young Mulhall improvise a fresh episode of his serial, not a word of which was ever written down."[49]

The miners' concern with the quality of life clashed with the operators' concern with profits. The question of who would decide the hours of labor and the level of productivity became a key bone of contention in the class struggle that unfolded between coal operators and miners. In the 1870s, the coal operator or his manager did not control what went on in his mine. Yet the miner's independence was under attack. In 1872, *The Working-*

man's Advocate reported, "The Kingswood colliers, near Bristol, received notice on Saturday that their services would not be required after Saturday next unless they abandon their 'stint' system, which restricts the production of coal to about 4 tons per man."[50]

Coal companies, especially the larger ones, began to introduce work rules to force contract miners to work longer hours, and to work when the company pleased instead of when they pleased. In August 1863, a large company near Pittston, Pennsylvania, imposed a rule specifying that "[w]hen any man is away from his work for 2 days, either from idleness, drinking or any other cause except sickness . . . such man shall be discharged."[51] In 1873, *The Workingman's Advocate* denounced "a villainous contract" that forced the employees of the Wabash Coal Company "to be ready for duty when the whistle blows every morning." To enforce this rule, the engineer who tended the cage was forbidden to lower any miner or laborer after 7:30 A.M.[52]

The contract itself became the operator's means of control, rather than an agreement between operator and independent miner, presumed equals. A miner's fumings to a labor paper as late as 1891 show the extent to which miners believed that their rights included control over their own time. "To our farmer friends we ask," the miner began, "how would you like to have to ask some boss for permission every time you took a day to go to town? We as miners have to sign a contract to do this very thing. . . . How would you like to ask a boss for permission to take two hours at noon instead of one, if you so desired? This is what we have to bind ourselves to do before we go to work in the mines."[53]

In theory, the contract miners were independent while the company men worked directly under the mine boss with no pretense to independence. In practice, by the 1870s the distinction between the two groups was growing hazy. Mine bosses regularly hired day laborers or company men out of the group of contract miners. The contract miner worked as such only part of every month. For one to six days a week, he did his work under the direct supervision of the mine boss.[54]

THE CRAFT OUTLOOK: MANAGERS

Despite the stirrings of change, the coal industry of the 1870s was still in a preindustrial, which is also to say a premanagement, stage of development. In the smaller mines particularly, management tasks received the minimum of attention. These included supervising production, bookkeeping and financial planning, sales, raising capital, control of supplies, and planning the layout of the mine and the overall manner in which the coal would be brought out.

In coal mining, the need for management had a special technical significance. As the nineteenth century unfolded, the developing science of geology, spearheaded by Sir Charles Lyell's *Principles of Geology* (1830), rendered the skilled miner's traditional knowledge increasingly inadequate, especially for opening mines. New, accurate knowledge about the origin of coal, how it lay in beds in the earth, and how the coal strata could be folded and faulted, superseded the amalgam of experience and superstition that made up the miner's skill.[55]

But professional management had its social and political dimension as well. It emerged, as Alfred Chandler, Jr., explains, out of the large corporation. The professional manager was a new man. He represented the interests of the owner, but he was not the owner, nor was he a production worker. He occupied a new rung on the socioeconomic ladder. Originally, he was hired from the ranks of the most skilled production workers, but gradually it became more common for him to have received professional training in a school, rather than traditional training on the job.[56]

The coal industry fueled the new industrial capitalism but brought up the rear in adopting its new organizational forms. Most coal mines, even quite large ones, represented a proliferation of small underground workshops. In the nineteenth century, professional management existed in the coal industry mainly as an ideal, and not a very widespread one at that. For superintendents and bosses, the contract miners' independence translated into lack of control. Production could be unpredictable, filling orders on time a problem.

Consider the exasperating experiences of the superintendent of the Buck Mountain Coal Company, a Pennsylvania anthracite firm. In the 1880s, his daily written reports to the company's Philadelphia office reveal that several times a month an insufficient number of miners showed up for work. "Payday made quite a number of our men have other business," he wrote in November 1886, "so they could not come to work today, we were short handed." After the January payday, he reported the same difficulty. "Our men are not in good trim yet. Payday had a bad effect on them, some of them painted Mahoney City red on Saturday night and had to pay for their fun. We will try to start work tomorrow."[57]

The miners took holidays when they pleased, according to their own traditions. "Today is called 2nd Christmas and is kept by the Huns and the Poles. That would have prevented us from working today," the beleaguered superintendent recorded on 6 January 1887. "The Irish Catholics are not observing this day any more as it has been stricken from their list of holidays." But the Irish did, of course, observe St. Patrick's day. "St. Patrick's Day may be the cause of many of our men not coming to work tomorrow, quite a number of them have gone to Hazleton to a large parade there."[58]

On some days, the superintendent could not say why the men failed to appear. One day in January 1887, "[e]verything ran hard . . . and went slowly. The men did not all come out."[59]

In addition to shunning work on traditional holidays and for a day or two after payday, anthracite miners customarily quit work on the occasion of any fatal accident, declining to resume until after the victim was buried. At the Buck Mountain Coal Company, this happened three times during the winter of 1887. In the anthracite region as a whole, in the eleven-year period from 1876 to 1887, 2,703 persons died, causing the survivors to miss numerous days of work.[60]

The arrival of the circus could also embarrass production. This entertainment could inspire the teenage mule drivers to "stampede." The boys would arrive at the mine from their separate homes, consult with one another and despite the best efforts of the boss to break them up, leave in a body. This left the miners and laborers with nothing to do, because they could not proceed without their underground transportation system. Work was null for the day.[61]

Finally, the age-old problem of "the morning after" made great inroads on the superintendent's available labor supply. It is impossible to quantify, but true enough that drinking was a serious and time-consuming activity in all coal towns. A former miner, reading over the diary he had kept in the 1870s of his youth, reflected, "I am rather startled to find the number of days work that I missed through drink."[62]

If the early manager did not regulate the miner's hours of work, neither did he regulate the use of his time at work. The men engaged freely in conversation during their working hours. To begin with, the long trip from the mine mouth to the face provided (as Michael Yarrow shows for present-day miners) exceptional opportunities for socializing. Then, during any lull in the workday, "from want of cars or other cause," (a journalist reported from the anthracite region in 1877), "the men will squat down, miner fashion, and tell stories and crack jokes."[63] Carter Goodrich, in his classic account of "the miner's freedom," points to the *the gob pile oration*—as one of the sources of militance in the coalfields. He suspected that underground speeches and discussions were important in promoting solidarity in the work force. Indeed, in 1873 *The Working-man's Advocate* reported that the Wabash Coal Company had "positively forbidden" suspensions of underground work for the purpose of holding meetings, as "such practices materially increase the Company's running expenses, and result in no practical good to any one."[69]

Outsiders freely entered the mines to discuss religious or other matters, or to sell subscriptions or insurance. In the early 1890s, a solicitor for a labor newspaper reported on his success after spending an entire day underground conversing with miners. On 6 December 1907 a mine in Monongah, West Virginia, blew up, killing 362 men, including an insurance salesman who had been going from room to underground room selling life insurance.[65]

If the bosses did little bossing, it was partly because both managers and skilled miners conceived of the work in similar ways. Superintendents and bosses sometimes gained their positions through family or social connections with the owner, but often they were drawn from the ranks of the contract miners. A boss-miner knew more than a boss with no experience. Still, he had no special training as a manager and continued to perceive the work much as he had as a miner. Neither bosses nor miners understood management activities as requiring a great allocation of time or resources. Particularly in small firms, managers devoted a significant number of hours to nonmanagement, production-oriented tasks.[66] In the 1870s, a mine owner employing eleven persons in western Pennsylvania reported that the man who weighed and dumped the coal and did all the machine work, also acted as pit boss.[67] At a mine employing thirty persons, "the mining overseer, or pitboss as we call him, does all the inside work [laying track, etc.] with a little assistance from the other day hands."[68] At a still larger colliery, the Rock Run Coal Works employing 122 men, the weigh master also acted as the company bookkeeper, and when he was not busy weighing coal or working on the accounts, he helped load coal into the crafts on the river.[69]

Like the craft miner he had been, the early manager conceptualized and carried out the whole task of running the mine, without seeking expertise beyond his own. As late as 1924, the industry journal *Coal Age* criticized the manager who tried "to be a universal mining genius." The article advocated the hiring of specialists, "for any man who attempts to cover every field thoroughly and to control everything is too busy to do anything well or fails utterly in a comprehension of some one or more of his many diverse duties. Thus we get plants which reflect the mind of one man and not the combined genius of several."[70] Specialization was being advocated in the 1920s; in the 1870s, the thought had not yet occurred.

The premanagement stage of the industry also revealed itself in the lack of financial planning. Financial planning rests fundamentally on information provided by bookkeeping, an activity entirely omitted by some coal firms. One proprietor reported, "I board what hands I generally employ and have these only for two or three months of the year, sometimes off, sometimes on. It is hard for me to determine the amount of daily wages, unless I commence and keep a daily memorandum of each particular, which I have not done."[71] Another firm employing eighty persons reported ignorance of the amount of capital employed in the business and of the cost of materials, because under the previous management "the books of the company were not kept proper."[72]

Keeping track of supplies, essential for keeping down costs, was another management task observed mainly in the breach. In this case, the labyrinthian caverns and the dark worked against keeping track. In 1885, a Wyoming mine manager reported to his superior regarding a large quantity of lost rails: "I think that some rails must have been overlooked in the last inventory," he confessed. "[I] do not know how to account for the shortage. It would be a very easy matter to miss them unless a person searched each mine, as nearly every miner when he can will have a rail or two hid away for emergency and unless dug for they will not be found."[73]

Planning the overall manner in which the coal would be removed and maintaining the good condition of the mine constituted important management tasks. Coal miners themselves complained that the failure of good management in these respects aggravated working conditions. A Pennsylvania miner testified in 1871 that the air and general safety of the mines could easily be put in better condition, if better mining bosses were employed. "In many cases some pet or favorite is employed in this position, whether he is competent from knowledge or experience or not."[74] Another miner testified that he had worked in a mine where the pit boss had never dug coal. All summer, white damp had troubled the men. "We told him we could not work in there, he would come in fresh from the open air, and after stopping about ten minutes would tell us the air was good enough, that there was no bad air in there. . . . A man who does not understand mining makes not only a bad pit boss for the men, but a bad manager for the operator, because through mismanagement he will lose at many times the coal that a good manager could take out."[75] This miner proposed a law requiring supervisory positions to be filled by skilled men.

For many coal operators, short-term leases negotiated with landowners prevented long-term planning. In Schuylkill County, as Clifton K. Yearly, Jr., notes, five- to ten-year leases were the rule, while some contracts provided the operator with exactly three years in which to get in, make a profit, and get out. Compare this with the British industry: "Among the many leases in the eighteenth and early nineteenth century that have come to light," reports Michael Flinn, "none was for less than 21 years." In the American industry, a shortage of capital combined with the short-term lease to produce dangerous and wasteful working methods. The leases themselves contributed to the shortage of capital in that many stipulated that the landowner would be paid by a royalty on tons mined. This allowed individuals with little more capital than the change in their pockets to go into mining, with the object of producing coal as fast as possible.[76]

Incompetent management could result in disaster. One management task was to supply the timber miners used to prop the roofs in their working places. Another was keeping the main entryway and other inside roads propped as they were constructed. An Illinois miner reported in 1875 that in the mine where he worked, after a roof fall had killed three men, the mine inspector "pronounced the mine in a most awful state, worse than any he had ever visited, and gave orders that the mines should be properly timbered before the men worked any more." At the time of the accident, the superintendent had been absent.[77]

Pennsylvania's coal mine inspectors, first appointed in 1869, worked to educate the industry by including in their reports maps of well laid-out coal mines, discussions of new methods, and diagrams of mechanical innovations.[78] While praising the occasional exception, the inspectors scolded the operators incessantly for poor management: "[I]n regard to the reckless manner in which the bituminous coal mines [are] worked and ventilated," chided one inspector, "it is lamentable that such a condition of things [is] allowed to exist in a state that boasts of its mines as its greatest wealth." He suggested that lack of training had caused this state of affairs:

Superintendents [are] chosen in many cases because he is a relative or friend of one of the company. . . . [H]is knowledge of conducting a mine is limited. . . . Such a superintendent makes another blunder by choosing an incompetent inside boss. The result can be imagined. If we examine the mine we find no system of mining adopted. *Chaos* can be the only word fitted to describe such a mine; no ventilation, no drainage; pillars, rooms or breasts worked into one another; thousands of tons of coal are lost . . . annually through this sort of management.[79]

The following year the inspector took up the same theme. He reported (in the context of the severe depression of the 1870s) that many superintendents had stopped maintenance work altogether. "Drains that should be kept open are left to fill up; bottom and top that should be blasted is left undone; airways that should be driven are postponed; pillars are mined out that should be left standing. . . . A company getting coal under such circumstances . . . will find, when it is too late, a ruined mine." He concluded "the sooner our coal operators discover the fact that to work their mines successfully, they [need to] employ competent men to act as superintendents and inside bosses, the better."[80]

Competent men meant "practical" or skilled miners, not mining engineers with professional training. In the 1870s in the United States, mining engineers who acted as coal mine superintendents were in the extreme minority. Most of these had come from across the Atlantic where professionalism had advanced further than in the United States. Of twenty-eight mine bosses and superintendents who testified to an Ohio mining commission in 1871, twenty-two had come up through the ranks as practical miners. Only two had received professional training, both in Germany, where before a man was permitted to take charge of a mine three years of scientific and engineering education were required, including studies in geology, chemistry, and mapping. (The German professional schools admitted practical miners only.) Of the twenty-eight Ohio managers, two had never been miners.[81]

THE CRAFT OUTLOOK: CHILD LABOR

The primitive state of management reflected the persistence of the craft tradition in the coal industry. Another indication was the particular form of child labor that flourished in the bituminous fields. In traditional craft production, tasks were carried out by all family members. Although Britain made women's underground work illegal in 1842 (and male miners emigrating to the United States were increasingly turning against it), the family system of production survived in America in the form of child labor. In the bituminous mines of the 1870s, young boys commonly worked alongside their fathers, a relic of the old way. So little did the superintendent involve himself with this form of labor that he often could not say how many children worked in the mine over which he presided. "The number of boys employed in mining cannot be estimated," an operator wrote in 1875. "According to the custom in this valley, every man having a boy able to walk to the pit may take him along whenever he pleases."[82] Another employer of 125 miners revealed, "Boys in mines not employed by me, but go in with their fathers or others. Not having any account with them am not able to give any figures as to numbers or earnings."[83]

On this matter, too, opinions were shifting. By the 1870s, some workers believed that children should stay out of the mines. A coal miner testifying before a commission in Pittsburgh denounced child labor:

> There is a practice of taking boys into the banks very young—from 8 to 9 years of age. They work from 10 to 12 hours per day, and hence are deprived of the advantage of schooling. They likewise suffer from the bad air of the mines. They sometimes work as "trappers" but a great many work with their fathers. . . . [F]athers take their boys in as soon as they can get a car allowed for them for the sake of earning a little more money. The little boys contract vicious habits. . . . A great many boys have been crippled. . . .[84]

Children were not only crippled but also killed; the persistence of youngsters on the fatality lists attests to their continuing presence in the mines. One day in 1878, a boy named James Burt was loading coal in a Pennsylvania mine when he was crushed by a fall of slate. His father took him out and obtained immediate medical attention, but the boy died the next day. He was thirteen.[85] That month, Sheridan West, a mule driver, jumped onto a passing tram of coal wagons, missed his step, and was dragged and crushed to death between the cars. He was fourteen. Of him the inspector said, "I am clearly of the opinion that the deceased was too young to be engaged in driving a mule."[86]

A decade later, when children under twelve were prohibited by Pennsylvania law from working in the mines, an inspector scolded, "The law

. . . is daily and hourly disregarded and evaded. . . . Boys really under ten
years of age, are asserted to be above twelve. . . . False representations are
made by parents and guardians, and acquiesced in . . . by the employer.
The child is not alone deprived of education, but oftentimes suffers from
coarse and even brutal treatment. . . ."[87]

Economic necessity prompted parents to lie about their child's age;
many families would have starved without the labor of their children.
Child labor laws failed to raise the father's wage to enable him to support
his family. But it was also true that many parents approved of their
children working in the mine. Most grown miners had themselves begun
work as children; it seemed natural to them that their sons would do the
same. They themselves had once fit the description given in 1907 by the
National Child Labor Committee, of the lad so small that he could hardly
carry his dinner bucket without dragging it.[88]

"Boys are employed upon the urgent demand of the miners," a mine
owner testified in 1871 to an Ohio commission. "[They] . . . want them
to be employed so as to help to support the family, to learn the mining
business, and to be their company." He further testified that prohibiting
boys would "clash with the general economy of mines," and that if
"boys were expelled from the mines, many of them would turn loafers,
and learn all kinds of bad practices. . . ." This operator's concern for the
morals of the boys fit nicely with his economic interests. But he was
right to suggest that the weight of tradition tipped the scales in favor of
child labor. Learning the mining business was not the least of it, as
miners wishing to pass on their trade to their sons believed that training
must begin in boyhood. It was a sign of change that most miners testifying before this commission favored raising the age of boys permitted to
work to 12. This must have seemed quite old enough to the typical
witness who, like Richard Grafton, could say, "I am a native of England; emigrated to this country in 1863; am 37 years old; have been a
miner 30 years."[89]

The situation of children working beside their fathers under prefactory
conditions was probably better than that of children caught in the mangle
of factory production. The typical boy working underground in the 1870s
and 1880s worked at his father's side, under his father's care, learning his
father's skill. The treatment he received depended on his father. A boy
could take pride in the fact that he was learning a hard-won skill and
contributing to the family income. In *A Miner's Life*, John Brophy

recalled his feeling of satisfaction when (around 1890) he was permitted to go underground:

> . . . on the eve of my twelfth birthday, I entered upon a man's job as a coal miner. I was childishly excited by the prospect, and pleased that I could do something to help the family through hard times. The boss, a good-hearted man, gave me a new mine cap and an oil lamp. I got a dinner bucket with a short enough handle so it wouldn't trail on the ground, a small shovel . . . a small pick, my first long pants, and I was equipped to go to work. I was even small for my age—I couldn't have weighed over seventy-five pounds—so for a while I was the subject of good-natured teasing by the other miners.[90]

Brophy's father took good care of his son, going to work early and working late to make up for the boy's limited strength and skill. With the father doing far more than half the work, the two could load their allotted number of coal cars and thus earn a subsistence income for the family. As an adult, the son reflected, "After a long ten hour day in the mine (eleven hours, including time out for lunch), I would go home a thoroughly tired child, with the task of washing up still facing me. It was a rigorous routine for a youngster, but I had the comfort of my father's unfailing kindness and of pride that I was learning to do a man's work."[91]

But he was not a man. Despite his continuing dependence on his mother, his apprenticeship required him to travel with his father from coal town to coal town searching for employment. During these times, they would leave the mother behind with the younger children. One day as they worked, the father mildly corrected his son for some minor offense. The boy sat down and began sobbing into his hands. The trouble was, it came out, that he "wanted his Mommy." Brophy was fortunate in his father. The two began the journey home the next day.[92]

Brophy's experiences in the bituminous industry contrasted sharply with those of children working in the anthracite industry. For it was in the hard coalfields, in the coal processing plants above ground, that the coal industry's first real factories came into existence. Unlike bituminous coal, anthracite was subjected to extensive processing before it was sent to the cellar bins of New York and Philadelphia. It was broken up, washed, separated into different uniform sizes, and cleaned of impurities, especially slate. This was done in the breaker, a high, windowless building at the mouth of the pit.

Anthracite breaker, 1905. Reproduction from the collections of the Library of Congress.

Breaker boys picking slate, 1911. Photo by Lewis Hine, courtesy National Archives.

The coal was poured into a chute at the top, and traveled down a conveyer. It was broken between rollers or by men wielding hammers, and then separated into different sizes by passing over screens—metal sheets with different sized holes cut out. During screening, the coal was washed with water. It then passed by the breaker boys, who picked the slate out of the coal by hand.[93]

That children should principally be engaged in going to school, or in play, is a twentieth-century view. Yet the employment of children in the anthracite breakers differed, both quantitatively and qualitatively, from their employment underground. In the British context, Sidney Pollard has written that the first factory workers were the most vulnerable members of society. They were paupers and, especially, children. Children made up three-fourths of all workers in the first British silk factories and half of all workers in the new cotton mills. In Pennsylvania of the 1870s, the burgeoning of child labor in the hard coalfields itself signaled the coming of the factory system. During that decade, the number of children working underground (in bituminous and anthracite) remained static, while the number of children employed in the anthracite breakers increased by seventy-eight percent. By 1880, twelve thousand boys worked as slate pickers in the anthracite counties of northeastern Pennsylvania.[94]

The little slate pickers did not work with their parents. Instead, they were crowded together with other children and a few men too old or crippled to work underground. Unlike any other workers in the coal industry, they worked under continuous surveillance (by the breaker boss). In 1877, a labor reporter observed a group of breaker boys working from seven in the morning until dark:

In a little room in this big black shed . . . forty boys are picking their lives away. The floor of the room is an inclined plane, and a stream of coal pours constantly in from some unseen place above, crosses the room, and pours out again into some unseen place below. Rough board seats stretch across the room, five or six rows of them, very low and very dirty, and on these the boys sit and separate the slate from the coal as it runs down an inclined plane. They work here, in this little black hole, all day and every day, trying to keep cool in the summer, trying to keep warm in the winter, picking away among the black coals, bending over until their little spines are curved, never saying a word all the live long day . . . the coal makes such a racket that they cannot hear anything a foot from their ears.

The children wore old coats, old shawls, old scarves, and ragged mittens to keep their hands from freezing. They looked like so many black dwarfs. "[N]ot three boys in this roomful could read and write. . . . They have no games. . . . [T]hey know nothing except the difference between coal and slate."[95] They were, in short, the most vulnerable, coerced members of the country's new industrial working class.

EXTRACTING PROFITS: MINING THE MINER

Underground, craft methods and attitudes persisted. For one thing, geography delayed steps toward factory methods of production. Above ground, the factory system placed work stations closer together to enable greater supervision and the more efficient passing of supplies from one work station to another. In the room-and-pillar method of mining, it was impossible to alter the physical space so as to move the individual work stations closer together. Quite the opposite. The older the mine, the farther apart the rooms or chambers became. They might be separated by a mile or more. Nevertheless, in many cases the physical barriers to more factorylike methods of production could be overcome by the longwall method of mining, a method largely ignored by American coal operators.[96]

Traditional mining procedures provided advantages to operators that were as critical as geography in preventing the introduction of a factory-like system of mining. Most importantly, the contract miner, not the operator, paid the price for inefficiency, whether or not he had caused it. For example, the time miners spent just waiting for items supplied by management—cars in which to load coal, timbers for propping the roof, and mining supplies—could be considerable, but this fact did not perturb the operator, who paid the miner by the ton, not by the hour. Operators who wanted to increase output usually increased the number of miners working rather than worrying about increasing individual productivity.

The manager exerted little control over underground procedures, but in recompense he shouldered little responsibility for his work force. Just as miners felt no obligation to work except when convenient, so operators felt none to pay except when convenient. Often miners would not see a paycheck from one month to the next.[97] Moreover, no one expected the operators to provide full-time work. The Pennsylvania miner of the 1870s worked an average of nine months out of the year. "Work around here is almost next to begging," a miner complained in 1884. Another confessed that he did not know what the miners were coming to, "but I am afraid there will be hundreds of families short of their daily bread at a very early date."[98] Year after year, poverty cast its dark shadow over the coalfields; its major cause was lack of full-time work.

Despite the operator's general lack of influence over his underground operation, he exerted overwhelming influence over one significant item: the amount he paid the miners for coal. Even when miners found themselves able to significantly influence the tonnage rate (usually only when both the economy and the union were strong), the operator had an impressive array of tools designed to undermine the agreed-upon rate.

The principal tool was the altered scale, operated by the weigh boss. No student of the history of coal mining can fail to be impressed with the widespread, extensive, perhaps at times nearly universal, use of cheating in weighing by employers to reduce the actual tonnage rate paid. A miner testified in 1873 that a fruitful source of irritation and strife at M'Kinney's Bank (at Saw Mill Run, Pennsylvania) was that miners thought "they were not getting just weight."[99] His complaint was one of a large production of the same which fill the annals of coal mining literature from 1860 through the 1920s. Short-weighing gave rise to dissatisfaction, strife, and strikes, and to the repeated demand for a check-weighman—a miner hired by other miners rather than by the employer to constantly check the weigh boss's weighing.

Operators frequently docked miners for sending up coal with too many impurities. Of course, a superintendent could legitimately complain about some carloads. Disputes over this erupted often. An anthracite miner grumbled, "[I]f there is any bone [carbonaceous shale—that is, coal with an insufficient amount of carbon], dirt, or slate found in a car, the miner loses the whole car." A West Virginia miner couldn't have agreed more. "The company has a thing on the cars they call a man and he does the docking," he wrote, ". . . a very genteel way of stealing is the opinion of your humble coal digger."[100]

Where the miner was paid by the carload, the operator could change the shape of a car to make it hold more coal. "The cars were formerly square in shape," said a miner, "which contained a surface on top of thirty square feet, but the shape of the cars have been changed to a taper from the bottom upwards, with a surface on top of forty square feet." Another miner pointedly observed that "the company gets all coal that is knocked off or falls off cars."[101]

Screens provided operators with another cost-cutting tool. Before the coal was weighed, it was passed over a metal sheet with holes cut out, or metal bars with spaces between the bars. The operator considered any that fell through too small to sell and did not pay for it. Miners often noticed that the holes (or the spaces between the bars) had grown larger.[102] One operator even contrived an "improved" version designed to further break up the coal as it was thrown down on the screen.[103] "The screen system," complained one miner, "robs me of nearly half the fruits of my labor."[104]

That small coal was supposed to be unsalable did not prevent the operators from selling it. In the early 1870s, miners around Pittsburgh accused operators of selling nut coal (the coal that fell through the screen) for two cents a bushel less than lump coal (the larger pieces). The miners themselves bought nut coal for five cents a bushel in order to heat their houses, although they had been paid nothing for mining it. One miner asserted that twenty to twenty-five percent of all coal mined was nut coal.[105]

The coal operator's control of the physical plant—the cars, the screens, and the scale—gave him significant control over the cost of labor as against the contract miner's lack of control over the real tonnage rate. Efficient production was not the operator's concern, because he was not paying for wasted time or for wasted coal.

Even more important, efficient production would have resulted in fewer miners bringing up the same amount of coal. This would have deprived many operators of an important source of profit. Coal owners did a lucrative business selling blasting powder and other mining supplies to miners, and renting houses and selling groceries to the mining community. An efficiently run mine offered the operator fewer mouths to feed.

The company coal town originated in the isolation of the coal camps. Of necessity, operators built towns for the workers in an industry conducted far from urban centers. But the company town became a significant business activity that many operators were loathe to dispense with even where alternatives existed. In 1885, an operator not in the grocery business lamented to a committee of the Ohio legislature: "As an operator I will say that at times I find it virtually impossible to run my mines when there is close competition [with] . . . parties who have stores. They sell their coal at a price which it costs to produce it and take their profit out of the store, which is generally a large one."[106]

Forcing the community to buy in the company store guaranteed the coal operator his profit on groceries. "As the company has gone to the expense and trouble of establishing a store, butcher shop and saloon for the accommodation and convenience of its employes," went Rule Number 3 of the Northern Pacific Coal Company (Timberline, Montana), "and as its employes derive their living from the company, all employes will be expected to patronize these places to the exclusion of all other similar establishments or peddlers."[107]

The store was no minor source of income. In 1891, the Union Pacific Coal Company purchased the (formerly independent) store at Hanna, Wyoming, for some 17,600 dollars. Within three months, gross sales had mounted to nearly 20,000 dollars. The profit was twenty percent. The company expected to pay for the entire investment in one year. Around 1900, the company's coal mine superintendent at Rock Springs, Wyoming, urged his superior to increase the salary of the manager of the company store: "A man that can and does make his store pay a clean profit of from eighteen to twenty thousand per year is worth more than $125 per month."[108]

The company town, and its store, enabled the operator to do quite well while ignoring the problem of increasing the productivity of the work force. It was to the operator's advantage to have a surplus of miners. "The thick coal mines were generally overcrowded with miners . . . who found ready employment," wrote Andrew Roy of the ten-foot seams of Ohio's Hocking Valley, "as most of the operators owned stores and gave employment to a surplus of miners for the sake of the store trade, which was very profitable." In 1886, the miners at an Ohio protest meeting declared that a strong and bitter feeling prevailed against the operators. These gentlemen, Roy explained, "charge exorbitant prices for their goods in the stores, and . . . keep the mines overcrowded with men, so that every dollar earned may be taken out at the stores in trade. No doubt the stores were where they intended to make their profits in the first place, and no wonder that the miners were opposed to them."[109]

Housing, too, was good business. At Rock Springs, Wyoming, the rent from 308 company houses (in 1898) brought a gross income of some twenty thousand dollars. Upkeep and expenses came to five thousand dollars, leaving more than fifteen thousand dollars in profit. Requesting permission in the 1890s to build more family-sized houses, the superintendent told his superior that the town had enough houses for single miners. But, he argued, the family house was a good investment, and "gives us a better class of men and more mouths to feed, so the company is a gainer all around."[110]

The manager lacked control over the work process, but he exerted it

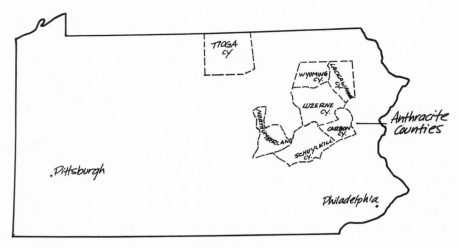

Anthracite coal fields of Pa. and Tioga Cy.

over the mining community. He used coercion both to profit directly from the town and to keep the union out; invariably, coal miners' unions threatened the operator's ability to profit in any way except through the efficient production of coal. Yet many forces, from the weight of tradition in one of the world's oldest industries to the profitable side effects of inefficiency, worked against the solution of efficient production. Given the operator's commitment to the old system, keeping the union out became imperative. Unions opposed forced buying in the company store, cheating in weighing, and other practices that for the operator constituted fundamental techniques for reducing costs.

Superintendents ruled company towns absolutely. During the course of investigating a strike in 1873, a Pennsylvania official graphically described three company towns located in Tioga County. Of one town, Fallbrook, a nearby resident wrote, "The superintendent makes all the laws and those who are in his employ must abide by them."[111] The superintendent himself confirmed the accuracy of this accusation, responding to inquiries with, "[M]aking laws . . . by the manager . . . always has been the custom in all this coal region. Were we not rigid in enforcing order, living in such a community would be out of the question."[112]

An investigator traveled to Morris Run, a pretty place with gardens, where, however, he found nothing, not so much as a pigpen, that didn't belong to the company: "There is not a road or path that is a highway for the public; the public roads of all the adjoining townships terminating when they strike the borders of this, so that if any person is found on its roads or streets, who are distasteful to the manager, he can compel him immediately to leave, or arrest him as a trespasser."[113]

A short distance away the Fall Brook Coal Company employed one thousand miners, mostly Swedes and Poles, with a few Irish and Welsh.[114] The superintendent at Fallbrook rented the houses and forced every miner he hired to sign an agreement permitting the company to evict him ten days after giving notice. (This contract provided a loophole to a state law to the contrary.) The power of eviction was no small matter in a town where the company owned all the houses. To miners' complaints about the matter, the superintendent replied, "If you are not satisfied you can go down the road."[115]

Superintendents may or may not have been expert in extracting coal, but they were expert in extracting money from the miners. Fallbrook miners complained that the company not only short-weighed the coal but also collected taxes and imposed arbitrary and unjust fines. A miner wrote to *The Workingman's Advocate*, "[I]f a man gets drunk and should happen to quarrel with any person, there is a heavy fine imposed, and if he refuse to pay he must leave." The miners claimed they were paid in scrip—money printed by the company (which was unlawful). The superintendent denied this, but the miners assembled in the investigator's hotel room turned out their pockets to demonstrate their lack of anything to show for their labors except "bogus money."[116]

The company's refusal to pay wages in lawful money forced the community to buy at the company store. "They have a company store here where [miners] can receive the necessities of life on presenting the so-called money, of course paying two prices for everything," grumbled a Fallbrook miner. "That is the object of issuing this money, to get it back into their own pockets."[117] Indeed, a survey done by the Pennsylvania Bureau of Statistics in the late 1870s showed that prices at coal company stores were consistently higher than those at nearby independent stores. Butter, for instance, sold for thirty-five cents at the company stores while it sold for thirty cents elsewhere.[118]

At Morris Run, the most frequent trespasser was not the union organizer (as might be expected) but the local farmer hoping to sell his produce. On one occasion, the superintendent confiscated a farmer's wagon loaded with butter, pork, and eggs, and drove it from the town. He threatened people buying from the farmer with discharge.[119] On another occasion, a mining boss informed a woman that she must return the pork she had just purchased to the peddler's wagon unless she wanted her husband discharged. Later, the boss informed the miner in question that "he did not want any of those damn peddlers to come here."[120] He ordered another farmer out for selling cheese at fourteen cents a pound, the identical item offered in the company store for seventeen cents. In the course of expelling yet another farmer from the town, he proclaimed, "You think this is a free country, but it ain't."[121]

In Arnot, the same lack of freedom prevailed. The superintendent informed one peddler that if he ever saw him again he would throw his pack in the river. He told another, this one furnishing "Yankee notions, dry goods, and jewelry," that if he returned he would take his pack and burn it.[122]

The company store illustrates one way in which the large corporation presented a common threat to coal miners, to nearby farming families, and to local merchants in adjacent independent towns. For small-scale producers and retailers of food and merchandise in the coalfields, miners and their families were important customers. Independent merchants often supported striking miners by extending credit for groceries, sometimes for months at a time. During the 1873 strike in Tioga County, the Business Men's League of Blossburg (a nearby independent town) joined miners' strike committees to report to the U.S. Treasury Department that the coal companies were illegally paying the miners in scrip. Nearby farmers and local businessmen supported the strike, an important cause of its ultimate victory.[123]

Underground, the skilled miner could mine coal how he pleased, when he pleased, and in the amount he pleased. But the coal operator expropriated the product to such a degree that frequently miners were paid less than nothing (expressed on payday by a slip of paper signifying the sum the miner owed the company). The operator could do this because the miner's legendary independence vanished when he emerged into daylight.

In the land of the free, the company town was isolated, remote, and anything but free. The working people of the community knew one industry; they served one employer; and they were ruled by one firm, usually in the person of one man—the superintendent. But their isolation, as one student has written, was a collective experience.[124] The company town gave its inhabitants something in common. The people were separated by sex, race, national origin, underground occupations and hierarchies, language, and religion. Coal dust covered them all. Company policies affected them all. Above ground, all were equal—in the eyes of the company.

4

THE HAVEN OF JUSTICE
AND RIGHT IS OUR DESTINATION:
COAL MINERS AND THE CIVIL WAR

This is a land where the government is based on the popular will.

Daniel Weaver, miners' union leader, 1861 [1]

Labor and capital should go hand in hand . . . for without labor, what would capital be worth to the capitalist?

Coal miner, 1863 [2]

The year Americans elected Abraham Lincoln president of the United States, the North's arsenal included ten of the fifteen coal producing states and most of the country's 36,500 coal miners. In 1860, southern coal states like Alabama and Tennessee could provide the incipient Confederacy with fewer than 2,500 mine workers. [3] Lincoln was elected on a platform of high tariffs (to protect manufacturing), opposition to the spread of slavery, federal financing of internal improvements, and homesteads in the West. His government supported the emerging manufacturing system that employed wage labor. It opposed a southern planting aristocracy that employed four million African slaves. Immediately after Lincoln's election, six states seceded from the union. Soon after he took office, on 4 March 1861, the country fell into the grip of war.

The war required coal, coal, and more coal. On 19 April Lincoln proclaimed a blockade of southern ports: 3,500 miles of Confederate coastline. For this grandiose scheme, the Union Navy could provide a few rickety wooden sailing ships; five steam frigates, all out of commission; and various other outdated, broken down, and totally inadequate boats. The Civil War ejected naval warfare into the steam age. Rapidly, the Navy converted its dilapidated collection of leaky vessels into a burgeoning fleet of gunboats, powered for the most part by coal-fired steam engines. On land, the Union Army had to be transported back and forth from one front to another. Railroads thrived. Munitions—guns, cannon, ammunition—had to be manufactured, only to be swallowed up in a war that required their immediate replacement. The iron industry flourished as did the anthracite coal industry, provenance of fuel for the blast furnace.[4] In 1864, the U.S. Census Bureau reported that the requirements of war "have vastly increased the production of coal in the loyal States, and rendered the mining interest unusually prosperous."[5]

Demand for coal was intense: Its price rose. In most sectors, inflation counteracted the stimulating effects of the war, but in mining, as David Montgomery observes, the value of output increased more during the Civil War than at any other time between 1839 and 1899. At the same time, the battlefields drained the coal industry of labor. As mine workers marched off to fight for the Union cause, a severe labor shortage developed. The war drove the economy, turning the coalfields into a union miner's dream.[6]

Such was the economic context of America's first national union of coal miners. But there was a historical context as well. Coal miners could look back to a myriad of local strikes beginning as early as the 1820s, and to a shifting scenario of local, state, and regional miners' committees and organizations.

THE AMERICAN MINERS' ASSOCIATION

The American Miners' Association emerged from a spontaneous strike that began in Abraham Lincoln's home state a few weeks after he was elected president.[7] In the autumn of 1860, according to Edward Wieck, Illinois coal miners refused a reduction in the tonnage rate and protested their employers' habit of cheating in the weighing and measuring of coal. Shortly after the miners walked out, they established a formal organization. Daniel Weaver, an English-born miner, and Thomas Lloyd, a Welsh miner fresh off the boat, assumed leadership of the new union.[8]

Immediately, the organization petitioned the Illinois Legislature for "protection against the rapacity of our employers" in regard to weighing.

The legislature enacted a Weights and Measures Act, and the operators rescinded the reduction. Miners returned to work victorious. (As they couldn't know at the time, the Act would never be enforced.)[9]

Success lifted hopes and made for more success. "We have launched our vessel, on the ocean of probation," Daniel Weaver declared in 1861, "with 500 passengers on board. . . . [T]he haven of justice and right is our destination."[10] By 1863, miners from Illinois, Ohio, and western Pennsylvania had crowded onto the ship of the American Miners' Association.

Did the new union reflect the emergence of class consciousness among coal mine workers? Were underground workers beginning to see their interests in common, and as opposed to the interests of their employers? Certainly, the Association represented a step in that direction. But its members were more craft conscious than class conscious. As skilled miners, they looked askance at laborers; it is certain that some branches barred laborers from membership. Our knowledge of the association's policy on this is ambiguous, but in 1863 the convention plainly resolved that "no more laborers be admitted as members of this Association, as we have found them to be an injury to us." On the other hand, at least one branch, in Tioga County, Pennsylvania, *did* include laborers.[11]

Despite the presence of laborers in some branches, the organization's program served mainly skilled contract miners. The two most frequent demands—an increase in the tonnage rate and the elimination of fraud in weighing and measuring—did not concern laborers who were paid not by operators according to the tonnage rate but by contract miners according to whim.

Still, the labor shortage created by the war undoubtedly benefited the laborer by reducing the time a man had to load coal before receiving a room of his own as a miner. In July 1865, David Morgan emigrated from Wales to Minersville, Ohio, where his eldest brother took him on as a laborer; within two months, he had received a room of his own.[12] Later in the decade, a Welsh miner from Scranton censured the hard work and low pay that was the lot of the laborer. "[I]t has become the custom for a man to labor first of all wherever he comes from," he explained. "Most labor for six to nine months before they get a place of their own."[13] This may have been difficult, but it was no lifetime. These experiences indicate that laboring in the 1860s (at least for the Welsh) was but a short sprint to mining.

The association favored the shorter workday, but its method of reducing hours—the voluntary restriction of production by miners—did nothing to reduce the hours of laborers. (The miner could bring down far more than a day's work of loading in far less than a day's work of mining.)[14] On the other hand, the miner's shorter workday did improve matters for his wife. "We used to work twelve to fifteen hours a day in this district," an Ohio couple informed their Welsh parents, "but this spring a resolution was passed to work only ten hours a day. . . . We do not start before seven in the morning." This meant the woman could sleep longer before rising to cook breakfast and prepare her husband's dinner bucket. "The women like this new system very much. They do not get up now before six o'clock where formerly they had to be down at four or five."[15]

Within the organization, cooperation among skilled craftsmen required the melioration of ethnic conflict. The association worked diligently to mend these rents in the fabric of union solidarity, in this way foreshadowing all American coal miners' unions to come. The intensely mixed ethnic coloration of the work force made this the only winning strategy. In 1861, Daniel Weaver lectured the miners on this point. "Does it not behoove us as miners," he wrote, "to use every means to elevate our position in society . . . by obliterating all personal animosities and frivolous nationalities . . . ?" Later in this *Address to the Miners,* he repeated, "[L]et there be no English, no Irish, Germans, Scotch, or Welsh. This is our country and 'All men are brethren—how the watch-words run! / And when men act as such is justice won.' "[16] From Ohio, where miners had formerly conducted their meetings in Welsh, a miner informed his parents in Wales that the association conducted its meetings in the English language.[17]

Initially, the Germans held back, but in 1861 Weaver reported, "the Germans now, are joining us by the thirties and forties. Confidence in them has taken root; some of the bosses have made it a point to discharge their old hands under various pretexts to make room for Germans who were not in the Union; but the incomers are as bad as the outgoers. They begin to see that our aims are right, our objects pure. We *think* as well as *work.* Our cause is theirs, our interests are identical."[18]

The association worked to overcome national differences in order to fight for shorter hours and higher wages (in the form of restricted production and a higher tonnage rate). But these concrete, material concerns formed strands in a larger web of issues. Coal miners strove for a more humane society in which human relations would be characterized by freedom and equality. For many of these foreign-born miners, America symbolized these very values. They, like the working people David Montgomery has described, were committed to "the world's only political democracy."[19] "We have a weekly paper of our own," an association member reported to a friend in Wales, "which . . . belongs to no one but us. We own the press and everything belonging to it and we can convey what we are thinking to every part of the country. There's a free country for you!" "It is wonderful to live in a free country where the rights of men are upheld," he continued, "one weighing as much as the next in the scales, with no difference between rich and poor and if you happen to meet the two on the street it would be difficult to say which was the gentleman."[20]

This miner relished his newfound sense of social equality. He basked in the notion that the working man had a right to be regarded with respect. The American Miners' Association went even further than this, proposing equal rights between labor and capital. Equality, an ideal quite different from its modern equivalent, equal opportunity, infused the thinking of the miners. It was a point of view reflecting the possibilities in a world in which individual firms were still quite small and personal contact set the tone of relations between employer and employees. The ideal of equal rights probably also reflected the relative independence of the contract miner, his view of himself as a contractor or craftsman rather than a worker.

In 1861, Daniel Weaver went so far as to say, "We must keep in constant recognition the rights of bosses as well as our own. Capital, as well as labor, must have its due."[21] Two years later, a member aired his opinion that the signers of the Declaration of Independence never intended the country to be governed by capitalists. "Labor and capital should go hand in hand, and not be as it is . . . with the coal operators and miners of Illinois." He went further, suggesting that labor should rule *over* capital, through the laws of the land, "for without labor, what would capital be worth to the capitalist? Without labor where would all our inventions be? Without labor where would all our mineral wealth be? . . . Without labor we would be without the power of steam . . . for without labor this great invention would not have been developed, for who can say that Watt, the inventor of the steam engine, was a capitalist? I think history will prove he was a working man."[22]

Two important influences colored the worldview of these men: their economic experience in a time of small-scale enterprise, and their recent British political experience. In Britain, the Chartist movement of the 1840s had struggled for universal manhood suffrage. Although the movement was destroyed, its ideas continued to flourish among British miners, including emigrants to the United States. Here, these miners could vote as soon as they became citizens. For them, the vote was a symbol, as E. P. Thompson has noted in the British context, "whose importance it is difficult for us to appreciate, our eyes dimmed by more than a century of the smog of two-party . . . politics."[23]

The right to vote inspired the miner with confidence in America. In 1863, an association member proclaimed, "[I]n our glorious country of adoption and armed with a powerful weapon—the Ballot Box—labor is, and ever will be, the ruling power of a nation like the one we live in."[24]

The right to vote elevated the miner's self-esteem. "Where is the man who does not feel the . . . ennobling influence of being recognized as a full, and a free man?" asked Daniel Weaver in 1861. "To a man who has escaped from the Old World's tyrannies and inequalities, where he was taxed heavily enough but not represented; such is a blessing incalculable. . . . This is a land where the government is based on the popular will."[25]

As for New World tyrannies and inequalities—in particular, slavery— Welsh, English, and Scottish miners objected. They supported Abraham Lincoln, volunteering for the Union Army in large numbers. Near Mount Savage, Maryland, a slave state, miners chalked the name Lincoln ("only a little ill-spelled") on the sides of their coal cars. Katherine Harvey reports that coal miners from Frostburg, Maryland, formed themselves into a company and marched off to war. A miner acquainted with them reflected that it "took nerve to be a Republican in a slave state." These miners supported the Republicans in part because they believed in the protective tariff.[26] But they also opposed slavery. One Welsh miner's racist views reveal that among his countrymen he was an exception. "Almost all the Welshmen are on the radical [Republican] ticket," he wrote from a Missouri coal camp in 1868, "and I would be so myself except for the niggar equality which does not agree with my views that the Creator never intended them to be equal to a white man. . . ."[27]

The American Miners' Association thrived in the context of a scarcity of labor but disintegrated under conditions of a surplus. During the war, operators understood their dependence on the skill possessed by miners. Unless they wanted to renounce this exceptional opportunity to sell coal at a high price, they had no choice but to grant concessions. To do otherwise would have been to invite strikes—and a lull in production.

In March 1864, the first organized group of coal operators, called the Pittsburgh Coal Exchange, conceived the idea of creating a surplus by importing miners. In a pioneering employer strategy, the exchange formed an immigrant company that proceeded to supply European miners to midwestern coal firms. Significantly, these were *skilled* miners, mostly from Belgium. The operators could not have mined coal without the skill monopolized by the miners. The new anti-union strategy worked; the following year, these operators forced the miners to accept a wage cut.[28]

But the end of the war did a better job for the operators than any conscious policy could have. In 1866, the American Miners' Association went into a steep decline. With the war's end came the end of government demand for coal and, with the return of mine workers from the "fields of blood," a huge labor surplus. In May 1865, an Ohio miner wrote to his people in Wales, "The miners of Ohio and Illinois and parts of Pennsylvania have not worked a month in the last six months."[29] Another miner reported from Pittston, Pennsylvania, that "there are dozens if not hundreds who have not had a day's work in five or six months."[30]

A devastating reverse came at the hand of the Fall Brook Coal Company, in Tioga County, Pennsylvania, where the American Miners' Association had organized a strong and successful branch. In August 1864, the company president, John Magee, asked his manager whether the company would be better off fighting the union, or "shall we submit to such further insult as the Devil chooses to impose on us?" His question was rhetorical. In early 1865, the Fall Brook Coal Company locked out its entire work force. The sheriff, aided by a company of Union Army troops just returned from the battlefield, evicted more than four thousand men, women, and children from their company-owned houses. During the proceedings, the men vacated the town, while the women and children jeered the soldiers as they moved from house to house removing furniture. One Mrs. Arrowsmith threw a handful of flour into the sheriff's face, whereupon he narrowly avoided death by choking. Ultimately, the union work force was defeated, but it is interesting to note that the effort at resistance included both the women of the community and the laborers, perhaps a reflection of the homogenizing influence of the company town.[31]

For management an important issue in the lockout was the question of who would determine production levels. After defeating the union, the company posted new work rules. One stated: "Any attempt by persons or organizations of persons to interfere with the progress of work, by coercing their fellows to limit the time or amount of work . . . will be rebuked and put down, no matter what the cost. Every person will be protected in his natural right to work as many hours per day as he chooses, and to earn as much as he can."[32]

During the recession following the Civil War, reverses also crippled the association farther west. Employer opposition combined with factional strife to wear the union down. One casualty of factionalism was the organization's newspaper, *The Weekly Miner*, an important unifying force for miners scattered throughout the bituminous coalfields. By 1867, only a few isolated locals existed to remind coal miners of their first national union.[33]

THE CASE OF IRISH MINE WORKERS

The American Miners' Association never penetrated the anthracite region, the one place inundated by workers new to the industry. Like a vacuum, the labor shortage drew unskilled Irish into the hard coal mines. From 1860 to 1870, the number of anthracite mine workers jumped from 25,000 to 53,000. Most of the newcomers were Irish. By 1870, they constituted thirty-two percent of the region's major ethnic groupings, nearly equaling the combined English and Welsh populations.[34] Unionism among these various groups took the form of local committees and local strikes, with one union near Scranton claiming nine hundred members in 1864.[35]

Ethnic and religious conflicts abounded both within and outside the union movement: Brawls between Welsh and Irish mine workers dated from the 1820s.[36] Such conflicts worked against class solidarity. Yet clashes based on opposing class interests also occurred. A complex mix of class and ethnicity had its effect on the relations among miners and laborers.

If the typical Welsh or English miner of the 1860s opposed slavery, favored the high tariff in company with his employer (generally a small-mine owner), and viewed the United States favorably in light of his ideals of freedom and equality, what of the Irish who flocked into the anthracite region? The typical Irish laborer, according to the conventional view, voted Democrat, despised blacks, and resisted the Union draft with riot and bloodshed. As one contemporary (though prejudiced) observer said of the Irish, "[T]he mining districts, where not one man in twenty can read and write, are the Democratic districts. . . . [T]hese men . . . were instructed by their Democratic leaders . . . not to go to the war, but to stay at home, go to the election and vote 'and put down this infernal Abolition Government. . . .' "[37]

In 1861, Lincoln's Republican Party lost heavily in the coal towns of Schuylkill County, Pennsylvania. Democratic newspapers "fanned the flames of bigotry," as David Montgomery notes, harping against emancipation of black slaves, picturing hordes of blacks coming north to compete for jobs. White racism played a strong role in the politics of the region.[38] But before concluding that Irish miners and laborers were racists to a man, we must consider other forces at work as well.

The Democrats did not represent working people any more than the Republicans did. As William Gudelunas, Jr., and William Shade have shown, affluent business and professional men dominated both parties. True, electoral politics revealed ethnic loyalties. The Irish tended to vote Democrat, the Welsh and English, Republican. But because both parties represented capitalists, merchants, professionals, and workers, electoral politics revealed nothing about class loyalties. Further, the parties were not precisely split along ethnic lines. Even Gudelunas and Shade, who use voting patterns to argue that ethnic conflict took precedence over class conflict in the anthracite region, show that forty-one percent of the Democratic leadership was not Irish, but Welsh or English.[39]

The outlook of Irish mine workers had to be influenced by Benjamin Bannan, the region's leading Radical Republican voice. The Welsh-American Bannan published an early and important industry newspaper, *Miners' Journal.* Not incidentally, he was also a coal operator, a supporter of the small-scale enterprise. The ink Bannan spent in lauding the high tariff was equaled only by the quantity poured out in loathing for Irish Catholics. As he advocated temperance, he noted the numbers of Irish saloons, the swarms of Irish drunks, and the proliferation of Irish charity cases, which emanated, he felt sure, from the Irish saloon. Adding insult to injury, Bannan advocated Sabbatarianism, a pointedly anti-Catholic movement favoring certain laws for keeping the Sabbath. What is more, the Republicans, with Bannan sounding the battle cry, condemned strikes and labor organizations as detrimental to the industrial system. In Bannan's mind and in his influential newspaper, a strike was no different from treason to the Union cause. It is hard to imagine an Irish miner or laborer warming up to this man, and few if any did. In contrast, the Democrats in the anthracite region preached tolerance for religious diversity, saving their insults for blacks. The *Democratic Standard*, popular among the Irish, oriented its editorials to working-class interests.[40]

IRISH MINE WORKERS AND THE DRAFT

The mix of ethnicity, race, and class in the anthracite region contained another important dynamic, this one introduced by the Civil War: an enormous increase in the power of the federal government. Miners and laborers of all backgrounds felt this new power in the form of the draft.[41]

The Conscription Act of 1863, the first military draft in United States history, was a class-conscious act. It excused the rich. A man could avoid service in one of two ways: He could provide (read pay) a substitute, or he could buy his way out for three hundred dollars. For a mine laborer, this amounted to the better part of a year's wages. In contrast, the best paid lawyer in Schuylkill County earned eight thousand dollars per year. This particular man paid the requisite three hundred dollars to avoid the battlefield.[42]

Conscription went along with suspension of the Writ of Habeas Corpus (the right of protection from arbitrary arrest without due process), and with an openly pro-employer attitude on the part of the Republican Party. In the anthracite region, the Provost Marshal charged with carrying out conscription was Charlemagne Tower, a Pottsville attorney with heavy investments in coal lands who was also a coal operator.[43]

The war provided Provost Marshal Tower with extensive and arbitrary power. He could draft any man he pleased, and arrest any who resisted or who did not appear when called. Once in Tower's net, a man had no recourse in the civil courts. In theory, aliens who had declared no intention to become citizens were exempt, but in practice Tower drafted aliens freely, claiming that affidavits from aliens could not be relied upon. (Large numbers of mine workers were of course, aliens.) Protests from various consulates that "aliens are hurried into the U. S. Army" did not faze him. He was also supposed to grant exemptions to the sole supporting sons of aged and infirm parents. Tower granted some such exemptions but denied others. To one Irish mine worker, he denied an exemption because the man's frail *mother* was not a citizen, an arbitrary interpretation of the rules. In another ruling, this one particularly heartless, he denied an exemption to the father of a motherless child.[44]

Understandably then, many miners and laborers in the predominantly Irish mining towns of Schuylkill County opposed the draft. In 1862 (when the states rather than the national government were attempting to carry out conscription), a group of anthracite mine workers stopped a train carrying unwilling recruits bound to Harrisburg, and sent them home.[45] The following year, the general in charge of conscription in Schuylkill County warned Washington that coal miners near Pottsville had organized to resist the draft, were armed, and conducted drills each evening. They numbered two or three thousand.[46] After Tower assumed his duties, he reported that draft officers were driven out of towns "by women throwing hot water and dirty water and other missiles at them," and by men and boys wielding "clubs, stones, and weapons."[47]

A military contingent enrolling men in the Irish-populated Cass Township found a crowd to be "cross and very saucy, even in the presence of the military." More seriously, in November 1863 a coal operator who had provided a draft officer with a list of his employees was murdered.[48]

Obviously, there was some truth to the observation that the Irish opposed the draft. Yet draft resistance erupted all over the United States, not just in the Irish coal towns of northeastern Pennsylvania. Enthusiasm for actually going into battle dwindled even among supporters of the North, as the war gradually became a horrendous bloodbath, with more than a million casualties (half of them killed) out of a total American population of thirty-five million. Even many Lincoln supporters regarded the conscription law as an affront to principles of freedom and equality. A newspaper in Easton, Pennsylvania, not a coal town, compared the popularity of the draft with that of a smallpox epidemic. In agricultural Bucks County, a tailor threatened the Provost Marshal with a club, and the employees at a tannery stoned the draft officer after a large dog had prevented him from entering. The New York City draft riot which took place in July 1863 and left some twelve hundred dead, is well known.[49]

Although draft resistance was notorious in the anthracite coal towns, even in these "receptacles of the worst classes of mining and laboring men," as Tower called them, there is a flip side to the standard historical account. The truth was that the Irish fought on both sides of the war, that whereas some Irishmen supported the Confederacy out of racism or for other reasons, others supported the Union. In the early months of the war, as Grace Palladino shows, more Irish from Schuylkill County volunteered to serve in the Union Army than did any other ethnic group. After the widely unpopular draft was established, certain coal towns populated mainly with Irish mine workers did no worse than the admittedly abysmal records of certain Republican strongholds.[50] "Many of us went into the war," reflected *The Irish World*, a newspaper popular among anthracite mine workers, "[because] the Union was for us the symbol of strength of [the] liberties of a free people."[51] Even Provost Marshal Tower exclaimed in the midst of his cavalry raids into coal towns, "I am myself astonished at the success with which I am carrying through the enrollment."[52] Although Tower attributed his success to the presence of military force, it is quite possible that many Irish mine workers opposed not the Union Army but the unfair draft law, the suspension of Habeas Corpus, and the arrogance of power represented by Charlemagne Tower.

The Provost Marshal's power extended far beyond the draft. Tower used the conscription law to fight the union activism of the Irish mine workers, activism having nothing to do with the war. Of Cass Township, Tower informed his superior that the military presence not only subdued the rebellious, and "prevented their making any hostile demonstrations, but even moved them to do more work and more quietly than before." Reporting on another occasion that "[t]he mines at several collieries in Schuylkill County have . . . stopped work," Tower urgently requested more troops and "three or four small howitzers" immediately. This was a strike, not draft resistance, but both amounted equally to treason in the mind of Charlemagne Tower.[53]

Captain Tower's habit of equating union activism with draft resistance had two effects. First, it undoubtedly increased the actual incidence of resistance by inflaming the resentment of pro-union miners and laborers. Secondly, it exaggerated the incidence of opposition to the draft. Because Tower was not authorized to arrest men for anything but draft resistance, he had no choice but to charge miners arrested for union activism with violating the conscription law. For example, in February 1864 he arrested seventeen "ringleaders" of a union calling itself "the Committee" in the heavily Irish mining district of Cass Township. He charged them with resisting the draft but could produce no evidence. In reality, he arrested them for attempting to collectively restrict production in order to shorten their hours and raise the tonnage rate.[54]

Upon closer inspection, then, it is apparent that supposed ethnic/political conflicts had a class component. It is also true that class bitterness and class-motivated violence haunted the anthracite region during and after the war. In 1862, a mine foreman was set upon by a crowd and beaten to death. In 1864, a miner working underground in Cass Township struck the mine boss in the head with a pick, nearly killing him.[55] Altogether, in Schuylkill County between 1863 and 1866 there were fifty-two murders. In the first three months of 1867, there were six murderous assaults and twenty-seven robberies.[56]

Looking back on this "reign of horrors never to be forgotten . . . by those who lived through them," a Pennsylvania Commissioner of Industrial Statistics brought out the elements of class conflict in the strife:

> At one time a superintendent murdered; at another, a boss; again, an attempt to rob a paymaster; and again, a farmer carrying produce, robbed and beaten . . . a long train of murders and attempted murders, of horrible beatings. . . . The outlook for the people . . . was gloomy enough. The cost of the necessaries of life was still high; wages very low and prospects of going still lower; man arrayed against man by their mutual necessities, underbidding each other for work; . . . angry and bitter contests with employers, with the fault as often on one side as on the other. . . .[57]

For the class war in Schuylkill County, there was no lack of volunteers. The direct violence of inferiors often met simple lack of interest in preserving human life on the part of superiors rushing madly forward to make a buck. A threat note pinned on a mine boss's office door in September 1863 read, "Mr. Snow—If you don't leave this neighborhood Glen Carbon inside of Monday September 23—remark the consequences—if written words nor writing will do we will try what virtue there is in cold lead—it is better one damned bugger should die than a whole crew."[58]

Whole crews dead were, of course, no exaggeration; underground fatalities resulting from indifference and greed probably seemed violent enough to survivors. In 1881, a commissioner recalled, "In the old time of pitiless oppression on the one hand, and violence on the other, objectionable rules or orders were often followed by violent threats and violent acts against the bosses. . . ."[59]

The violence subsided with the rise in 1868 of the second important coal miners' union in the United States.

5

BEATEN ALL TO SMASH: THE RISE AND FALL OF THE WORKINGMEN'S BENEVOLENT ASSOCIATION

I have always endeavored to impress upon Mr. Siney the fact that the way to benefit his [miners'] association was to increase the consumption of coal. His idea is to make it scarce on the market. That is not the way.

> Franklin B. Gowen, President of the Philadelphia and Reading
> Railroad, 1871[1]

When have the operators ever treated the miners so bad as the miners have treated the laborers at Ashley?

> Mine laborer, anthracite region, 1875[2]

During the 1870s, Americans experienced their first monopoly of a natural resource, when John D. Rockefeller's Standard Oil took control of ninety percent of the country's oil refining business. They experienced their first modern industrial depression, a cataclysm that began in the panic of 1873 and did not release its grip until 1878. They saw the first nationwide strike, the great railroad strike of 1877 that spread like wildfire from Baltimore to Pittsburgh.

For the most part, capitalism was still competitive. But as businessmen widened their markets in an expanding economy, they invaded one another's territories. Profitability carried with it the seed of its own destruction—increased competition.[3] The seed sprouted in the 1870s. The depression ruined hundreds if not thousands of businesses. The strong absorbed the weak in the coal industry as elsewhere, and new corporations expanded by swallowing smaller firms.

In the midst of these changes, coal miners established their second and third important unions. The Workingmen's Benevolent Association (WBA), founded in 1868, quickly became one of the two largest labor organizations in the United States.[4] Some believed the new union was behind the noticeable decrease in the violence that had been plaguing the anthracite region.[5]

In the early 1870s, WBA organizers set out from the anthracite region to organize bituminous miners. Their efforts resulted in the National Miners' Association, an association of local and regional unions that ostensibly included the WBA, but that in practice became a national union for bituminous miners. The WBA continued to serve anthracite miners exclusively.[6] Because the two unions coexisted and because their histories are highly similar, this chapter will focus on the Workingmen's Benevolent Association only.

Miners formed the WBA, as they had formed the American Miners' Association, in the midst of a spontaneous strike. The Pennsylvania Legislature inadvertently sparked the walkout in 1868 by enacting an eight-hour-day law. Radical Republicans had pressed for the law, but the fly in the ointment was their insistence upon freedom of contract. Thereby, "a loop-hole big enough for a cow to pass through was provided," as a miner remarked of a similar piece of legislation, "and the operators could stand and laugh it up." In both cases, the loophole excused employers who contracted otherwise with their employees. After the law was enacted, coal operators made such contracts a condition of employment.[7]

The law did the mine workers no practical good, but it validated their aspirations and spurred them to action. At the noon hour of the day before it was to go into effect, laborers and company men gathered around the mouth of a Mahoney Valley mine decided to demand the eight-hour day. They rose in a body and proceeded to confront their superintendent. He would not hear of it. At this, according to the state Commissioner of Industrial Statistics,

> . . . the hands all quit work and moved to the next colliery, making the same demand, and the reply being the same, were joined by the hands there, and proceeded to the next, and the next, and so on until every colliery in the valley was stopped, and the moving mass being joined at each colliery by all the men who were able to endure the fatigue of walking, it soon

Anthracite coal towns

comprised nearly the whole working population of the valley, gathering enthusiasm and determination in proportion as their numbers increased.[8]

The eight-hour strike of 1868 pervaded the entire hard coal region and outlasted the summer. In September, operators and dealers depleted their stockpiled coal. With cold weather coming on, the operators compromised, offering miners a ten percent wage increase but not the shorter workday. Miners returned to work with a partial success. This would have resembled a hundred other strikes except that a week before the strike ended, an estimated twelve thousand to twenty thousand miners assembled at Mahoney City to establish the Workingmen's Benevolent Association.[9] The new union embraced unskilled laborers along with skilled miners, making it America's first unambiguously industrial union. John Siney (1831–1880), an Irish immigrant from England, became the leading spokesman for the WBA, a union whose members were Irish, Welsh, English, and Scottish.[10]

Siney, the most important coal miners' leader of the decade, was the eldest of seven children, the son of Irish peasants evicted from their

*John Siney. Courtesy of
the Schuylkill County
Historical Society.*

traditional lands.[11] The dislocated family moved to England when John
was five. There he worked as a child in a textile mill and as an adult in
a brickyard. He grew to be an unusually tall man with strongly pro-union
views and extensive early experience in union organizing. Yet by age thirty
he had not mastered the rudiments of spelling and writing.

A turning point in his life came at the death of his wife in 1862. Leaving
his infant daughter with his mother, he emigrated to Pennsylvania in the
midst of the Civil War and found work in St. Clair (Schuylkill County)
as a common laborer doing repairs in a coal mine. He was one of a new
breed: an adult entering the mines with no prior training. Within a year,
he had received his own chamber as a miner and began earning two to
five dollars per day. He sent for his mother and child, and also began
taking writing lessons from a miner's wife who kept him as a boarder.

In 1868, five years after going underground as a laborer, Siney became
the prime mover in the new WBA. In April 1869, he was elected Chairman of the Board of the all-important Schuylkill County branch. His
1,500-dollar per year salary enabled him to leave the mines; he remained
in office until 1873, at which time he left the anthracite union to become

president of the National Miners' Association. (The main difference between the two was that the National Miners' Association, and now Siney,
opposed strikes.) By 1876, coal operators had succeeded in smashing both
unions. Siney remarried that year, and returned to St. Clair without funds.
He became a small-time saloon keeper, but two years later he developed
"miner's consumption." During his last illness, he sank into destitution.
He died of black lung on 10 April 1880, at the age of forty-nine.

John Siney represented both the growing importance of the Irish in the
anthracite field, and the growing importance of laborers in the union
movement. One indication of the laborers' participation was the WBA's
attention to the issue of the shorter workday, expressed as such. This
explicit demand for the eight-hour day also signaled the deteriorating
independence of the contract miner. Miners and laborers were coming to
have more in common.

In other ways also, the WBA took the concerns of laborers into account. The union worked for legislation against forced buying in company
stores, and for safety legislation. It accumulated a fund to care for its sick
and injured. "If a member is sick he is supported and visited by brother
members," an officer testified to the Pennsylvania Legislature. "If a member of the Association dies he is buried by the Association whether his
folks are in good circumstances or not." The WBA supported widows of
slain mine workers and sent their children to school.[12]

But these efforts paled before the union's devotion to issues affecting
contract miners exclusively. Its chief goal, pursued in numerous strikes,
was to tie the contract miner's tonnage rate to the operator's selling price
of coal, a system known as the sliding scale.[13]

THE MINER AND THE INDIVIDUAL OWNER

Contract miners and individual (noncorporate) owners fought over ventilation and weighing, but they shared the goal of high-priced coal. Both
miners and operators believed that the road to prosperity lay in limiting
production in order to make coal scarce in the market. One district's
charter explained that the WBA's purpose was "to make such arrangements as will enable the miner and laborer and operator to protect and
promote their mutual interests."[14] This notion of mutual interests of
employer and employee is a far cry from E. P. Thompson's notion of class
consciousness: "the identity of interests as between themselves and as
against their employers." Another district held that the union's purpose

was "to make such arrangements as will enable the operator and the miner to rule the coal market."[15] Here was the miner identifying with the coal operator, and viewing himself as a small contractor, not as an industrial worker.

The WBA blamed the low tonnage rate on the low prices operators received for coal, and blamed low prices on overproduction. Reflecting this philosophy, the union announced a strike in the spring of 1869 for the purpose of reducing stockpiles of coal. The small-mine operators actually supported this self-inflicted layoff, which lasted for most of the summer. The union's General Council permitted the districts to return to work when the price of coal rose to five dollars per ton at one distribution center and to three dollars per ton at another. Yet the back-to-work movement began before these prices were reached. Women strike supporters stoned some of these defecting miners. In other districts, mine workers went back with a partial victory—the tonnage rate tied to the price of coal. In order to avoid another glut on the market, the union set strict production limits for contract miners—one fewer car or wagonload of coal per miner per day.[16]

In 1871, a miner elaborated on the common interests of the operator and the miner. "We are all on suspension at the moment and not on strike," he explained to a friend. "The name suspension is more 'genteel' than strike; suspension does not create bad feeling between masters and workmen because on both sides they are suspending work until the coal on the market has been bought up. . . ."[17]

Like skilled miners and small-mine operators, the owners of local retail stores favored high prices. The merchants supported the miners, providing them with groceries on credit during strikes. On one such occasion in 1871, the WBA resolved "to tender our sincere thanks to our friends, the merchants, for their assistance, pecuniary and otherwise." Any miner who did not repay a merchant for credit extended would be expelled from the organization.[18]

If the world of the coal industry had been populated exclusively with contract miners and the owners of small collieries, the strategy of restricting production might have worked. But in the real world, cracks and fissures marred even the "genteel" relations of these two groups. The owners of small mines wholeheartedly approved of high-priced coal gained by reducing the supply on the market. But as soon as their goals diverged from those of the contract miners, they withdrew their support. When the miners began to agitate for a closed shop (in which only union miners would be hired) and for the formation of miners' committees to settle grievances, the small-mine operators refused to go along.[19]

THE MINER AND THE LABORER

If contract miners failed to get along perfectly well with the small-scale mine owners, neither did they see eye to eye with the laborers. Disagreements erupted frequently between miners and laborers, as Harold Aurand has shown, although both groups now belonged to the same union.[20] It was no wonder. "Look again at the unfairness of the system to the laborer who has to fill from six to seven cars a day with coal and he gets but one third of the wages of the miner," wrote a mine worker from the anthracite region:

> The miner and laborer go to work at seven o'clock in the morning and probably the miner will cut enough coal by ten or twelve o'clock. Then he will go out leaving the poor laborer up to his waist in water and he will have to pile the lumps and fill three or four cars with coal after the gentleman [the miner] has left. He [the miner] will wash, put on a shirt, and a white collar and will go to dinner boasting that he has cut enough coal for the laborer. . . . He calls for his cigarbox and enjoys himself for an hour or two and because he is a religious man he says that it is nearly time for him to go to a prayermeeting. Between five and six o'clock the laborer, poor thing, arrives home wet as a fish, and after eating his supper, in spite of his weariness, goes to the prayermeeting, and who should be praying at the time but the man he works for.[21]

In 1870, the WBA attempted to ease the tension between miners and laborers by calling for a more equitable ratio between the tonnage rate the operator paid to the miner and the wages the miner subtracted from this to pay the laborer. Further, the General Council resolved that the miner could no longer pay his laborer directly. Hereafter, the company would pay the laborer by deducting the amount due him from the money owed to the contract miner.[22]

These steps probably did some good, but they did not prevent conflict between miners and laborers from breaking out again in 1871. The occasion was a WBA-led strike against reductions in the tonnage rate. The miners favored the strike, as did small-scale mine operators facing falling prices and an oversupply of coal on the market.[23] But the laborers questioned whether they would gain anything by a victory, and in Wyoming County they met among themselves to discuss returning to work. Miners attended their meeting and broke it up. At this, the laborers withdrew from the WBA with the idea of forming their own union. Their grievances focused not on the operators but on the miners. They demanded: "That the miner pay to the laborer one third of the whole amount received by him each month, together with the price of any cars lost through the neglect of the miner . . . and that in case the miner is not capable of cutting his coal . . . the laborer will not be bound to assist him."[24]

Another of the laborers' demands shows how far from a true apprenticeship system (in which the laborer expected to learn the miner's trade and eventually to become a miner himself) the relations between miners and laborers had strayed. The laborers demanded that "a laborer be entitled to a chamber in his turn if he is capable of working said chamber and that work shall be divided among all nationalities for the future, as has not been done in the past." Apparently, the miners were blocking qualified laborers from becoming miners on account of nationality. Ethnic loyalties and antagonisms were feeding into the existing class relations.[25]

Miners controlled the conditions of laborers. In 1875, a laborer complained to a local newspaper, "When have the operators ever treated the miners so bad as the miners have treated the laborers at Ashley?"[26] The following year, in the midst of the devastating depression, miners working at a shaft near Scranton reduced the wages of their laborers to twelve cents a day. The laborers struck. The miners replaced them with new laborers. Great excitement and several fights resulted. It was officers of the company who called a meeting between miners and laborers, and conducted negotiations between them until the matter was settled amicably.[27]

Laborers had no reason to support miners in their struggle to maintain their accustomed independence, and miners could not rely upon them. In 1873, the Philadelphia and Reading Coal Company established new work rules requiring miners to leave the mine on foot if they wanted to go home before the company-designated quitting time. The miners protested this attempt to force them into a more factorylike discipline but concluded that they would be unable to win a strike over it because, as Harold Aurand reports, "they were afraid the laborers would not back them in a strike."[28]

Conflicts between miners and laborers in the coal industry show one way in which the industrial working class was only partially formed. Despite their shared membership in the WBA, the contract miners' independence and the laborers' dependence on the contract miner made it difficult for the two groups to experience an identity of interests. The miners exploited the laborers directly. For their part, the laborers had no interest in preserving the miners' traditional control over many aspects of the job. Ethnic divisions intensified the problems between them, because often the miners were Welsh or English while the laborers were Irish.

THE PROBLEM OF NATIONALITY

In the anthracite region, conflicts based on class tensions were not necessarily separable from those based on ethnicity. If an Irish laborer attacked a Welsh mine boss, who can say whether class or national antagonism fueled his hostility? Yet it is certainly true that within the WBA ethnic conflicts simmered and threatened the organization. In 1871, a group of English miners in Schuylkill County countered rumors that they were disloyal to the union by publicly denouncing "those men who have circulated false reports concerning our actions."[29] In another incident, three coal companies fired the Welsh miners and replaced them with Irish and Germans. Threats followed insults. The Irish and German interlopers formed committees for self-protection. Whatever the Welsh did to express themselves has not come down to us, but the English, Scottish, and Irish agreed to shun the Welsh, because, as one miner explained, "in their latest murderous outrages they have shown us they are a class of beings who should never be allowed to associate with peaceable and law abiding citizens."[30]

Yet the union worked to assuage these conflicts, and it made significant progress to that end. In 1871, a spokesman told the Pennsylvania legislature that the WBA comprised five nationalities.[31] The Pennsylvania Commissioner of Industrial Statistics commented that "[a] great thing had been achieved by the gathering together of all the different nationalities and clans together in the new society, removing the bitterness of factions, and uniting all in a common union for the good of all."[32]

One member recalled his initiation into the WBA (in 1869), noting especially the interethnic cooperation that characterized the union: "The president of the miner's union in the district . . . was a young Pennsylvania

Dutchman. The Secretary was a blunt old Englishman, honest and rugged, but stubborn as a mine mule. When he was right, so much better
for the right; but when he was wrong, nothing could move him from his
position. The treasurer was a Welshman. The next leading officer . . . was
an Irishman, cautious and conservative."[33]

THE RISE OF THE LARGE CORPORATION

Although the WBA worked hard to repair rifts in the work force, the
union was destroyed before its achievements in these directions could be
put to a full test. The intervening force was the "large, greedy, cruel, and
unfeeling companies who want everything for themselves," as one miner
called the large corporations.[34] In 1869, three large firms (which were
anthracite railroads) and numerous small ones mined the region's hard
coal. The managers of the large firms were conscious of their class interests
if the miners were not. During the 1869 strike, the president of one
expressed himself on the mutuality of interest between the miner and the
operator: "We are not prepared to take in new partners."[35]

How different was the self-concept of the contract miner. As small
contractors, miners believed it was their prerogative to set the price of
their own labor. During a strike, the WBA asked the large-scale operators how they would reply if miners presumed to set the freight rate. "If
you were to deign to reply," the miners argued, "it would be that we
had no business to interfere with your business. Have you any business,
therefore, to interfere with our business? We name the price of our
labor, you may have the price at which you sell the coal. There we
stand." To these miners, the concept of freedom was central to the
question of who would determine the tonnage rate. "We are free men,"
they announced. "[W]e wear no collars; we are not slaves; we will labor
at the prices we choose to establish for ourselves and it is not in your
power to coerce us. . . ."[36]

The small, individually-operated or partnership-operated coal firms
were concentrated in the southernmost part of the anthracite region—
Schuylkill County. They were soon to pass into history. The engineer of
change was Franklin Benjamin Gowen, the District Attorney for Schuylkill County from 1862 to 1864, a failed individual coal operator who had
climbed onto and now drove the corporate bandwagon. Gowen became
acting president of the Philadelphia & Reading Railroad in 1869.[37] That
same year the railroad acquired the Schuylkill Canal, which ran along the

Franklin Benjamin Gowen.
Courtesy of the
Library of Congress.

Schuylkill River from Pottsville to Philadelphia. Owning both the canal and the railroad gave the company a monopoly of coal transportation out of that important anthracite county. Because transportation costs could amount to as much as fifty percent of the final price of coal, Gowen could ruin any operator by raising the freight rate.[38] The company could now dictate production levels, wages paid to miners, and anything else it pleased to the remaining "independent" operators.

But Gowen went further, aggressively buying coal lands in order to centralize control of the anthracite industry. Several independent operators quickly became superintendents of mines they had once leased or owned. By 1875, the Philadelphia & Reading Railroad had spent twenty-five million dollars to purchase 100,000 acres of land. The company controlled eighty-five percent of the collieries in Schuylkill and Northumberland counties. The company's executives, one scholar writes, "had taken complete control of the economic basis of every city, town, and industrial village in the region."[39]

Gowen's goals diverged sharply from those of the small colliery owners. His idea was to lower the price of coal and expand its markets. "I have always endeavored to impress upon Mr. Siney the fact that the way to benefit his association was to increase the consumption of coal," he testified before the Pennsylvania Legislature in 1871. "His idea is to make it scarce on the market. That is not the way. One month of stoppage induces people to burn bituminous coal and other fuels."[40]

Another corporate manager noted the high fixed expenses of running a railroad. "Our aim," said Asa Packer, president of the Lehigh Railroad Company, "is to get the largest tonnage we can carry at the lowest rates."[41] This made sense. The anthracite railroads profited as much from carrying coal as from selling it. The more freight the coal trains carried, the greater the profit, no matter what it sold for at the end of the line. Further, the locomotives burned coal for fuel: Railroad corporations wanted it cheap. Finally, low prices enabled the large corporation to undersell the competing small operator and drive him into some other endeavor.

It was Gowen, therefore, who in 1870 led the operators to sign a contract with the Workingmen's Benevolent Association. The "Gowen Compromise" tied the tonnage rate to the price of coal on the market, although it omitted the *minimum* price the miners had desired. As the price of coal fell, the Gowen Compromise resulted in a wage cut, exactly as he had planned.[42]

In the end, however, the depression outdid Gowen in destroying the WBA. Gowen insured that it would never return. The depression lasted from 1873 to 1878 and brought near starvation to thousands of working people; few labor unions survived it.[43] The occasion of the WBA's demise was the "long strike" of 1875. It was doomed from the start.

Determined to smash the union, Gowen announced a suspension of work in the fall of 1874. By January, the union realized that the suspension (which involved the majority of anthracite firms) was not a temporary layoff, but a lockout. In response, the WBA called a strike. Grossly underestimating the strength of the operators, mine workers announced that if operators had not ended the lockout by 1 March they would return to work only if granted an eight-percent raise.

The operators did not respond. In Lehigh County, they took the highly unusual step of hoisting the mules up out of the mines, thus revealing their expectation of an extremely long suspension. In Schuylkill County, Gowen's Philadelphia & Reading Coal and Iron Company began hiring both gunmen and strikebreakers. Strikers responded by burning down breakers and by attacking the locomotives and cars of the Philadelphia & Reading Railroad. Union leaders, believing violence played into the hands of the operators by creating bad publicity, tried to stop it. They failed. They had lost control of the membership.[44]

Ultimately, the strikers were starved into submission. Andrew Roy recorded that, in the last days of the long strike, "[h]undreds of families rose in the morning to breakfast on a crust of bread and a glass of water. . . . Day after day men, women, and children went to the adjoining woods to dig roots and pick up herbs to keep body and soul together. . . ." The anthracite mine workers returned to work on 14 June 1875, on the operators' terms. One superintendent found his returning employees too weak to work, for want of food. He urgently requested emergency provisions from his superior.[45]

A coal miner's song about the long strike of 1875 preserves the fate of the WBA:

Well, we've been beaten, beaten all to smash,
And now, sir, we've begun to feel the lash,
As wielded by a gigantic corporation,
Which runs the Commonwealth, and ruins the nation.[46]

"It was a terrible thing to submit to a twenty percent reduction on contract work," the union's last secretary recalled. "Evil days had come. We went to work, but with iron in our souls."[47]

After the strike collapsed, the Philadelphia & Reading Coal and Iron Company posted new work rules at each of its forty collieries. From now on, the company would furnish empty wagons for the purpose of letting men into the shafts or slopes only between six and seven in the morning, between noon and one in the afternoon, and again after five-thirty in the evening. In short, all employees, whether contract miners or not, would be required to work the ten-hour day.[48] For the firm's nine thousand employees, factory discipline became a distinct and looming cloud on the horizon.

In light of the depression, perhaps the ten-hour day was a small matter. "Reports . . . from the coal regions are truly distressing," a labor paper revealed in August 1876. "Many families have been subsisting on the flesh of domestic animals which they were compelled to eat to prevent starvation. Numerous instances of *actual starvation* have been recorded in Wyoming and Lackawanna Counties."[49]

Again, the level of violence rose. Coal operators armed vigilante committees for the purpose of harassing union suspects, and furnished guns to managers. On two occasions, mine bosses fired into crowds of miners. Superintendents and bosses began to receive notes threatening their lives. Murders and beatings of those on the company side were again on the rise. There were also numerous street brawls, frequent highway robberies, and many other crimes committed.[50]

THE MOLLY MAGUIRES

Such was the desperate and murky context in which the Molly Maguire episode took place. Allegedly, the Molly Maguires was a secret terrorist organization of Irish mine workers.[51] Presumably, its members operated in the anthracite field during the 1860s and 1870s, threatening, assaulting, and occasionally murdering mine guards, foremen, superintendents, and owners. Certainly, a legend existed about such a secret organization. According to Franklin B. Gowen, the "Mollies" formed a clandestine cell within the WBA as well as a cell of the Ancient Order of Hibernians, an Irish fraternal organization. Gowen orchestrated the campaign against the Molly Maguires, and in so doing successfully linked unionism with the breakdown of law and order in the public mind.

Possibly the first mention of such a secret society was made in 1857 by *Miners' Journal* editor Benjamin Bannan, an influential pro-operator voice who was also notoriously hostile to Irish Catholics.[52] Statements by such deeply prejudiced men are as close as the contemporary sources come to hard facts in the case of the Molly Maguires.

In 1871, Franklin Benjamin Gowen accused the union before the Pennsylvania Legislature: "I say there is an association which votes in secret, at night, that men's lives shall be taken, and that they shall be shot before their wives, murdered in cold blood, for daring to work against the [strike] order [of the WBA]. . . . *I do not blame the Association* but I blame another association . . . and it happens that the only men who are shot are men who dare to disobey the mandates of the Workingmen's Benevolent Association." To this, union spokesmen took extreme umbrage. One miner said, "You charged us with being a secret society of murderers?" Gowen sidestepped, saying that it was not the WBA but another organization. John Siney rose to testify that he had heard rumors about the Molly Maguires but added, "I wish to be placed upon my oath. As workingmen we are stigmatized as a band of assassins; anything coming from our lips is supposed not to be believed; I know of none; I have heard say that such a thing is in it; I do not know a solitary man belonging to it." James Kealy, WBA president, argued, "When he refers to outrages in the coal region he should know that the coal region has been more quiet since the organization of the association than before. At each of their meetings they must speak respectfully of their employers. If a committee is appointed on wages they are instructed how they shall act and speak. Yet we are told we are conspirators. We intend to put the saddle on the other horse. The conspiracy rests on the other side."[53]

Not until 1873 did Gowen contract with the Pinkerton Detective Agency to send agents into the anthracite towns to infiltrate the secret society and inform on its doings. The chronology is significant because some of the crimes for which twenty "Mollies" were eventually executed had been committed in 1863, when Gowen himself, as District Attorney of Schuylkill County, had presided over the county's law enforcement apparatus. During that period of violence, Gowen had failed to bring charges against anyone.[54]

The first Pinkerton agent sent into the anthracite coalfields was a miner from Illinois named P. M. Cummings. His name has not become familiar in the context of this story for the simple reason that he found no substantial evidence whatever for the existence of the Molly Maguires. James McParlan's name, however, has been remembered.[55] In 1873, this Pinkerton detective agent entered a local saloon as James McKenna, a jig-stepping, frequently drunk, loquacious, and amusing fellow with an elaborately constructed persona and a convincing past.

Largely as a result of McParlan's reports and trial testimony, twenty Irishmen, all of them members of the Ancient Order of Hibernians, most of them activists in the WBA, were accused of membership in the Molly Maguires, and of murder. The murdered included a mine owner, two superintendents, two foremen, a mine clerk, a paymaster, a mine guard (watchman), a miner, and one policeman. Several had been killed in the context of a riotous crowd scene, two of them as long ago as 1862 and 1863. The twenty accused were tried and convicted in 1876 and 1877 in an atmosphere of rising public hysteria. The prosecuting attorney was none other than Franklin Gowen. A careful student of the episode writes, "[I]n the current state of public opinion, it would have been dangerous for a witness to contradict the Commonwealth testimony even when he was telling the truth. So firmly had Gowen planted his version of the Molly Maguires in Schuylkill County that juries automatically accepted the Commonwealth's evidence when it disagreed with the testimony offered by the defense."[56] Accordingly, each of the men, after protesting his innocence, was hanged.

Not only the prosecution's evidence at the trials but essentially all historical accounts of the Molly Maguires rest on testimony and evidence supplied by one man: James McParlan (alias James McKenna). Ann Lane, and J. Walter Coleman before her, are two scholars who question whether so much should rest on so dubious a source. While there is no evidence that McParlan, in his reports and testimony, constructed an elaborate hoax, his capacity for doing so is demonstrated by his ingenious invention of James McKenna, a complex creation that he kept going for more than two years. Alternatively, as Melvyn Dubofsky has suggested, McParlan may have been an *agent provocateur,* instigating violent crimes before reporting them. (Defense lawyers at the time asserted that this was the case.) Another possibility is that the Molly Maguires existed as one of the several gangs that operated in the anthracite region before the formation of the WBA and after its demise. It is also possible, as the much-cited Wayne Broehl assumes but never really argues, that McParlan was reporting the truth.[57]

The convictions in the Molly Maguire cases were based on flimsy and highly contradictory evidence. True, in one trial witness after witness got up to give consistent evidence against the accused. This foreshadowed a famous trial thirty years later in which McParlan (now known as McParland) was exposed as having tutored the key witness. There is no proof that he did the same in the Molly Maguire trials but, again, later experience shows that he was capable of it.

In the later instance, the Moyer, Haywood, and Pettibone case (1906–1907), McParlan[d] persuaded Harry Orchard, the murderer of former Governor Frank Steunenberg of Idaho, to tell the court that he was ordered to commit the deed by a secret inner circle of a miners' union, the Western Federation of Miners. This story was strangely reminiscent of the Molly Maguires, alleged to be an inner circle of the WBA. McParlan[d] was thinking of the Molly Maguires too, telling Orchard that some of the prosecution witnesses in the Molly Maguire trials had gotten off scot-free from their own violent crimes in this manner.

But, unlike the Molly Maguires of the 1870s, the accused leaders of the Western Federation of Miners had Clarence Darrow, one of America's great lawyers, to defend them. Darrow won his case. The defendants were acquitted after a brilliant, eight-hour summation, in which Darrow had this to say about McParlan[d]: ". . . this Pinkerton detective . . . never did anything in his life but lie, cheat and scheme (for the life of a detective is a living lie, that is his business; he lives one from the time he gets up in the morning to the time he goes to bed; he is deceiving people and trapping people and lying to people and imposing on people; that is his trade. . . ."[58] As Broehl notes, the Mollies were tried in the 1870s and Clarence Darrow was not there.

An extremely plausible account of the Molly Maguires was presented by a contemporary newspaper, *The Irish World,* a source largely ignored at the time:

> Acts of violence were done by desperate men. We do not defend their methods, but we are satisfied that the number and nature of such acts have been grossly exaggerated, while the provocation under which those men labored is a factor usually eliminated altogether from the discussion. There have been in the mining regions Americans, Irishmen, Englishmen, and others who took part in the disturbances. Small batches of them, here and there, united to avenge the wrongs inflicted on them. After a while when any midnight deed was done, by an isolated miner or special policeman, it was attributed to an organized band of conspirators who received the name "Molly Maguires." Doubtless the men themselves helped to give color to this belief, hoping thus to terrify the capitalists into a more equitable frame of mind. But there is no convincing evidence that any regular or extensive society, such as the "Mollies" are alleged to be, ever existed in Pennsylvania.[59]

Any convincing history of the episode would have to depend upon evidence to corroborate or contradict James McParlan[d]'s reports and testimony. Undoubtedly, such will never be forthcoming. Whatever the truth of the matter, the Molly Maguire incident represents an early use of the idea of "terrorism" to inflame the public against a group, the Irish mine workers, and their organization, the Workingmen's Benevolent Association, by pinning upon it the entire burden of all the violence that had occurred.

By the end of the 1870s, the coal industry had changed. The decade had seen the rise and fall of the country's first industrial union. Further, the large corporation was an immutable presence on the industrial landscape.

Something else, also of great moment, was changing. Increasingly, employers understood the benefit of a surplus of labor. Not only did hiring more miners than necessary benefit the housing and grocery business; it also made it easier to cut wages, on the principle that the more a man needs the less he will insist upon. As early as 1869, the editor of *The Workingman's Advocate* landed in Liverpool and saw billboards advertising for miners to come to parts of the United States where thousands were out of work.[60]

In the 1870s, coal operators began to introduce several new sources of labor into the mines. Regarding an 1873 strike in Tioga County, John Siney wrote, "They may import Swedes, Arabs, or who they please, but *practical miners keep away from Tioga.*"[61] The following year, a miner reported from a strike in western Pennsylvania, "They have got the Harrisville colliery . . . filled with our old friends, the Swedes."[62] Italians, eastern Europeans, and blacks also began to enter coal mining, sometimes as strikebreakers, sometimes not. What the new workers had in common was lack of experience. They had no preconceived notions about how a thing should be done; they had no strong sense of their traditional rights.

In the context of a new labor force, the coal operators could introduce certain changes. In 1874, coal operators near Pittsburgh engaged an Italian labor company to import two hundred Italians from New York to work as strikebreakers. They were hired on a new basis. An operator explained, "We are bound to keep them all the time, whether they work or not. . . . [W]e would accept no sliding scale [the tonnage rate determined as a percentage of the operator's selling price], nothing of the kind. We take all the risks off the laborers, and of course if we get steady employment we make something by it."[63]

These operators were class conscious. They acted explicitly in their own class interests. One spoke of the "war between capital and labor," and another mentioned that it was the frequent strikes of the local miners that brought "these swarthy foreigners among us." It made no difference to them *which* racial or national group worked in the mines. "We are determined to put an end to strikes," explained an operator. "[I]f the Italians fail we intend to import Swedes to do the work for us, and if they do not meet our expectations then we shall bring a lot of Negroes from the south."[64]

The following year, a Pennsylvania coal operator faced with a strike imported three carloads of Italians to work in his mine. The strikers' wives joined their husbands in forcing their way into the strikebreakers' houses. They "dragged the men therefrom, threw their baggage out the window, and took possession of their provisions."[65]

In most cases, the capitalists of the coal industry did not import the new workers into the industry. They came of their own accord, spurred by conditions in Europe or in the South. Nevertheless, this perpetual stream of new players innocent of the rules of the game enabled employers to change the terms of the struggle. For coal mine workers, the question of whether class solidarity could transcend racial and national divisions would become even more critical than it had been before.

6

PENNSYLVANIA IS SWARMING WITH FOREIGNERS: OWNERS, WORKERS, AND METHODS IN TRANSFORMATION, 1880-1900

It was the beginning of a new era. The conditions existing between employers and employes were changed radically, and the relations of the workmen to each other were also of quite a different character. Society took on new forms. . . .

Editorial, United Mine Workers Journal, *1891* [1]

The large corporations . . . are, slowly but surely, riveting the chains of slavery around us.

Illinois coal miner, *1873* [2]

The new workers came with lower levels of skill, or with no coal-mining skills. To the artisans of the industry, they represented a red flag signaling defeat, hard times, a change for the worse. They were one sign that industrial capitalism, fueled by coal, had begun to transform the coal industry.

Between 1880 and 1900, the impact of coal on the nation would be profound. The development of coal, the U.S. Census Bureau reported in 1902, "has been coincident with the rapid advancement of this country . . . to the front rank among the industrial nations of the world. Indeed, the country's progress has been due largely to the abundance and cheapness of its mineral fuels, chief among which is coal. Most of this development has taken place during the last two decades and has far exceeded the growth in population, indicating a rapid change from an agricultural to a manufacturing nation."[3]

While coal fueled industrial capitalism, the new economic system began slowly to make its mark on the coal industry. The most striking change was the sheer, staggering quantity of coal dug. Decade after decade, coal production more than doubled, rising from 37 million tons in 1870 to more than 350 million tons in 1900. During the same period, the number of coal mine workers increased from 186,000 to some 677,000 men and boys.[4]

During the 1880s, the map of the American coalfields assumed its familiar aspect. West Virginia took its place there; by the turn of the century, the Mountain State would be the third coal producer after Pennsylvania and Illinois. (Not until the 1930s did West Virginia surpass Pennsylvania to take first place.) In the 1880s, the Rocky Mountain coalfields, with Colorado at the center, also began a rapid development. At the same time, the expanding industry in the older coalfields kept Ohio, Illinois, and especially Pennsylvania among the country's top coal producing states.[5]

In the mid-1880s, Old King Coal finally toppled wood as the nation's principal source of energy. His sooty reign was short but significant. By 1910, the organic rock supplied seventy-seven percent of the country's energy. Coal remained the quintessential fuel until the 1940s, when fuel oil and natural gas worked their energy coup.[6]

The infant steel industry was one of the pistons driving coal. A transforming innovation, the Bessemer process, made it possible to mass-produce steel, that strong, flexible material that had been manufactured in small crucibles a bit at a time. Steel changed the face of the cities. In 1883, skyscrapers began to renovate the skyline of Chicago, a slate wiped clean by the fire of 1871. Steel beams and the elevator, the familiar cage of the mine shaft put to a new use, released city buildings from their six-story limit. The railroads, having recovered from the depression of the 1870s, began to replace hundreds of miles of deteriorating iron rails. The manufacture of iron, steel's principal ingredient, continued to require anthracite or coke as a reducing agent.[7]

The discovery of sulphur-free bituminous coal suitable for coking spurred the expansion of industrial capitalism. In the late 1870s, coke began to replace anthracite coal as the chief metallurgical fuel. Coke served the purpose better, in part because the good coking coals of western Pennsylvania occurred near deposits of iron ore. This eliminated the costly

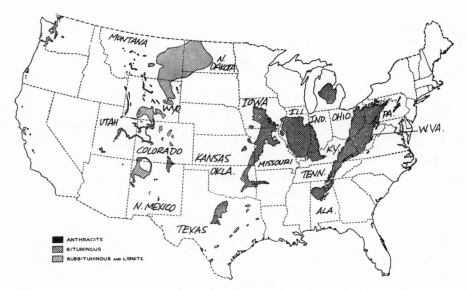

Coal Fields of the United States, 1900

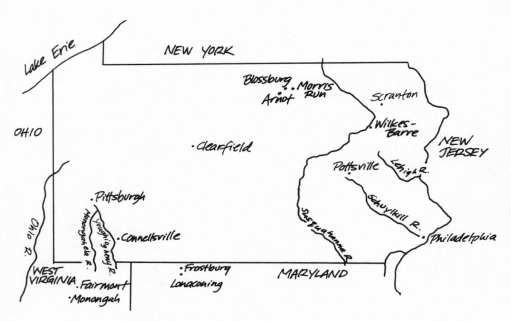

A few coal towns, including Connellsville

haul of iron ore to the anthracite region. Secondly, coke, unlike the dense anthracite, was full of air holes left when the volatile matter escaped as gas. Engineers found it easier to get the oxygen from the iron ore to combine with the carbon of the coke, because the air holes permitted greater contact between the two elements.[8]

The Connellsville coke region, a strip three miles wide and fifty miles long just south of Pittsburgh, sprang into activity as a major new coalfield. A Pennsylvania official reported that the coke region had expanded from 3,600 coke ovens in 1876, into "a vast furnace of 13,000 ovens, while the poorly-paid, helpless band of workmen of ten years ago has grown into a vast army of 13,000 cokers. . . ."[9]

Quantity told only part of the story. Equally as significant, the individual worker began producing more coal, an increase in productivity that became especially noticeable in the 1890s. At the beginning of the decade, one man could bring up about 476 tons per year. Ten years later, the same man was producing 565 tons per year.[10]

At the root of increased productivity was the undercutting machine, introduced into the largest mines to mechanize the task requiring the most skill. Even more fundamental, managers attempted to reorganize the work so that unskilled men could do it, an effort not always accompanied by mechanization. On their part, the new workers brought new cultures and viewpoints with them into the coal industry.

Finally, fewer owners began to control much more. Gradually, the coal industry passed into the hands of financial capitalists, men whose attention, in Thorstein Veblen's words, had "shifted from the old-fashioned surveillance and regulation of a given industrial process . . . to an alert redistribution of investments. . . ."[11] Some financial capitalists, men like Jay Gould, did take an interest in the day-to-day management of their vast holdings, but of necessity they had to rely on distant managers to carry out general policies. Others, like J. P. Morgan, prime mover in the concentration of capital in the anthracite industry, contrived profits from panics, or from combinations that increased neither efficiency nor production.[12] Either way, the financial capitalist was a new kind of owner, who might silently enter into proprietorship and just as silently depart. At the beginning of the twentieth century, he was coming into his own.

He was the subject of voluminous commentary. The articles on him contained frequent denunciations of his greed. Even commentators who did not disapprove frankly discussed his money, the portion of industry he controlled, and the realities of an emerging class system. The explicitness of the discussion is faintly shocking to modern ears accustomed to a journalism that barges into any bedroom but blushes at the specific mention of a person's financial condition. At the century's turn, euphemism on such matters was absent. An associate editor of the *Wall Street Journal* published a typical article titled "Our Financial Oligarchy." He peered into the pocketbooks of J. P. Morgan and seven other capitalists "who control two-thirds of the mileage and $9,000,000,000 of the $13,000,000,000 of capitalization of American Railroads. . . ." Illustrating his point were pictured seventy-six gentlemen from J. P. Morgan to John D. Rockefeller.[13]

THE MINE OWNERS

The coal industry was by no means the country's most concentrated industry, yet the trend toward concentration thoroughly affected it. "The spirit of combination of interests, which has affected the industry enterprises," we read in the 1901 volume of the *Mineral Resources of the United States,* "has to a greater or lesser extent also pervaded the coal mining regions. . . ."[14]

That year, a corporate director described the movement from competition to concentration to the United States Industrial Commission. Around Pittsburgh, there had once been 140 different coal companies, "with that many proprietors." Competition among these owners had "gradually reduced the price of coal to a point which was in most cases below the cost of production." After failing at numerous attempts to fix a unified price structure, the operators solved the problem of competition in 1899 by merging all 140 companies into two: the Pittsburg Coal Company and the Monongahela River Coal and Coke Company.[15]

Consolidations like these were occurring everywhere, yet from a national perspective, no one coal firm succeeded in cornering more than three percent of the U. S. market.[16] In 1905, 665 coal operators produced fifteen percent of the nation's coal. Compare this to oil: By then, one operator, John D. Rockefeller's Standard Oil Trust, controlled nearly ninety percent of the nation's oil production.[17]

Still, giant coal firms dominated by financial capitalists could be counted among the corporate goliaths. A modern (1974) manual listing America's hundred largest industrial and mining corporations around 1900 includes twelve coal companies. Among them were the Westmoreland Coal Company, located in western Pennsylvania, and the Maryland-based Consolidation Coal. In 1903, one of the twelve, the Mellon-controlled Pittsburg Coal Company, absorbed another, the Monongahela River Consolidated Coal and Coke Company. Yet nowhere in the manual do we see the giant anthracite firms. The author omitted them because he could not separate their financial data from that of the railroads that controlled them.[18]

Neither could the Census Bureau, reporting on the ownership of the coal industry in 1902, extract coal firms from the railroads. "The production of some minerals is probably controlled by a mutual understanding between operators . . . and transportation companies, but it is impracticable to identify operations which are controlled by arrangements of this character."[19] The railroads were of course the country's largest enterprises, overshadowing most mining or industrial companies.[20] By 1909, railroads controlled more than one-fourth of the country's coal output, and coal mines were their most important nonrailroad enterprise.[21] Coal's entanglement with railroads caused the degree of concentration in the industry to be understated, though by how much it is impossible to say.

The "biography" of Consolidation Coal reveals one type of development, in which the coal firm retained its own shape as a financially recognizable entity. The company began "life" in 1860 in the Georges Creek region of Maryland and began producing coal in 1864. Between 1876 and 1906, the firm was wedded to the Baltimore & Ohio Railroad; together, the two companies monopolized the production and distribution of Maryland coal. Between 1899 and 1901, the Pennsylvania Railroad, America's largest firm, acquired both the coal company and the railroad. Under this influence, Consolidation Coal began an aggressive expansion, acquiring the controlling interest in the Fairmont Coal Company of West Virginia (itself created by the merger of twenty smaller companies in 1901), and of the Somerset Coal Company, a Pennsylvania firm. During a 1906 Interstate Commerce Commission investigation of the relationships between railroads and coal companies, the Baltimore & Ohio Railroad sold its fifty-two percent of Consolidation Coal securities back to Consolidation Coal. Three years later, Consolidation Coal absorbed its West Virginia and Pennsylvania firms, becoming the largest coal firm in the United States. The company also acquired 100,000 acres of Kentucky coal land. In 1915, the Rockefellers purchased thirty-eight percent of the securities of Consolidation Coal to acquire the controlling interest. The depression of the 1930s brought setbacks and reorganization, after which the firm recovered and, in November 1945, merged with the Pittsburg Coal Company, by then a giant in its own right. In 1966, Continental Oil, reflecting a new trend, absorbed Consolidation Coal.[22]

Prior to World War I, Consolidation was an extremely profitable company. Katherine Harvey explains: "To appreciate the size of Maryland operators' profits at this time [around 1903], we need only consider the example of the George's Creek Coal and Iron Co. [a Consolidation Company], which on February 10, 1903 declared a one hundred percent stock dividend, 'to distribute among the stockholders a large surplus of net earnings that had been accumulating for some time.' "[23] This was astonishing but not unusual. The company had shown substantial profits back to 1887 and before.

Other coal firms also dazzled investors with high profits. In the Pocahontas-New River field of southeastern West Virginia, operators who shipped coal to the eastern seaboard over the Norfolk & Western Railroad in the late 1880s earned twenty-five percent on capital invested. In more than one instance, according to Joseph Lambie, they had been able to double their operations *out of profits* and to continue earning this huge percentage on the new capitalization. Profitability led to a rush to open new mines, which proliferated from four to thirty-eight in the eight years following 1886. Another highly profitable firm was the Pittsburg Coal Company, whose business was sufficiently lucrative in the early 1900s to enable the firm to take three million dollars out of net earnings (which totaled five million dollars) to acquire new coal lands.[24] It was not for nothing that the black rock was called "black diamonds."

This does not go without saying because it has become conventional for scholars to extrapolate backward from the 1920s—a time when alternative fuels were making deep inroads into coal markets—to describe a sick, unprofitable, "backward" coal industry wracked by overproduction and excessive competition.[25]

In 1909, the U.S. Census Bureau undertook to answer the question of whether or not the bituminous industry was profitable. The figures showed the industry to be extremely unprofitable. Indeed, in four states operating expenses exceeded the value of the product reported, leading the curious to wonder what motivated the operators to continue in business.

The Census Bureau concluded that the statistics did not provide conclusive evidence as to profits. For one thing, railroads and industrial concerns such as steel controlled fully half of the country's entire coal output. For these captive mines, the value of coal placed on the books was arbitrary and governed by bookkeeping considerations rather than by any need to accurately reflect the real world.[26]

For another thing, many coal company profits did not come from the coal itself: "It should also be noted that many mine operators make a considerable profit by renting houses and selling merchandise to their employees. The Bureau . . . corresponded with many operators whose returns showed an excess of expense over the value of products, and not a few of them stated that, while there was a loss in their coal mining business proper, this was more than counterbalanced by profits from selling merchandise and renting houses."[27]

Like other industries, bituminous coal suffered setbacks during depressions and recessions. For the smaller companies especially, competition reduced sales and interfered with profits. But to project the description of "sick" back into the industry's robust heyday is to forget what the patient was like prior to succumbing.

In anthracite (usually described as a separate industry but arguably just one of several distinctive coal regions), the concentration of capital was unambiguous.[28] "Concentration of interests has been the policy for the last few years in connection with the mining and preparation of coal," reported the U. S. Geological Survey in 1901. "Large central plants are rapidly replacing isolated smaller plants. One large breaker now does the work formerly done by a number of smaller ones. . . . This concentration naturally tends to better management, and to a saving in fixed expenses and in labor." In the anthracite region, "none but a lawyer" could tell where the railroads left off and the coal companies began.[29]

Through the 1880s and 1890s, the Philadelphia and Reading, then the largest corporation in the region, joined other railroad-anthracite concerns in attempting to restrict production to raise the price of coal, just as the old WBA had. Franklin B. Gowen's philosophy of lowering prices to expand markets failed during depressions when markets were contracting. This strategy also depended on the ability to *control* prices, lowering them to undercut competitors and to sell more coal, raising them when necessary to maintain or increase the percentage of profit over costs. In the last two decades of the nineteenth century, the large competing anthracite firms sought to gain control through pooling arrangements in which each firm agreed to supply only a certain percentage of the market. But the operators couldn't stop themselves from cheating: When prices started rising, they scrambled to produce more than their agreed-upon allotment. The system was constantly breaking down.[30]

By the end of the century, the consolidation of the anthracite carriers by means of overlapping directorates, a process directed by J. P. Morgan, ended the competition among them. Eliot Jones, a student of the region, described the results:

> Four out of the nine directors of the Reading Company constituted four of the nine directors of the Central of New Jersey. Six out of the thirteen directors of the Lehigh Valley were directors either of the Reading Company or of the Central of New Jersey. Four out of the fourteen directors of the Lackawanna were directors of the Reading Company or of the Central of New Jersey, three of whom were directors of the Lehigh Valley also. Three of the directors of the Erie were directors of the Reading Co., three were directors of the Central of New Jersey, three of the Lehigh Valley, and one was a director of the Delaware and Hudson. One of the directors of the Delaware and Hudson was, therefore, a director of the Erie, and another was a director of the Ontario, and also of the New York Central. . . ."[31]

The concentration of capital in the anthracite region is well known. But the merger movement was also conspicuous in the bituminous regions of Ohio, Illinois, and Indiana. In 1905, the *Mineral Resources of the United States* reported on "Consolidations in 1905." A typical entry read: "The Southern Coal Co. and Mining Company, of St. Louis, Mo. was formed by the consolidation of the Germantown Coal and Mining Co., the Muren Coal and Ice Co., the Oak Hill Coal Co., the Dutch Hollow Coal Co., the Glendale Coal and Mining Co., the Tower Grove Coal Co., the Dutch Hill Coal Co., the Walnut Hill Coal Co., most of the mines thus taken over being located in St. Clair County, Illinois."[32]

In some regions, a railroad exerted so much control through contracts and leases over the "independent" coal operators along its route that for the railroad to acquire the companies merely would have increased the carrier's risk without improving its control. In the 1880s, the Norfolk & Western Railroad became the dominant economic force in the Pocahontas-New River region of southeastern West Virginia. The railroad set the price it paid operators for the coal, as well as the price at which it sold the coal in the eastern market. (The railroad took seventy percent of the price and gave thirty percent to the operators.) In addition, the railroad distributed coal cars to the operators, and in this way could restrict production with precision, if deemed necessary. The railroad weighed the coal, and employed an inspector at each mine to insure its quality before loading. This system of dominance lasted in this important field until the depression of the 1890s. During that time, the contracts enriched the operators as well as the carrier.[33]

These "independent" operators prospered, but they were hardly independent. In other regions, the "independent" operators found their relationships with dominant railroad corporations less than mutually advantageous. In Rock Springs, Wyoming, a Union Pacific coal town, the three independent operators remaining in 1906 complained bitterly to the Interstate Commerce Commission that they regularly declined half their orders because the Union Pacific Railroad refused to provide them with enough coal cars.[34]

By the twentieth century, then, large corporations dominated the mining of coal, bituminous and anthracite. But, unlike the oil industry, no one company came even close to mastering the entire industry. Although railroad and steel giants divided coal among themselves, and the industry claimed a few giants of its own, the small firm did not disappear. The resource was sufficiently scattered, and the costs of entering the industry sufficiently low, at least at first, that the large companies could never entirely eliminate their smaller competitors. Many of these small operators were nevertheless under the influence or even control of a large corporation. And although no one corporation dominated the entire industry, several bituminous *regions* were as controlled by one group of large interlocked corporations as was the anthracite region.

THE MINE WORKERS

As the new industrial system brought an increased concentration in capital, so it brought a growing influx of new workers. The demographic changes affecting the coal industry mirrored those of the country as a whole. In the 1880s, with the economy on the upswing, the number of persons entering the country in pursuit of a job leapt from the usual two million per decade (since 1850) to more than five million. The depression of the 1890s caused a drop to three and a half million, but after 1900 the numbers leapt again, this time to nearly a million per year.[35]

Gradually, the composition of the incoming tide of humanity changed. Until the 1880s, eastern and southern Europeans had constituted fewer than ten percent of the immigrants. But in Europe, as industrialization spread south, it disrupted and uprooted populations. In the 1880s, sixteen percent of new arrivals came from southern and eastern Europe. In the

1890s, their numbers rose to nearly half. In the first decade of the twentieth century, seventy-one percent of all immigrants to the United States came from Italy or eastern Europe.[36]

The "new immigrants," or "Slavs" as they were often called (though some, such as the Lithuanians, were not Slavs) began entering the coal industry in huge numbers. (The Jews were the exception; they remained virtually absent from the industry.) In the 1880s, Pennsylvania's annual list of coal industry victims began to include men like John Koccollock, killed on 5 July 1882 in an anthracite mine. He was Hungarian, "said to have a wife and family in that country."[37]

Probably half of the eastern and southern Europeans came to industrial work from the farms of Europe. Others had known industrial labor, but not in mining. Others still were artisans skilled in trades other than mining. A few came from the European coalfields, such as those in Hungary.[38]

Hungarians crowded into the Connellsville coke region. There, the ties of the new industrial workers to a more traditional economic system, where all family members shared the work of production on a farm, were revealed in the way the women worked alongside their husbands doing "severe manual labor" at the coke ovens. The coal and coke operators hired the men and paid them by the oven. "Their sturdy wives," wrote the Commissioner of Industrial Statistics in 1885, "in their native land, were accustomed to out-door labor. Naturally to them, most unnaturally to the American view, they assisted their husbands in preparing the ovens, and in drawing the coke therefrom."[39] One condition, as John Bodnar notes, remained constant from one side of the Atlantic to the other: This was the difficult struggle for survival; in solving it day after day, the women were equal partners with the men.[40]

By the early 1890s, the work force in some regions had virtually been replaced. "In the mines surrounding Scranton," reported *The United Mine Workers Journal* in 1892, "it was found that nine-tenths of the miners at present employed are Hungarians, Italians, and Slavs. Five years ago the miners were nearly all Americans."[41]

The new immigrants crowded into the coalfields of Pennsylvania, Ohio, and Illinois. In Illinois, they increased during the 1880s from 604 eastern or southern Europeans to more than 3,000, accounting for virtually all of the industry's expansion.[42] In contrast, when West Virginia's industry began its rapid development, it was manned by white mountaineers along with a growing minority of blacks. But after 1900, more and more new immigrants set their sights on the Mountain State, eventually making up half the work force. In all American coalfields, by 1905 fifty-one percent of mine workers reporting their nationality were Slavs, Hungarians, or Italians.[43]

Another important group entering the coal industry were blacks. Blacks had mined coal in colonial Virginia, and they were also old hands in the Alabama coal mines. But the development of new coalfields in southern Appalachia (Virginia, West Virginia, Kentucky, Alabama, and Tennessee) represented a new kind of opportunity for many black farmworkers who had remained in the South after the Civil War. These new industrial workers shared a similar agricultural past with the new immigrants. By 1900, more than half of all Alabama miners were black. West Virginia's coal mine work force was one-quarter black; eventually, the state would claim more black mine workers than any other. Blacks also worked in other coalfields, but as a tiny minority. Nationwide, by 1910 some nine percent of all coal mine workers were black.[44]

Conventional opinion accused the new workers of pushing the traditional work force out of the industry, claiming that operators were hiring the new workers consciously in order to replace skilled or unionized workers.[45] There was something to this, but it was also true that the new workers came at a time of intense industrial expansion. As coal production doubled, and doubled again, each newly employed Italian or black mine worker could not possibly have replaced an Irish, English, or native-born white miner. Although the new workers contributed to a surplus of labor, the traditional work force simply could not reproduce itself fast enough to fill all the new jobs.

Although employers often replaced strikers with inexperienced strikebreakers, most new workers did not enter the industry by this route. Pennsylvania's Commissioner of Industrial Statistics investigated the coke region and reported, "[W]e believe the [Connellsville] coke operators have been wrongfully charged with the bringing of this nationality [the Hungarians] for the purpose of breaking strikes. . . ." Hungarians first entered the Connellsville region during the boom of 1879, "when everything was at high tide. Work was plenty and labor was too scarce to supply the demand. An effort was made to supply workmen from the cities . . . from New York there came a large number of Germans and Poles . . . here were also some Hungarians and these came when no strike was in existence. Word was sent by them to their friends at home and from that time forth they have come in a steady stream."[46]

Some new workers refused jobs as strikebreakers, particularly when strikers managed to get close enough to talk to them.[47] To give one of many examples, in the early 1880s a Pennsylvania official reported on a coal strike in western Pennsylvania sparked by a reduction. Operators promptly sent agents to Ohio, who returned in short order with "colored strangers." But these blacks, "after receiving more light on the subject," refused to work at the reduced rate. Next, fifteen Hungarians were brought, but they, too, were "easily persuaded to accept the views of the miners on the situation." It took the operators several tries to finally defeat the strikers with a new installment of black strikebreakers.[48]

Whether strikebreaking or not, however, the new employees worked to the advantage of the operators and to the disadvantage of the established work force, by virtue of their sheer numbers. In many coalfields, the new groups contributed to a large surplus of labor, even in the context of industry expansion. In 1892, a prosperous year, coal miners worked on the average just under four days a week, whereas they required five or six days to meet their expenses. In more typical years, many coal miners worked slightly more than half time.[49] Employers had an ample pool from which to draw strikebreakers, or new workers with weaker notions of what constituted their rights. "Labor is so plentiful that the operators can do just what they please," a Welsh anthracite miner complained in 1895. "Pennsylvania is swarming with foreigners—Poles, Hungarians, Slavish, Swedes, and Italians . . . who are fast driving English, Welsh, and Scotch miners out of competition."[50]

Mine workers believed that operators consciously introduced ever more workers in the face of a profound surplus. Peter Roberts, the first scholar to study the anthracite industry, concurred, and noted that the operators pursued this policy in order to "keep laborers in due subjection. They acted on the idea that employes could be better controlled, their tendency to combination more effectually frustrated, and industrial friction more successfully stayed, if they were kept near the starvation point."[51]

On the other hand, matters improved for many experienced miners who were pushed up, rather than out. Miners of English, Scottish, and Welsh extraction, the Immigration Commission reported in 1909, earned the highest wages in the industry.[52] Some moved to the newly developed coalfields of the West and Southwest, where operators offered higher wages in order to attract labor. Others became mine bosses, superintendents, or mine owners. In 1901, a Kansas official observed, "The different classes of labor have the same characteristics here that they have else-

where. The Anglo-Saxons are the best miners. They have a better comprehension of the work to be done . . . and almost all the bosses are chosen
from their ranks."[53]

A Pennsylvania mine inspector listed the nationalities of newly certified
mine bosses in one part of the anthracite field. Of fifty-two bosses certified
in 1886, twenty-six were Welsh, English, or Scottish, seventeen were
American, six were Irish, and three were German.[54] Other British miners
succeeded in becoming coal operators. In the anthracite region, the
Welsh, Scottish, or English operator could find a number of his countrymen in the same fortunate position.[55]

Far from being pushed out, some Britons rushed out of the industry
into related, professional jobs. They became the first state-appointed coal
mine inspectors or found positions in the state Bureaus of Labor Statistics.
In 1903, the Illinois Bureau printed portraits of the state's seven coal mine
inspectors. Caption biographical sketches disclosed their birthplaces: Five
were born in England, one in Scotland, and one in Wales.[56]

The life of W. B. Wilson (1862–1934), America's first Secretary of
Labor, reflected this pattern of upward mobility. Born in Scotland, he
emigrated to the United States with his parents and began working
underground at the age of nine. An early union activist, in 1890 he helped
to found the United Mine Workers of America, serving as its secretary-
treasurer from 1900 to 1908. In 1913, President Woodrow Wilson appointed him Secretary of Labor.[57]

Less prominent British and Irish miners simply left the dirty and
dangerous occupation for cleaner, safer, and better-paid pursuits, or at
least they encouraged their sons to stay in school rather than go into the
mines. Early in the twentieth century, the Immigration Commission
found that in South Fork, Pennsylvania, a coal town of over two thousand
inhabitants, the majority of business and professional people in town were
former coal miners.[58] According to a long-time resident of the anthracite
region, by 1903 most lawyers and some judges around Wilkes-Barre were
the sons of Irish miners.[59]

By 1900, the work force in the coal industry could be divided into two
distinct groups. Those in the first group had more experience both as
workers and as union activists. They faced, as we shall see, a losing battle
to prevent the erosion of their craft, but they had access to the individual
escape route of upward mobility. Those in the second group had little
experience in either industrial work or unionism. Their mobility was only
lateral: They could move from one coalfield to another, or from coal

mining into some other heavy industry such as steel. They had fewer alternatives to working collectively to improve conditions. Within the union movement, the fate of each of these groups became inexorably tied to that of the other.

Because different experiences lead to different views of the world, the radical change in the demographics of the coal industry inevitably affected the consciousness of miners. The new immigrants, naturally enough, had no attachment to the traditional ways of craft mining. Furthermore, as Gerald Rosenblum suggests, most came to the United States in pursuit of specific goals relating to their lives and communities across the sea.[60] They came with the idea of earning a given sum with which to return home.

The more provident of the Hungarian coke workers, according to a Pennsylvania official, saved "from $600 to $1000 with which they either pay off the mortgages held on their property in Hungary by the Jewish, Greek, or Armenian creditors, or they buy a tract of ten or fifteen acres in their native country, build a house and become comparatively independent."[61] Between 1900 and 1906, the Bureau of Labor noted, immigrants sent money orders worth more than nineteen million dollars home to Italy, Austria-Hungary, and Russia. This represented but a small portion of the whole; most immigrants carried their funds home in person.[62]

Many of the new immigrants did not think of the United States as their new country because they did not plan to stay. Neither did many view coal mining as a lifelong occupation. The United States did not keep records of returns, but in 1907 the Bureau of Labor Statistics estimated that twenty-eight to forty percent of all immigrants eventually made their way back to their country of origin.[63] Many were young single men, or married men who had left wife and children behind in the old country. They wanted to earn as much as possible, quickly, in order to return, and so tried to live cheaply, and to work every possible day for as long as possible. "The nine hour question does not strike the foreign element very well," a miner wrote to a labor paper in 1891. "They work 12 to 15 hours per day."[64]

Those who planned to stay had an equally pressing need to save: to bring their families over to join them. "He had only passage money for himself so he had to leave his wife and daughter behind," reflected the daughter, Bruna Pieracci, whose father emigrated from Italy to the Iowa coalfields in 1907. It took him two years to save enough for the passage of his wife and child.[65] Though the persistent cry against them was their willingness to work more hours for less money, most new immigrants had little choice but to work themselves to the bone.

The women of these families also worked harder. A contemporary wrote: "[T]he family income of the slav is . . . increased in ways the English-speaking miner would not think of. The foreign woman does manual work, such as picking coal from the culm banks, carrying driftwood from the forest nearby, and in a score of other ways lessens the cost of living to the slav family. In one or two cases these women have been known to work in the mine as laborers."[66] He might have added the important economic activity of keeping boarders. In 1900, one-third of the foreign-born wives (half of the southern Italians) in a typical anthracite community contributed to the family income by keeping boarders. Arriving in an Iowa coal camp one day in 1914, one Italian miner's wife, Paulina Biondi, took in four boarders that same day. For the next forty years, she kept a continuous minimum of four boarders in a four-room company house while also raising five children.[67]

A variety of factors—the need to save at all costs, the desire to return home combined with little attachment to the occupation of mining, the drive to improve the standard of living, and above all, lack of choice— enabled the new immigrants to tolerate poor living conditions. For these reasons, they were at first difficult to organize.

The established work force aggravated the situation, often, by greeting the new immigrants with open hostility, and not only when they arrived as strikebreakers. In 1884, 150 Irish and English miners in the Connellsville region struck H. C. Frick and Company's coke works because the management had hired seven Hungarians, not strikebreakers. Further east, anthracite miners stoned the houses of recently employed Hungarian mine laborers. There, miners formed societies not to recruit the new immigrants into the labor movement but to expel them from the region.[68]

In 1904, Peter Roberts commented in *Anthracite Coal Communities* upon the treatment the new immigrants received: "They were abused in the press and on the platform, maltreated in the works and pelted on the streets, cuffed by jealous workmen and clubbed by greedy constables, exorbitantly fined by justices of the peace and unjustly imprisoned by petty officials, cheated of their wages and denied the rights of civilized men, driven to caves for shelter and housed in rickety shanties not fit to shelter cattle. . . ."[69]

Even government officials did not hesitate to malign the new immigrants. For instance, in 1884, Pennsylvania's Commissioner of Industrial Statistics had this to say about the Italians coming into the coalfields: "The illiteracy, turpitude, and degraded habits of this class of immigrants, innate and lasting as they are, stamp them as a most undesirable set."[70]

Chief among the new immigrant's abusers was that bastion of unionism, the skilled contract miner. Peter Roberts reported that contract miners in the anthracite region were known to pay the newly arrived laborer $1.57 per day rather than the going rate of $1.75.[71] A black organizer working in Alabama reported that in the 1890s black laborers hired by white miners, "being ignorant . . . allowed themselves to become the servant of a servant, working for 75 cents to $1 per day . . . making for his servant employer fairly good wages, while he himself only earned a pittance."[72] Established miners may have resented the new workers, but they themselves were hiring them and, in the process, undermining current wage rates.

As a group, the new immigrants were no more homogeneous than the established work force had been. Vast differences separated individuals even of the same national group. An Italian who emigrated to the American coalfields recalled that when he first arrived in 1914 he often didn't have to speak Italian, much less English, because in his community people "got along" in the dialects associated with particular villages and regions.[73] If Italians were so divided, how much more so an Italian from a Greek, or a Serb from a Pole. Their common experiences first occurred in the coal camps of the United States.

The new work force profoundly affected the coal industry. Perhaps its greatest impact was on the progress of mechanization. On the one hand, the new mine workers enabled the operators to eliminate or to mechanize the most skilled part of the work—undercutting—and thus to purge a large portion of the most highly paid and most recalcitrant workers. On the other hand, the new abundance of cheap labor enabled the operators to forgo a radical overhaul of all work practices, to forgo the mechanization of loading, done by poorly paid workers, and to forgo the idea of improving efficiency overall, so that fewer workers could produce more coal.

TECHNOLOGY

Technology is the creation of human minds; its evolution is shaped on the anvil of human goals. The history of mechanization and the reorganization of work methods in the coal industry are comprehensible only in the context of the goals and values of contending groups in the industry.

Profit was the overriding goal of the owners, operators, and managers. But the means to this end varied. The owner of a small mine lacked the capital to mechanize, fought to limit production, and hoped for high prices. The corporate operator, usually a railroad, sought to control his fuel supply. Profits lay in the direction of freight charges and full trains, large-scale production and low prices. For this type of employer, extreme control over an extremely cheap labor force could act as a force against mechanization. For a third group, the financial capitalists, profiting could involve withdrawing capital from the coal industry to invest it more profitably elsewhere. Especially for them, mining coal was not an end in itself; what was good for the owner was not necessarily good for the industry.

In contrast, the skilled craft miner was emotionally invested in a coal industry that he had been born into. His skills formed part of a family heritage passed down from father to son. He felt at home underground as nowhere else, for, as one miner wrote, "[W]e seem to develop a mole-like sense which causes us to feel out of our element when above ground."[74] The craft miner's self-esteem was completely tied up with his technical mastery of coal mining. He held strong views on the correct methods of mining, and his life, more than anyone else's life, was profoundly altered by changes in the industry.

The unskilled worker, *when he first arrived*, focused his aspirations outside the industry. He worked to survive, and to save if possible. His work was not interesting. He did not expect to get satisfaction from it, but from the use of the money earned, from his family, and from his community. He was the prototype of the alienated worker of the twentieth century.

This mix of contending values and goals (with those of the operator predominating) provided the context for the limping pace of mechanization in the coal industry. The practical miner with his skill at undercutting and blasting was the first to go, but he was not replaced by an efficient, factorylike system of mining. Operators continued paying miners by the ton until the 1930s; this payment system guaranteed that the miners would bear the costs of inefficiency.

This system helps to explain the long duration of highly wasteful practices. A prime example was the long daily wait for coal cars in which to load coal. Carter Goodrich writes in his classic *Miner's Freedom* that in the 1920s mine workers spent twenty to forty percent of the day waiting for cars: "[J]ust as the greeting "Good Morning" fitted the economic life of our farming ancestors, and as "Low Prices" has been suggested as its proper equivalent for modern housewives on their way to market, so the question, "How are the cars running?" has become a frequent and significant how-goes-it in the mines."[75] Providing coal cars to the mines was a complex logistical problem for railroads and for coal operators. The problem had several dimensions, but the sluggish progress toward solving it stemmed partly from the fact that for the operator who paid the miner by the ton the miner's endless waiting did not present a problem.

Perhaps another disincentive was the fact that the scarcity of cars gave the pit boss an important lever of control over the work force. This lever was, as a Pennsylvania miner called it in 1886, "the accursed free turn system." A free turn was an extra coal car distributed to company favorites. A West Virginia miner, claiming it was practiced in nearly every mine, dubbed it "the abominable robbing system called the free turn." Because the number of cars available for a miner to load had a direct bearing on his typically insufficient income, the free turn was a tasty carrot indeed.[76]

This nonrational system of distributing coal cars did nothing for efficiency but everything for controlling the work force. The same was true of the first important change introduced into the work process of mining bituminous coal. The change involved no new tools or machines. It eliminated the most exacting and time-consuming task—undercutting— but did not replace it with anything. Although undercutting typically took about fifty-one percent of a miner's time, the new procedure was so wasteful of coal that it did not increase efficiency. Mining coal without making the long horizontal cut at the base of the seam was known as blowing the coal off the solid. The coal was drilled and blasted down with no preparatory undercutting.[77] In soft coal mines, blowing off the solid caused the coal to shatter into smaller pieces or into dust—a deterioration in quality. For the workers, "blasting the coal entirely from the solid breasts," as an 1882 mine inspector described it, produced large volumes of smoke and left the mine in an unhealthy condition.[78] Shooting coal off the solid provided one and only one advantage: It enabled operators to hire inexperienced workmen over whom they could exert considerably more control.

In 1900, a Kansas official explained the relation between the new labor force and this new work method: "With, or perhaps as a result of,

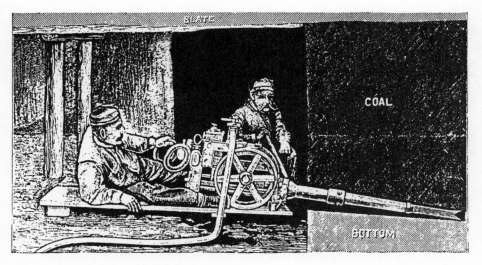

Harrison coal mining machine in operation.

the gradual introduction of other classes of labor the coal has been less and less undercut, and the practice of shooting from the solid has become more common. The result is that poorer grade coal is now marketed. . . . In such mining, where quantity and not quality of work counts, the higher paid laborer is at a disadvantage."[79]

Shooting the coal off the solid eliminated the need for skilled labor but required no capital to install. It is not surprising, therefore, that by the end of the century, the practice had become commonplace. In Illinois, for example, more than seventy-five percent of all coal mined in 1898 was shot off the solid.[80]

Improved productivity awaited mechanization in the form of the undercutting machine, first introduced by large bituminous operators in the late 1870s. There were two types. One, the Harrison Machine, cost only about three hundred dollars in 1890. It worked something like a horizontal jack hammer, its "pick" dealing two hundred strokes per minute to the base of the seam. The other, the Lechner Machine, worked by a revolving bar into which sharp steel points were inserted. The bar was driven into the bottom of the seam, making a cut three feet wide and six feet deep. The machine "runner" then backed it up and moved it over to begin the next cut. Both types were powered by compressed air produced by a steam engine above ground and conducted through the mine in pipes.[81]

In 1891, the undercutting machine was statistically insignificant: Only five percent of bituminous coal was machine mined. By 1900, twenty-five percent of the output of bituminous coal was so mined. Fourteen years later, the percentage had risen to one-half.[82]

The undercutting machine elevated the productivity of the individual miner, increasing his average daily output from 2.57 tons in 1891 to 3.71 tons by 1914. At the time, the U.S. Geological Survey attributed the change entirely to coal-cutting machinery, but Keith Dix notes that improvements in haulage also helped to increase productivity.[83] The important point is that in the last two decades of the century operators who desired to increase efficiency had gained the means for doing so.

The machine enabled the operator to reduce the work force, especially the skilled miners who alone knew how to undercut the coal properly. This lowered the cost of production. In one Ohio mine, the introduction of machinery in 1904 resulted in a reduction of the work force from 140 to 40 men.[84]

The undercutting machine benefited the coal operator only. Recognizing this, one of its inventors specified that it be made sturdy enough to withstand being clobbered with a sledge hammer.[85] The machine became the means to political ends, inasmuch as the skilled workers it eliminated were also the most pro-union segment of the work force.

Geological variations made installing the coal cutters less than perfectly routine, but the human reluctance to see them succeed proved to be an even more intractable difficulty. In 1886, a Wyoming coal superintendent reported to his superior that, although the machinery was first class, "we met with difficulty in getting men to load after them and do the other necessary day work. All parties wanted to see them a failure. We changed the entire outfit of men several times, and think that we now have a good lot of men, but the old outfits used the machinery up pretty well." "All our mishaps" notwithstanding, this manager reported, the work was accomplished cheaper than it could have been done by hand.[86]

Why should the miners have liked the machine? When introduced into mines where undercutting was still done, it increased the danger of mining.[87] It increased the fine dust suspended in the air, and sped up the release of gas from the coal, worsening breathing conditions. It was so loud that even the industry journal *Mines and Minerals* admitted its tendency to produce deafness. The noise drowned out the sound of the roof creaking and cracking—the traditional miner's warning of an impending roof fall. Finally, the machine vibrated so violently that machine operators frequently suffered from internal injuries.[88]

The machine undercut the miners' skill as surely as it did the coal, turning their hard-won knowledge into a worthless commodity. A coal mine inspector in the early 1880s couldn't comprehend why the miners objected to the machines, given that undercutting was "the most laborious part of the work." He supposed it would be natural for the miners to look upon the machine with some degree of satisfaction, but he found quite the opposite to be the case. The miners felt the machine would throw them out of work, reduce wages, and create a surplus of labor—in short, that "all of the profits derived from the machine would go into the pockets of the operator."[89]

A decade later, one coal miner rebutted an advocate of the undercutting machine thus: "He does not say what could or should be done with the thousands of practical miners that would be thrown out of work who depend upon and know no other craft but the art of digging coal to support themselves, wives and little ones."[90] "[T]he mining machine is the natural enemy of the coal miner," we read in an 1888 report of the Illinois Bureau of Labor Statistics. "It destroys the value of his skill and experience, obliterates his trade, and reduces him to the rank of a common laborer or machine driver. . . ."[91]

The relationship between the new labor force and the new machine was plain. "The character of the men employed in the Illinois coal mines has changed," an industry observer reported in 1895:

Formerly English, Scotch, Welsh, and Irish. At present there are some Germans, majority are Slavonians, Russians, Italians. This change has resulted partly from the prejudices among old hand miners. When on account of competition and frequent labor troubles, a cheap and reliable system became necessary and machines were introduced, there was a great deal of bitterness and the result was that nearly all the old miners left the business. . . . Many operators think it doubtful whether machine mining would have become practicable without this introduction of foreign labor. The fact is that under the present system coal mining in Illinois has either ceased or is fast ceasing to be a skilled occupation.[92]

A similar innovation was the "driller," a mechanized drill that enabled a man to bore the three-and-one-half-foot hole required for blasting in two minutes.[93] Another change was the employment of one shotfirer to do the blasting throughout an entire mine. In this case, however, the new practice did more than provide management with a means of reducing wages. It unquestionably improved safety in the mine, especially if the shotfirer did his work at night when the mine was relatively empty. By the twentieth century, the employment of a shotfirer working alone at night had become a trade union demand.[94]

The elimination of hand undercutting, hand drilling, and individual blasting occurred sporadically over many years. Even with these changes, the disintegration of the craft miner's job did not translate into full-scale automation. Loading by hand, for instance, persisted into the 1930s. But mechanization, gradual as it was, carried with it a corresponding increase in the division of labor. In 1885, a Pennsylvania coal mine inspector offered a glimpse of the new work organization. A large mine operating thirty-two undercutting machines "requires one man to handle each machine, and a boy to scrape the dust away from the cutter. One man and a boy can [undercut] about 35 tons per day. Another set of men blasts the coal down and loads it into wagons."[95]

In the traditional style of work, every third person working underground thoroughly comprehended the process of coal mining, and was presumably engaged in passing his knowledge on to someone else. The splintering of the whole process into smaller, more repetitive jobs caused the disintegration of the body of knowledge that previously had been passed along. New workers coming in could no longer expect to learn the procedures of coal mining from beginning to end.

At the same time, management grew ever more important. This was accomplished in part through an amendment to the Pennsylvania mining law enacted in 1885, requiring inside bosses to pass a competency examination administered by the Commonwealth before they could be so employed. The law also required every coal operator to employ a certified mining boss in every colliery.[96] The law had the effect of increasing the number of managers in the industry. It accelerated the shift in the location of information about the process of mining, from the production worker to the manager. Finally, it pushed open a little wider the door through which the skilled craft miner could pass from production work into management.

The increased division of labor in the coal industry, and its social meaning, was plainly described in the 1888 Report of the Illinois Bureau of Labor Statistics. The skilled miner "takes his own tools into the pit and undertakes to deliver from the wall of mineral before him certain tons of coal for a certain sum per ton. He mines and drills and blasts and loads his own coal, timbers his own roof, takes care of his own tools, and is responsible mainly to himself for his personal safety and the amount of his output." This was the old method. The miner was responsible for mining. He autonomously made dozens of decisions required for carrying it out. In the machine mine, on the other hand, "[I]t takes 7 or 8 men to perform these various functions. . . . [First] the coal is undercut. . . . A blaster follows with tools and explosives, loosening the mass; the loaders reduce it and shovel it into pit cars; the timbermen follow and prop the roof, which no longer has the mineral to rest upon. Laborers assist in every process, and a machinist is retained for repairs. Each one does his own certain portion of the work and no more."

The value of the system to the owner, the Report explained, was that it relieved him to a large extent of skilled labor "and of all the restraints which that implies. It opens to him the whole labor market from which to recruit his force."[97]

7

A CLOSING UP OF OUR RANKS: MINERS AND THEIR UNIONS IN THE NEW ERA, 1880-1900

Our Order contemplates a radical change in the existing industrial system . . . while the trade unions accept the industrial system as it is and endeavor to adopt themselves to it. The attitude of our Order to the existing industrial system is necessarily one of war.

Officer of the Knights of Labor, *1880s* [1]

The introduction of the factory system added greatly to the progress of trade unionism. Trade unions became a necessity, and . . . a possibility. The bringing together in the industrial centers of large numbers of workmen created new opportunities, and the . . . hardships accompanying every transition from hard labor to production by machinery made them, more than ever, a necessity. Philosophically considered, trade unions are an essential part of our industrial system.

Editorial, United Mine Workers Journal, *1891* [2]

Few of the changes turning the coal industry upside down and inside out benefited the mine workers. Growing concentration of ownership increased the power of the employer, and the mining machine attacked the value of the miner's skill. The new workers in their horrifying numbers eliminated any hope for that important lever in the union cause—a scarcity of labor.

But in the 1880s the changes seemed anything but inevitable. Wage labor, it is true, was firmly established: More than half of all American workers now worked for wages. Yet nearly as many still occupied themselves in some sort of independent craft. As late as 1900, only sixty-one percent of the work force worked for a wage or salary, a sharp contrast to the more than ninety percent who do so today.[3]

Neither, in 1880, did the large corporation dominate the economic landscape as it would twenty years later. And mechanization, though present, was an oddity even in large coal firms. The sight of an Italian or Slav laborer was only slightly more common.

It was not irrational or even backward-looking for the skilled craft miner of 1880 to oppose the large corporation, the mining machine, and the new worker. The new worker faced the blatant racism of the skilled miner, and this touches upon a seamy and yet unresolved side of American history. In the context of his battles with his employers, however, the miner's antagonism was understandable. The newcomer was a human being with intelligence, hopes, aspirations, fears, and the dignity of his humanity, no less than the established worker. His very history, both personal and collective, was entirely similar to that of the established worker. Yet, in the view of both the skilled miner and the coal operator, he was also a tool in the class struggle.

Coal miners' unions of the 1880s and 1890s at first opposed the changes of industrial capitalism. Later, miners fought over whether to oppose the system or to reform it. Finally, the success of the new system spelled the failure of those who had opposed it absolutely. Of necessity, miners moved their struggle onto a new terrain, one dominated by corporate employers.

THE KNIGHTS OF LABOR

The Knights of Labor, the most important working-class organization of the 1880s, opposed industrial capitalism. It was a national organization composed of regional and local assemblies made up of men and women from various occupations. Coal miners began joining in the late 1870s, after coal operators had destroyed their alternatives (the WBA and the National Miners' Association).[4] The Knights, founded by Philadelphia garment cutters a decade before, at first conducted its affairs in extreme secrecy. It dropped this policy in the late 1870s, in part to avoid being accused of "Molly Maguirism." From then on, one could read in place of the customary five stars, "the Noble Order of the Knights of Labor."[5] In July 1879, the newly public organization gave a picnic in the anthracite coal town of Shenandoah, and it was attended by more than 10,000 persons.[6]

The Knights welcomed farmers, small-scale merchants, artisans, and other types of workers. The glue binding this eclectic membership together was their common antipathy to the large corporate enterprise that threatened their way of life. For this reason, the Knights shut their doors to three groups representing the new industrial system—bankers, lawyers, and stockbrokers.[7]

The organization worked for a humanistic society in which equality was the leading value. A Declaration of Principles formulated in 1878 held that the "alarming development and aggressiveness of great capitalists and corporations, unless checked, will inevitably lead to the pauperization and hopeless degradation of the toiling masses."[8] The Declaration proposed legal equality between capital and labor, envisioning a society in which the small-scale enterprise formed the economic basis for equality. The purpose of the Knights, stated its nationally prominent leader, a machinist named Terence V. Powderly, was "to make each man his own employer."[9]

Indeed, in a system of small-scale enterprise, the apprentice could realistically look forward to becoming his own employer. Reflecting this wish, throughout the 1880s the Knights advocated the formation of cooperatives—businesses owned by workers on a cooperative basis. A group of striking coal miners in Illinois embarked on such a venture but found their efforts thwarted when railroads refused to haul their coal. To the Knights, the small cooperative enterprise seemed like a good alternative to the large corporation. The looming reality of the large firm, not to mention its uncanny ability to pervade the economic environment of the small firm, makes this program seem utopian. But hindsight fails to take into account that in the early 1880s the coming predominance of the large-scale corporate enterprise did not seem inevitable.[10]

The Knights' vision of a society composed of equal small enterprises was also the source of its organizational structure, a distinctly nonmodern arrangement in which people from various occupations belonged to the same local assembly. The structure reflected the Knights' goal of reforming all of society (not just the workplace) and its base in the community. It also reflected a conviction about the common interests of a polyglot group comprising merchants, craftspeople, farmers, and others. All opposed the coming corporate society.

But the Knights were not Luddites. In 1880, their national newspaper editorialized that no sensible man would "turn back the hands upon the dial of human progress by abolishing machinery." Machinery could be made to lighten the burden of labor. Yet, in hideous fact,

> *the first result* of the introduction of labor saving machinery . . . is the degradation of labor through the subdivision of labor. Man . . . becomes a part of a machine, and because a piece of machinery can be run as well by a child as by a man, his labor is brought into constant competition with child labor. The cradle is robbed to satisfy the insatiate greed of the soulless corporation *that owns the machine. . . . .* From dull monotony a man's labor becomes perpetual torture, and his mind narrowed within circumscribed limits of toil becomes dwarfed, thus extinguishing the manly independence so essential in all who exercise a freeman's right.[11]

Egalitarian ideals caused the Knights to welcome women as members along with men. The Declaration of Principles supported equal pay for equal work, and women participated actively in the organization. The Knights also favored equality between blacks and whites, although a condescending tone crept into pronouncements on the subject. "[T]he (outside) color of a candidate shall not debar him from admission," Powderly ruled in 1880. "[R]ather, let the coloring of the mind and heart be the test."[12] In places like West Virginia, black and white miners belonged to segregated Knights assemblies, but worked together on joint regional committees.[13]

As for southern and eastern Europeans, generally speaking the Knights simply excluded them from consideration. Judging from the organization's newspaper, in the early 1880s such immigrants did not exist. A philosophical difficulty may have caused the blind spot: A society conceptualized as an aggregation of small enterprises had no room for hordes of unskilled workers. As these workers became more apparent in the real world, the Knights became hostile. In 1885, Powderly vilified Hungarians as moral lepers and described in gross detail the filthy living conditions of one group living in Scranton.[14] But he had nothing to suggest in terms of dealing with them in a realistic manner. The organization seemed powerless against this new reality; in the latter part of the decade, immigrant strikebreakers played a decisive role in breaking its back.[15]

What the Knights finally did do about the new immigrants was as unrealistic (in retrospect) as their opposition to the large-scale corporate enterprise. They launched a campaign for legislation against the importation of contract labor, a system in which employers' agents hired workers in Europe, advancing them transportation money. Contracts brought America only a minute percentage of the immigrants, mostly skilled glass workers. Nevertheless, the Knights wholeheartedly put their energies into the campaign for the Foran Act (1885) to prohibit contract labor. In lobbying for the bill, they did not mention glass workers but denounced Italians, Hungarians, and others who came without the benefit of a contract.[16]

The Knights' cardinal ideal of an egalitarian society prevented them from coming to terms with either the new industrial system or its new industrial workers. But their lack of empathy with the Europeans was mild indeed compared with the sheer loathing they turned against the Chinese. One would think that egalitarian values would preclude such attitudes and, indeed, at first the organization carefully articulated an antiracist position toward all peoples, including Asians.[17] But in the context of an epidemic of anti-Asian racism then sweeping the nation, the Knights did an about-face and acquired a raging case of the disease. In 1880, the governing General Assembly voted down overwhelmingly a point of order that would have welcomed the Chinese as members.[18]

In 1882, the U. S. Congress legitimized such sentiments by enacting the first exclusion law, which prohibited entry into the United States of "convicts, lunatics, idiots, persons suffering from particular diseases, paupers, and the Chinese."[19] The curious demographic context of Chinese exclusion was that the number of Asian entrants to the United States was positively minuscule compared to any other group. Only 400,000 of the thirty-two million immigrants to arrive between 1820 and 1930 were Chinese.[20] Perhaps their very lack of numbers made them ideal scapegoats for the new industrial system.

The Knights' firm belief in the small-scale enterprise as the foundation of society influenced the leadership to favor a certain prescription for resolving industrial conflict—the Board of Arbitration and Conciliation.[21] Such a board consisted of three to five members chosen by both employers and workers involved in a dispute. Conceived as a neutral third party, the board heard both sides before determining the terms of a settlement. Supposedly, its decisions were fair and binding. Arbitration and Conciliation differed radically from the trade union idea of a Joint Wage Agreement negotiated directly between opposing sides. The Joint Wage Agreement (which came into existence in the late 1880s) assumed that the opposing sides had opposing interests. The outcome was heavily influenced by the relative power of the participants, with the power brought to the table on the miners' side being the threat of a strike.

In contrast, Terence V. Powderly and the leadership of the Knights of Labor opposed strikes. In 1879, Powderly called them a "relic of barbarism."[22] The following year, the governing General Assembly resolved that strikes produced "more injury than benefit to working people, consequently all attempts to foment strikes will be discouraged."[23] The organization's national newspaper railed against strikes in virtually every issue.

But history, along with the membership of the Knights of Labor, was passing the leadership by. Between 1881 and 1886, the U.S. Bureau of Labor counted more than 22,000 strikes across the country, including hundreds initiated by local assemblies of Knights. More than half were either successful or partially successful, demonstrating that in fact strikes could be a valuable instrument for making changes in the face of intense employer opposition.[24]

Coal miners in particular stuck with their long tradition of striking to maintain or increase the tonnage rate.[25] In Pennsylvania from 1881 to 1886, they embarked on eight hundred different strikes.[26] "[Coal] strikes are of almost constant occurrence in one part of the country or another," we read in the 1889 volume of the *Mineral Resources of the United States*, "and no annual report of the industry can be written without mentioning some section which has been seriously injured from this cause."[27]

Powderly struggled against the coal miners' propensity to strike. In 1882, he admonished an assembly of striking Maryland miners that if instead they had kept working they could have purchased a mine of their own and been free of the bosses.[28]

There may have been more than the vision of a cooperative society of artisans behind the reluctance of the Knights of Labor to support the strike weapon. The organization welcomed both merchants and managers into its ranks. Coal mine superintendents and pit bosses as well as the men they supervised were called upon to decide on issues affecting the miners. In Alabama, a coal mine superintendent also functioned as a key statewide leader of the Knights; many other leaders were entrepreneurs or managers, not workers.[29]

For these and other reasons, the leadership worked diligently to stop coal miners from striking, but to little avail. In 1882, Clearfield, Pennsylvania, coal miners asked their employers for a fifteen cent raise. "The announcement that a mass meeting would be held . . . created grave apprehension," according to a state official, "for it was feared that in the event of unfavorable replies . . . a general strike would at once be declared." At the mass meeting, an official of the Knights "pointed out in a forcible manner, the folly of labor, in a disorganized state, entering into conflicts with organized capital." On this organizer's advice, the miners made three extended attempts to persuade operators to submit to an Arbitration and Conciliation Board. The operators declined, in the meantime discharging those making the requests.[30]

This action prompted the miners to write to Powderly requesting aid. He fumed in reply that he wished the men would "have patience and become organized before launching out on strikes, but it is the old story over again. Strike in a hurry and repent at leisure." Powderly's antipathy for strikes blinded him to the fact that the Clearfield miners, far from striking, were doing exactly as he had recommended. They were making polite requests of the operators, who were discharging them for their trouble.[31]

The strike waves rippling through the American coalfields began to include mine workers who lived outside the purview of the Knights of Labor. In 1883, a local union in the Connellsville coke region, now dominated by the gigantic H. C. Frick Company, began organizing the coal miners and coke workers. The attempt failed, but the significant fact was that one of the organizers was a Hungarian.[32]

In 1886, the Connellsville coal and coke workers became agitated over low wages, the more so because operators were enjoying a period of prosperity. "The Huns had not been saving money as they longed to do," explained the state Commissioner of Industrial Statistics, "and it was very easy to convince them that they were being made, with their fellow-laborers, the victims of oppression." This feeling soon gave rise to "self-constituted representatives . . . walking about spreading views antagonistic to the coke operators."[33]

This "walking about spreading views" fertilized "one of the most stubborn and bitter strikes" yet to occur in western Pennsylvania. Hungarians inaugurated the strike on 18 January 1886, demanding a wage increase and the right to hire their own check-weighman at every scale. "Grown bolder by repeated successes and liberal supplies of Hungarian whiskey," the commissioner reported, "the fiery Huns resolved upon war in earnest. . . . A mob of 400 men, armed with clubs, knives, coke forks, and revolvers, marched from Mt. Pleasant to Stonerville, stopping at various plants along the route, driving away the men at work, break-

ing down oven fronts, demolishing wheelbarrows, and throwing loose
tools and supplies . . . into the burning ovens. At some points resistance
was attempted, but the mob swept on, beating a number of bosses almost to death." To some extent contradicting himself, the commissioner added, "The mob consisted largely of Hungarian women, who
brandished knives and other warlike implements in the air."[34]

In the context of prosperity and high prices, the operators met the
strikers' demands, and they returned to work victorious. This startled a
few English-speaking miners into revising their opinions on the "docile"
Hungarians. A miner from Sutersville, Pennsylvania, humorously expressed the hope that "some of the Hungarians from the coke county will
come down and organize us 'poor degraded' English speaking people
. . . as we are now at the mercy of the bosses."[35]

Increasingly, mine workers like those in the coke region worked in the
shadow of a gigantic firm. Yet the leaders of the Knights would not work
for tangible improvements within this context. Trade unions, organized
according to occupation, began to spring up alongside Knights of Labor
assemblies to discuss how best to work for higher wages and shorter hours.
An officer of the Knights explained the difference between his organization and these trade unions: "[O]ur Order contemplates a radical change
in the existing industrial system . . . while the trade unions accept the
industrial system as it is and endeavor to adopt themselves to it. The
attitude of our Order to the existing industrial system is necessarily one
of war."[36] Another officer stated, "We do not believe that the emancipation of labor will come with increased wages and a reduction in the hours
of labor. We must go deeper than that. . . . [T]his matter will not be
settled until the wage system is abolished."[37]

The new immigrants of Connellsville faced a daily grind of extremely
long hours and a life of grueling poverty with no end in sight. Neither
they nor the more established workers could minimize the importance
of hours and wages. Only men who had never experienced such conditions could decide they were unimportant. The very organizational form
of the Knights—the mixed-occupation assembly—made it difficult to
deal with concrete matters even when the members so desired. "The
idea of trade unionism," a coal miner revealed in 1887, "is stalking
through the Knights of Labor with monstrous strides. . . . A miner does
not want his wages adjusted by an axe-maker, nor a glass-blower by a
stone cutter."[38] Often a local trade union open to workers sharing one
occupation coexisted with a local Knights of Labor assembly. Many peo-

ple belonged to both. Chris Evans, a miner who eventually broke with the Knights, recalled, "There was always something lacking in the mixed local where the doctor, the grocer, and the business men were called upon to act on questions in which the mine workers alone were directly interested. The commercial man rarely looks upon a strike in the same light a coal miner does. . . ."[39]

Miners also resented the habit their officers had of sending representatives who knew nothing of the inner workings to negotiate with employers of the mines.[40] Geological variations made adjusting miners' grievances a complicated, technical affair. Nonminers did not know enough to negotiate effectively.

For years, coal miners within the Knights agitated for their own separate division, an organizational change opposed by the leadership. Finally, in May 1886, after a lengthy internal battle, they got what they wanted. The Knights allowed the formation of a new division (called National Trade Assembly 135) controlled exclusively by coal miners.[41] It was in a very real sense a national union of coal miners, though it remained part of its multioccupational parent organization. Within the Knights of Labor, it represented a shift toward trade unionism.

But the timing was interesting. Coal miners formed their own division of the Knights in the wake of the founding convention of another national trade union for coal miners, the National Federation of Miners and Mine Laborers. For the previous several years, coal miners had done without a national union. Now they had two. As many diggers would have occasion to remark, two was worse than none at all.

THE NATIONAL FEDERATION OF MINERS AND MINE LABORERS

The National Federation, founded in September 1885, included members of the Knights as well as nonmembers.[42] Many mine workers did not consider the two organizations to be intrinsically hostile to each other. Nevertheless, the coal miners' movement proceeded to lose five years in the quarrel that developed between them. According to the National Federation version of events, just as the new coal miner's union was making excellent progress, "[t]o our utmost surprise, a rival steps in between us, emanating from the noble order of the Knights of Labor, who form a similar organization . . . following our footsteps in almost every particular, and as a firebrand thrown in among us is only calculated to create discord and disunite the miners and mine laborers of this country, and bring ruinous and disastrous effects upon the members of our trade."[43]

The National Federation's objectives open a window on the values and the realities of the 1880s. Wages, the union announced, were no longer regulated by the miners' skill as workmen, nor by the value of the coal produced, but by competition with cheap foreign labor. Excellent workmanship, which involved not breaking the coal into overly small pieces, was made irrelevant by the practice of screening the coal before it was weighed.[44]

The National Federation advocated the "principles of arbitration and restriction [of production]," in order to remove the causes of strikes. It hoped mine workers would become citizens, "that we might secure by the use of the ballot the services of men friendly to the cause of labor." The traditional labor force still included numerous noncitizens. It advocated the eight-hour day, a law requiring weighing *before* screening, payment every two weeks, the abolition of company stores, and legislation for "more efficient management of the mines so lives and health of our members might be better preserved."[45]

In a dramatic move toward trade unionism, the National Federation pioneered the Joint Wage Convention, a meeting between representatives of operators and of union miners for the purpose of negotiating a contract. The first Joint Convention was held in the region known to the coal industry as the Central Competitive District—western Pennsylvania, Ohio, Indiana, and Illinois. It became the model upon which union contracts have been negotiated in the coalfields to this day.

This element was important because in many other ways it was impossible to distinguish the National Federation from its rival. For instance, the Federation and its constituent trade unions looked down on new immigrants no less than the Knights did. Consider this comment on Hungarian, Polish, and black strikebreakers working in Ohio's Hocking Valley in 1885: "They are living under all conditions, except the conditions in which semi-civilized or even tolerably moral people would live. At many places large numbers are packed into houses like hogs in a pen and apparently with aspirations no higher than that of the quadruped mentioned."[46]

The following year, the Federation's newspaper favorably reprinted a sermon asserting that "free American laborers" were being driven out of the coke region to make way for "half savage Hungarians."[47]

The trade unions making up the National Federation opposed contract labor, believing it "offered a menace to the stability of American institutions." An Ohio union opposed this type of labor "in the name of the American artisan." The attack against contract labor was no more relevant in trade union clothes than it had been in the garb of the Knights of Labor.[48] This Ohio union also favored uniting with the independent operators for the purpose of "hastening the departure of the soulless corporation," a sentiment that could have come straight from the mouth of any loyal member of the Knights of Labor.[49]

Both the National Federation and the Knights of Labor had, after all, emerged from the same cultural soil—that of the British craft miner. Locally, the two unions were not even always separate. In 1885, a miner from Dixon, Indiana, became destitute after being fired for organizing miners into both the Knights and a local trade union.[50] A miner from Ohio's Hocking Valley reported that the Knights had taken up the local trade union "in its infancy, reared and fostered it as an auxiliary, and when the Federation started we held the same views and did what we could to make it a success."[51]

Despite their similarities and even an overlap in membership, the escalating rivalry between them made it increasingly difficult for either to function. The Joint Conferences between operators and union representatives began to stagger under the weight of the feud. (Although the Federation had initiated the Joint Conferences, the Knights participated.) At the negotiating table, operators witnessed public disputes between the feuding parties; in 1889, the whole apparatus of the Joint Convention collapsed because of disunity on the miners' side.[52]

The dual union controversy raged through the workplace as well. "In every mine there were angry discussions between members of the respective organizations," wrote Andrew Roy, "and when arguments failed to convince the debaters, they frequently resorted to blows."[53]

Union organizing, difficult at best due to employer opposition, became embroiled in the controversy. In 1886, two different campaigns sank in the mire. In the anthracite region, two Federation organizers held "successful meetings at all points visited. . . . [M]any new recruits were made members of the already organized local unions." A short time later, organizers from the coal miners' assembly of the Knights of Labor made their way over the same territory and "the fight for supremacy was so furiously waged that the power and influence of organized effort was wasted. . . ."[54] Both unions embarked on a strike the following year, which was largely supported by the new immigrants. But, as Victor R. Greene notes, the reluctance of the English-speaking miners combined with a bickering leadership brought the effort to a disappointing end.[55]

The same scenario was replayed in the Connellsville coke region. In 1886, the trade unionists sent one of their best speakers "and from that day lodges . . . multiplied throughout the coke region, gathering together of thousands of employees of all nationalities."[56] The Hungarians joined the new trade union movement in large numbers. They worked for the H. C. Frick Coal and Coke Company and urgently required what officers of the Knights were loathe to fight for: improvements in wages and hours. Yet others in the coke region, mostly the more established workers, adhered to the Knights of Labor. "The great house of cokers is divided against itself," declared a Pennsylvania official. "The jealousies of leaders have arrayed the two labor organizations in the coke region against each other in a hopeless war for supremacy."[57]

THE UNITED MINE WORKERS OF AMERICA

This chaos resolved itself when the two unions, faced with the utter futility of doing otherwise, merged.[58] Some 240 coal miner delegates accomplished this astonishing feat of diplomacy in Columbus, Ohio, in January 1890. After several days of deliberation, they founded the United Mine Workers of America.

The merger achieved a delicate balance between opposing positions and antagonistic parties. No miner was asked to sever his relationship with his parent organization. The new union was affiliated equally with two umbrella associations: the Knights of Labor and the American Federation of Labor (the federation of trade unions). Within the UMWA, Knights of Labor assemblies and trade union lodges had equal rights. Knights and trade unionists were equally represented on all committees and in all official positions.[59] Within a few years, the Knights would lose influence, but at the time no other organizational form would have survived.

From this most unpromising beginning, the United Mine Workers of America grew into the largest labor union in the United States. Within one decade, it was the largest constituent union of the American Federation of Labor, with more than 91,000 paid-up members.[60] Although the AFL consisted mostly of craft unions, the UMWA was an industrial union, one that embraced both skilled and unskilled workers.

The founding resolutions bared old, unresolved issues. They advocated the eight-hour day and payment in legal money. They opposed compulsory buying in company stores, short-weighing, the employment of children under age fourteen, and the use of hired gunmen to enforce company policies. They advocated improved safety procedures "to reduce to the lowest possible minimum the awful catastrophes which have been sweeping our fellow craftsmen to untimely graves by the thousands. . . ."[61] As if to emphasize this last point, two separate disasters obliterated two of the new union's local assemblies in the first year.[62]

The UMWA's position on strikes was a delicately composed amalgamation of opposing viewpoints. Mine workers would ameliorate their conditions by means of "conciliation, arbitration, or strikes."[63] The difference was that the new union made it a high priority to accumulate a defense fund for use in the event of a strike.

The discussion did not end there of course. In 1892, the president of the Ohio district articulated the trade union view against "the enchanting phrase," arbitration and conciliation: "It tends at times to allure those to whom it is proposed, yet, notwithstanding this I am an implicit believer in the potency of a strong labor organization. A closing up of our ranks will do more to secure justice and fair play than all the arbitration boards that ever existed to take advantage of the laborers' weakness while ostensibly proposing justice."[64] On the other side, proponents of arbitration and conciliation also continued to influence the organization for years to come.

The UMWA's stance toward industrial capitalism itself contrasted with that of the Knights of Labor. "The wage system," announced President John Rae, "is a natural and necessary part of our industrial system. The solution to the many problems now confronting us is to be found, not in the arbitrary overthrow of the system, but in maturing and perfecting it."[65]

Mechanization, too, seemed inevitable, if not exactly pleasing. In Ohio, a UMWA officer told his district convention:

The tendency of the age is to substitute machinery for hand labor. This is noticeably so in our business. . . . [E]very day machines are increasing in numbers and improving in efficiency. The displacement of hand labor by this method has only begun. It is matterless if we approve or disapprove. My judgment is that mining machinery is but in its infancy and hence we should prepare ourselves to yield with as much grace as possible to the inevitable and turn our attention to securing the best terms possible.[66]

A significant change was occurring. Yet mine workers had not merely abandoned a vision of a more equitable society in favor of "bread and butter" unionism. For instance, a Cincinnati suffragist persuaded the delegates of the first annual convention to pass a resolution in support of women's suffrage. The following year, the convention advocated the abolition of the Senate, or failing that, the direct election of senators. (Until 1913, senators were elected by state legislatures, not by popular vote.) In 1894, with the People's Party in its heyday, the miners' convention endorsed not only government ownership of railroads, telegraphs, and mines, but also the collective ownership of the means of production.[67]

The UMWA started out with 17,000 members, mostly white Anglo-Saxon Protestants or Irish Catholics. Within a short time, however, the recent ethnic entrants to the industry again demonstrated their readiness for collective action. In 1891, the union announced a campaign to introduce the eight-hour day into the coal industry. By way of response, the H. C. Frick Company cut the wages of nine thousand men in the Connellsville coke region and announced that the nine-hour day was on for three years to come. The miners, who were not union members, struck. After the fact, the UMWA decided to back their strike. The new immigrants thus led the way in the union's first officially sanctioned strike.[68]

Obviously, they were prepared to stand up and fight, and the UMWA welcomed them into the union. Still, the typical miner probably looked down on his new fellow worker, whether consciously or not. The highest praise could reek with condescension. "The great majority of the Connellsville coke and mine workers were Slavs, Huns, and Poles, and until this suspension occurred, were regarded by the American miners as a servile people," recalled Andrew Roy, an erstwhile miner born in Scotland. "But during the strike they displayed the haughty and fierce spirit of the Anglo-Saxon race. . . . Even the Slav women showed fight, shaking their fists in the faces of the sheriff's deputies."[69]

The Connellsville strikers put up a good struggle but lost the battle. Some "new immigrants" were newer than others. The more experienced Europeans were defeated, a labor paper reported, by the importation of "green ones to take their places."[70]

Nevertheless, the infant UMWA fared rather well until the stock market collapsed in 1893. The ensuing economic storm carried off hundreds of members and eroded the finances. Then, in 1894, in the depths of the depression, the leadership made a disastrous decision to call a nationwide strike. Across the nation, coal miners struck for the purpose of reducing stockpiles of coal in the hope of raising the tonnage rate. Here was a startling reproduction of the long and ill-advised strike of 1875. The defeat of 1894 was nearly as bad. At the end of the year, the secretary was obliged to report that "charity, sweet though humiliating, has saved thousands of our miners from absolute starvation."[71] Two years later, President Phil Penna listed the union's chief possessions as "disappointment, low wages, adverse conditions, and an empty treasury."[72]

The union struggled on, and managed to survive the depression. Sometime during the 1890s, it gained an organizer whose name was to become a household word, first in the coalfields and then across the country. John Brophy recalled being introduced to the Irish-born "Mother" Mary Jones in his underground workroom. "She came into the mine one day and talked to us in our workplace in the vernacular of the mines. How she got in I don't know; probably just walked in and defied anyone to stop her." She was then in late middle age, "conventionally dressed," of medium height, and "sturdily built but not fat." Brophy recalled her at a mass meeting: "Her voice was low and pleasant, with great carrying power. She didn't become shrill when she got excited; instead her voice dropped in pitch and the intensity of it became something you could almost feel physically."[73]

By 1897, coal miners had withstood a series of wage reductions, to the point where, as an economist commented, "No one at all familiar with . . . [these] conditions will deny that the miners' earnings had been reduced way below the living point. Everywhere poverty and degradation were manifest." He concluded that the last straw had been put on the camel's back.[74]

UMWA leaders called another nationwide strike. Seizing upon the symbolism of Independence Day, they announced it was to begin on 4 July 1897. In answer to the call, 150,000 miners, the vast majority of whom were not union members, lay down their tools. This time, the fight was over wages, not prices, and this time the recovering economy lined up on the side of the miners. The three-month strike turned the wheel of fortune for the UMWA. It succeeded beyond anyone's wildest dreams.[75]

The anthracite region had remained more or less unorganized since the 1870s, and in July the predominantly Slavic miners who now worked there failed to respond to the strike call. But in August the employees of the immense Lehigh and Wilkesbarre Coal Company struck to protest a new superintendent hired "to restore discipline."[76] The strike spread quickly.

'Mother' Mary Jones. Photo courtesy the Colorado Historical Society.

Strikers soon discovered that their designated UMWA organizer, an Irish miner named John Fahey, had gone to Harrisburg to lobby for a stronger anti-immigration bill. They demanded his immediate return. He obliged and organized them into locals, but then went back to Harrisburg. Again, the strikers insisted that he return. And so the cart led the horse.

The anthracite strikers added some demands of their own to the list. Besides protesting wage cuts, they objected to unfair pay discrimination between English-speaking and non–English-speaking mine workers. They demanded an end to the company store and the right to choose their own physician. (Companies commonly docked wages to pay for a company doctor, and miners commonly wanted out of this arrangement.)

The strikers organized immense demonstrations throughout the region. One of these ended in tragedy. On 10 September 1897, five hundred demonstrators, peaceful and unarmed, were marching down a street in Lattimer when the sheriff's deputies turned on the parade and fired at point-blank range. Nineteen persons died in the Lattimer massacre, and at least thirty-nine were wounded. The next week, massive funeral processions accompanied the dead to their burials. Afterward, the strike surged forward, stronger than ever. Approximately eleven thousand anthracite mine workers were on strike, and now they were clamoring to join the UMWA.[77]

The following year, President Michael Ratchford stood triumphantly before a miners' convention whose delegates represented a membership that had increased by three hundred percent. "Our trade history shines brilliantly with the victories and achievements of 1897," he proclaimed. In the anthracite region, many operators had granted concessions, while in the Midwest the UMWA had signed its first Joint Agreement with coal operators in the Central Competitive District—western Pennsylvania, Ohio, Illinois, and Indiana. Everywhere the victory had boosted the validity of the strike weapon. "Facts without force," Ratchford asserted, "have never righted a single wrong."[78]

Women had proved themselves a strong force in the strike. Ratchford saluted them as militant demonstrators under the union banner: "Noble women, mothers to whom our success is due in no small measure, led the marchers under the burning summer sun, with their babe on one arm and the flag or banner in the other. They bore their share of that memorable struggle without a murmur of complaint and at its close were found in the forefront battling for home and dear ones."[79] This concept of ideal womanhood—militance combined with motherhood—ran wide and deep in the coalfields. It would be put forward repeatedly in the next twenty years, for it formed a central core of the worldview of the people of the coalfields.

The victory of 1897 marked the accommodation of the UMWA to its Italian and eastern European constituency. Ratchford proposed that the union begin printing parts of its newspaper in "the popular foreign languages." In 1899, the union hired a Slovenian organizer "to work among other eastern Europeans in their own languages."[80]

Organizer John Fahey had seen the light in 1897 and had already begun hiring organizers out of the new ethnic groups. One of the best was Paul Pulasky, born in 1869 in the United States to Polish parents. Like his British counterparts, he started working underground at age nine. He spoke fluent English in addition to five other languages. His genial personality and drive to succeed made him the ideal organizer, and in 1899 he was elected vice president of one of the union's three anthracite districts. He stands as a reminder that an increasing number of "new immigrants" were not new and they were not immigrants, though people everywhere kept on calling them "foreigners."[81]

In the anthracite region, the strike of 1897 had been a partial success, but much remained to be done. From that year until 1900, organizers like Fahey and Pulasky persuaded eight thousand anthracite mine workers to join the union. Most were Slavs—it had become difficult to organize the more privileged English and Irish. Anxious to see some results from their activism, these unionists agitated within their union for a district UMWA convention. Finally, the national office authorized one to be held in August 1900. The delegates drew up a list of demands including a wage increase and asked the operators to negotiate. The operators refused. The national leadership then authorized a strike to begin on 17 September 1900. On that day, the 8,000 union members who stayed home were joined by 80,000 non-union anthracite mine workers. A week later, the strike force had reached 127,000—ninety-seven percent of the work force.[82]

The 1900 anthracite strike was the third important coal miners' strike of the decade. It was during this strike that Mother Jones emerged in newspaper accounts as a conspicuous and colorful organizer, particularly of the wives. One evening at McAdoo, a town solidly behind the strike, she advised the women "to leave the men at home to take care of the family. I asked them to put on their kitchen clothes and bring mops and brooms with them and a couple of tin pans." In the middle of the night, the women marched over a mountain road toward nonstriking Coaldale, beating on tin pans. They reached the mine ahead of the morning shift and greeted miners arriving for work with a din of clattering pans and "Join the Union! Join the Union!" Coaldale joined the strike.[83]

The next year, Mother Jones stood before the United Mine Workers Convention and argued for the participation of women: "My friends, it is often asked, 'Why should a woman be out talking about miners' affairs?' Why shouldn't she? Who has a better right? Has she not given you birth? Has she not raised you and cared for you? Has she not struggled along for you? Does she not today, when you come home covered with corporation soot, have hot water and soap and towels ready for you? Does she not have your supper ready for you, and your clean clothes ready for you?"[84] A woman's work in the family gave her rights in the struggle.

The United Mine Workers of America took on the twentieth century with a new president, John Mitchell, at the helm, and with Mother Jones carrying the flag. By now, the new immigrants, like the undercutting machine, had become a fact of life for the coal miners' union. The membership included eastern and southern Europeans, blacks, Britons (and many American-born sons of Britons), and other native-born whites.

In contrast, the leadership remained largely of British or Irish birth or descent. This lent a new twist to the dynamics of nationality within the union. For most officers, success in the union unlocked the exit door out of the working class. In this early period, virtually every national officer had begun working in the mines as a child. These men were self-educated; their letters and reports, as well as verbatim transcriptions of meetings, provide eloquent testimony of their mastery of speaking, writing, and debating. These skills prompted their election to office. After serving the union, many were appointed to government positions; others were hired by operators; and a conspicuous number ended up serving employers as well or better than they had served the mine workers.

Phil Penna, elected UMWA president in 1895, resigned in 1896 "for family reasons." Following a period of obscurity, he emerged as a commissioner for the Indiana coal operators. For many years thereafter, he represented the operating interests of the Central Competitive Field.[85]

Michael Ratchford, elected president in 1897, shifted to the operators' side at a more dignified pace. In 1898, he resigned the presidency to accept an appointment on the United States Industrial Commission. In 1900, he became Commissioner of Labor Statistics for Ohio, a position he held until 1908. He rose from there to serve as a commissioner for the Ohio coal operators, ending his career as a commissioner for the Illinois coal operators.[86]

Few national or even district officers ever returned to the pit after holding office. Some succeeded in elevating their economic status without compromising their principles; they left the industry entirely or spent their lives in the union movement. Others passed through their job working for the union on their way to a better one working against it. Consider John Nugent, a West Virginia district officer who became "Immigration Commissioner" for the state. Entirely financed by two large coal operators, the job required Nugent to bring mine labor into the state, "enticing men to work," one newspaper wrote, "under the very conditions that he once bitterly opposed."[87] It is impossible to say to what extent such opportunities colored the conservative stance of the union's leadership— its pattern of embracing the still relatively new industrial system—but it is certain that the personal interests of the union leaders and those of working miners were far from being strictly identical.

This drift was not lost on union members. As early as 1891, a miner complained of :

. . . the deliberate action of nearly, if not all ex-officials, in construing the influence of the organization in making good their escape from being the so-called hewers of wood and drawers of water. We have yet to hear of a single instance where any returned to the coal mines and took up his kit of tools whereby he might coach his successor with the experience that was bought and paid for by his fellow craftsmen. All that goes over to the enemy, the monied power.[88]

As the distance between leadership and membership widened, that between skilled and unskilled members narrowed. The proof was in the increasing relevance of the eight-hour day for contract miners. In 1891, an Iowa miner explained:

People who are not familiar with the conditions of the miner while he is performing his labor, often say that the miner need not ask for an 8 hour day, because he works by the bushel, and he can quit work whenever he feels like doing so. Let me tell you how much you are mistaken. This very same contract which lies before me as I write this article, says: "That every employee of the company must be ready for duty when the whistle blows at 7:00 every morning and will be expected to perform a full day's work unless the foreman orders less time to be worked."[89]

To the extent that miners were Anglo Saxon, Irish, or German, while laborers were blacks or Slavs, the breakdown in the distinction between the two types of jobs must also have reduced one source of conflict among nationalities. As miners and laborers drew closer together in the conditions they experienced, a strong union policy against racism further alleviated conflicts between them. The UMWA founding convention resolved that "no local union or assembly is justified in discriminating against any person in securing or retaining work, because of their African descent."[90]

At the 1892 convention, a black miner and dedicated union organizer, Richard L. Davis (1864–1900) offered a resolution designed to confront racism in the membership.[91] It stipulated that any Local Union or Assembly preventing a member in good standing from obtaining employment on account of race, creed, or nationality, would be suspended for three months, and on the second offense, for six months or longer.[92]

Black miners became a strong force for the union.[93] "We are made up of all races and colors," a West Virginia miner declared in 1891. "The colored element are in the majority and be it to their credit are mostly the best men to stick up for their rights."[94] The whites were more absent than present and on one occasion when they did show up to a meeting in Pocahontas, Virginia, the district president felt "agreeably surprised to find a large number of the white miners present . . . this being the first time in six years that any white men have attended a meeting when organization was the topic under discussion. The cause for their absence in the past I do not wish to discuss in a newspaper, as it would be no benefit to our cause."[95]

Official policy notwithstanding, the racism of white miners hindered organizing while increasing the difficulties of black miners. In 1892, William R. Riley, black secretary-treasurer of District 19 (Tennessee), reported that in Oliver Springs the white miners had attempted "to run off all the colored from that place. . . ." He reassured his readers that he didn't think all whites were hopeless. "I do hope that . . . readers . . . will not for a moment think that I meant that there were no honest hearted white men in the south, for there are as good white men in the south as ever breathed the breath of life, but they are so far in the minority that they cannot do the good in the way of justice that they would like to do."[96]

A Welsh miner writing from West Virginia hit harder. The white miner of West Virginia, he proclaimed, "is about the most contemptible person on the face of God's earth. He is unbearably ignorant and does not know it. He has generally been brought up on the mountains, hog fashion, and when they come to the mines and earn a lot of money, they swell out and don't know themselves. . . . These detestable cranks seem to think the poor niggar was made to receive their insults and brutality."[97]

The advantages of racial dissension accrued to the operators. "[A]llow me to say," a West Virginia miner commented in 1891, "that the operators do not fail to try and keep up race dissensions whenever they see the slightest opportunity."[98]

But national policy transcended racist local practices, and this was important. Blacks were noticeably present in the union leadership. In the 1890s, Richard L. Davis served on the national executive board. By 1904, the union employed three black organizers; mine workers in Alabama, Tennessee, West Virginia, and Kentucky had elected black district officials, including a district president in West Virginia.[99] Nationally, miners aired the race question at length in the letters columns of the *United Mine Workers Journal,* and blacks were prominent among the correspondents.[100] Yet in the early twentieth century, some local unions still barred black mine workers from membership. Resolutions passed at the national conventions condemned the practice but would not go so far as to expel these racist locals.[101]

Also on the down side, the racial composition of the leadership did not reflect the number of blacks in the rank and file. The black delegates to the 1906 convention argued that "our race of people will easily be estimated to constitute at least one fourth of the entire membership of the organization." They proposed a change in the constitution to make the proportion of "colored brothers" on the district and national staffs consistent with the number of black members. The convention would not go along with this pioneering proposal for rectifying the wrongs of discrimination with a quota system.[102]

Furthermore, black leaders did not experience the upward mobility of their white counterparts. The case of Richard Davis provides a sad illustration. He acted as a committed and insightful organizer and officer for several years, for which the coal operators blacklisted him. He lived the last two years of his life in extreme economic distress (which apparently the union did nothing to alleviate). He died of "lung fever" at the age of thirty-five, leaving his wife and children in a state of desperate poverty.[103]

Nevertheless, as Herbert Gutman first emphasized, the United Mine Workers of America was a force against racism, and in Gutman's words, "quite possibly ranked as the most thoroughly integrated voluntary association in the United States of 1900."[104] Black and white organizers traveled together through segregated areas of West Virginia and Alabama to promote the union cause. In 1892, Richard Davis accompanied a white organizer into southern West Virginia, where the two encountered only segregated hotel accommodations. At one rooming house, Davis refused to sleep in a filthy cabin in the back. The white organizer joined him in pressuring the proprietor to integrate. As a result, Davis became "the first negro to eat at a table in that man's dining room." He slept in the best bed in the house.[105]

Alabama was a different world. There, in 1903, a coal company superintendent and his cronies attacked a pair of organizers, shot the white man in the arm, beat the black man, and then forced the one to kiss the other.[106] In Alabama, where more than forty-five percent of the mine workers (in 1889) were black, coal operators and their allies attacked the biracial coal miners' movement on explicitly racist grounds. They succeeded in smashing the movement in 1908 during a two-month strike opposed by the governor, the state militia, and citizen groups who felt that "the people of Alabama" will never "tolerate the organization and striking of Negroes along with the white man."[107]

The UMWA's stance toward the darker-skinned Europeans was also positive, if also far from perfect. The new immigrants, as we have seen, were in the forefront of the union's first strike in 1891. Again in 1897, they positively rushed into the union. From the beginning, the union understood the importance of organizing them: The founding convention designated the *Hungarian Journal* of Pittsburgh as one of its several official newspapers.[108] But, at first, ideas on how to proceed were quite scant. In 1891, a solicitor for the *United Mine Workers Journal*, going from mine to mine selling subscriptions, reported helplessly from a mine near Pittsburgh that his visit had been fruitless because of the lack of English-speaking people "and I have not yet learned to talk to the Hungarian, Bohemian and many other different races of people whom I meet in the mines . . . and if I could I guess it would not do much good, for they would not be able to read our papers."[109]

Some Anglo-American coal miners used the newer groups to excuse their own passivity. In 1892, an Illinois activist sneered at non-union miners who complained of the impossibility of organizing because " 'there are so many nationalities,' 'so many different languages,' 'we cannot understand each other,' etc., etc., etc. These are the stock arguments the poor fellows gasp out as they hold their backs and spit and cough to relieve their ill used lungs and livers. Alas! They have already forgotten that in the five months strike here in 1889 the 'many nationalities' stood as firmly as the English-speaking miners. . . ."[110]

In truth, some groups of white Americans were more difficult to organize than the new immigrants. Farmer-miners presented a particular problem. Those in southern Illinois went back and forth from their farms to the mines and showed little interest in the union. They worked underground for ready cash (as most new immigrants did at first) and felt no long-term commitment to the industry.[111] The same organizing problem occurred in the farming country around Meyersdale, Pennsylvania. A frustrated activist reported: "A great many of these fellows here will say . . . when asked about starting a branch of the miners' union: 'No use, boys, the d--d "wheats" will go back on us and every one of us will get the bounce.' A 'wheat' dear readers, is a name some of those weak-spined miners give to the small farmer who has not farm enough to support him without working some in the mines."[112]

The mountaineers of West Virginia also earned a reputation for resisting the overtures of organizers. They too had a deep, pre–coal-mining connection with their region, and for them coal mining provided more cash than they had ever seen in the backwoods. In 1901, native-born whites still made up fifty-seven percent of the coal mine work force in West Virginia. Mountaineers were on a par with "wheats" when it came to enthusiasm for the union.[113]

Despite the noticeable activism of the European nationalities, prejudice against them plagued the union's early years. Somehow, the UMWA's antiracist policies failed to screen grossly racist sentiments from the pages of the *United Mine Workers Journal.* In May 1891, subscribers read that in Ohio the "men and their families [were] turned out upon the roads to make way for . . . the docile Hun and the gentle Pole, with a sprinkling of Dagoes. . . ."[114] The following month, the *Journal* printed this: "Every day thousands of Italians are landing on our shores, steeped to the lips in ignorance and superstition, unscrupulous, treacherous, revengeful; bound by secret oaths to their societies; brigands at home, prepared to resume their infamy in their new country."[115]

Prejudices among the members were apparently rampant, causing President John McBride to raise the problem in his report to the 1893 convention:

[T]he internecine strife occasioned by religion and nationality has . . . prevented your officers from effecting local organization [and] . . . in many instances has disrupted those already established. Coal operators have not been slow to avail themselves of the opportunity thus afforded them to create animosities between employees and to impose upon them oppressive conditions which, under other circumstances, would not be tolerated. Men who originate and circulate stories that stir up national prejudices and create religious fanaticism presume upon the mine workers ignorance. . . . [Y]ou cannot afford to harbor or countenance such a spirit of bigotry and intolerance. . . .[116]

How well did blacks and new immigrants get along in the coal industry? Were their relationships ridden with conflict or did they find bonds and common interests? There is no definitive answer, but it is suggestive that the black organizer Richard Davis found it easy to relate to "the nationalities." In 1897, he spoke to an outdoor meeting attended by a large number of foreign miners:

[T]hey were very attentive to the business of the meeting and especially when one of their own number was speaking. I will just here make this plainer. The checkweighman at this mine is a Polander, but can speak the English language quite fluently. After Vice President Miller and myself got through speaking this gentleman got up and interpreted it to the Polanders, Huns and Slavs in a very able manner. It was quite interesting to notice how they would flock around him. . . . Although the meeting was an out-door one, one could almost hear a pin drop while he was talking.

This entire body of men joined the union with loud cheers.[117]

The UMWA achieved an imperfect but generally positive record of welcoming black and European mine workers. In contrast, the union sank in the mire of racism toward Asians. The 1892 convention put it on the record: "We are unalterably opposed to the landing of a single Chinese in this country. . . ."[118]

In 1901, the UMWA expressed loathing toward the "numberless hordes of Chinese" coming to America and urged Congress to reenact the Chinese exclusion law, soon to expire.[119] Prodigious quantities of organizing energy slid down the exclusion drain. An anthracite official instructed every local to adopt anti-Chinese resolutions, and organized mass meetings in every coal town to this end.[120]

The 1902 convention depicted the Chinese as "a vicious and evil class . . . particularly degrading to American labor."[121] They multiplied like rodents and lived like rats, the *Journal* ranted. They were always ready to take a job at half-price.[122] Two years later, the UMWA convention traduced the Japanese as "morally and intellectually a curse to the nation."[123]

Anticorporate sentiment fused with anti-Asian racism. A typical diatribe stated that the defeat of the Exclusion Act would put millions of dollars into corporate pockets while millions of American workmen would starve.[124]

Even UMWA President John Mitchell, considered a champion of racial harmony, repudiated the Asians. Mitchell welcomed to the United States "the best men and women from every country," with the exception of the Chinese and Japanese "because these two races always work for low wages and can not be assimilated."[125]

In 1906, coal miners embraced the rhetoric of the Eugenics Movement in resolutions declaring that "the social incompatibility as between people of the Orient and the United States presents a problem of race preservation which it is our imperative duty to solve in our favor."[126]

The curious thing about all this was that Chinese and Japanese immigrants posed no threat to the jobs of most coal mine workers. There were never more than one thousand Asian individuals in the entire coal industry.[127] Most miners voting to ban the Chinese had seldom if ever seen the object of their disdain. Only in the West, in the Rocky Mountain region, did miners actually encounter Asians. It was there that Asian mine workers withstood the most vicious attacks, and there also, years later, that Caucasian mine workers finally accepted them as brothers in the struggle.

II

COAL IN THE AMERICAN WEST

CATALYST OF CONFLICT AND CHANGE

PROLOGUE:
REFLECTIONS
AT MIDPOINT

We have arrived at the midpoint of our particular journey into the American past. Before turning to the West—to the heart of American mythology—it may be well to cast a glance backward, to evaluate and summarize where we have been. Looking back, we can see how vital immigration was to the American experience and how strong the value of equality, though it struggled and shrank in the face of the growing inequalities of the new industrial system. The new industrial system increased productivity and lowered prices. Its main casualty was skilled work; those who fought it most bitterly were skilled workers. Skilled miners were fighting not so much to maintain certain privileges they held vis-à-vis unskilled workers as to stave off the destruction of their whole way of life, their self-esteem, their manner of surviving and living in the world. In the coal industry, new methods of management evolved slowly both because the owners profited by the old ways and because mine managers were often as not little more than skilled miners, in a better position, to be sure, but schooled nonetheless in the ways of their fathers.

We have glimpsed the women of the coalfields and seen how their ideas of appropriate feminine behavior included some rather aggressive activities. We have also seen that in general the coal-mining people held democracy dear, but that in many coal camps it existed as an ideal rather than as an experience. Yet, the acid rain constantly falling on such values was that of racism; the idea that the harsh new system was the fault of the new workers and grew out of their leading innate characteristic—subservience.

As we turn to the West, we are prepared to bring to the second leg of our journey a set of questions. Was the West really different from the East and if so how? Was the West the land of opportunity, a national safety valve against a steam of oppression and discontent that could burst forth into revolution? Was the "Wild West" of the coalfields unique culturally, or was it a variant of the industrial development that had occurred farther east?

The region designated as District 15 by the United Mine Workers of America—Wyoming, Utah, New Mexico, and especially Colorado—embraced the most important coalfields of the West. Although Colorado was the nation's eighth largest coal-producing state, the true national significance of its coalfields was their strategic regional position as a source of fuel to the railroads and to the industry of the West.[1]

In 1859, the high, grassy plains east of the Rockies seemed untouched by industrial capitalism. Less than half a lifetime later, the wilderness had been transformed into a center of industrial civilization based on coal for energy. This process had a dark side: the devastation of the Indians, a procedure concluded in the West between 1860 and 1880 in tandem with the rise of industrialization. Our national history includes, even rests upon, the genocidal destruction of the Indians; certainly the losses Indians sustained form part and parcel of the industrial history of the West.

Not that the Indians were peaceful, but they viewed themselves as part of nature. To feel superior to nature was a point of view intrinsic to the developing capitalism. Even more than other whites, the nineteenth-century capitalist looked down upon the rest of the world, which he saw in terms of its utility to himself. Superiority was the lens through which he saw the "savages," the buffalo he shot from moving trains, the coal resource that seemed inexhaustible, and the men he set to digging out the rock that burned.

The capitalist was the fabricator of a new world, but his plans did not unfold without resistance. The last of the Indian wars coincided with the first of the labor wars. The latter reached their height in the Colorado Fuel and Iron strike of 1913–1914. The Ludlow massacre of 20 April 1914 gave rise to an armed rebellion of strikers that lasted ten days and ended only when President Woodrow Wilson ordered the United States Army into the southern Colorado coalfield. "This rebellion," wrote a federal investigator, "constituted perhaps one of the nearest approaches to civil war and revolution ever known in this country in connection with an industrial conflict."[2]

The Colorado Fuel and Iron strike involved deeply opposed ways of looking at the world. The outcome of the struggle had far-reaching consequences for all future conflicts of this kind. The coal operators were victorious, but theirs was a hollow, ideological victory that cost them very dearly. Open class conflict wreaked havoc on the ability of coal firms to make money. This spurred an important group of business leaders, led by John D. Rockefeller, Jr., to move toward new attitudes and policies that, taken together, have come to be known as corporate liberalism. The national importance of the Colorado Fuel and Iron strike and the national reaction to it will lead us back to a national perspective in the last chapter of this book.

The strikers were defeated, yet the history of the strike, and of the long evolution of conflict leading up to it, opens a window onto one of the "other Americas" of our national past, in terms of values as well as economic conditions. By the time of the great coalfield war, the strikers saw themselves not only as Italians, Slavs, or Greeks, but as working people and as Americans. They flew the stars and stripes proudly. The flag symbolized freedom, something they could only dream about in the coal camps of the West. For them, the class struggle was real, and it was inseparable from the struggle for democracy.

8

CAPITALISTS, RAILROADS, AND COAL: AGENTS IN THE TRANSFORMATION OF THE WEST

We can examine almost any hill and find evidence of this treasure—coal here, coal there, coal everywhere.

Trinidad Enterprise, *13 December 1873*[1]

Nothing lives long except the earth and the mountains.

White Antelope, *a southern Cheyenne*[2]

Far from the Civil War raging in the East, huge deposits of iron ore and coal lay undisturbed under a wild terrain—present-day Wyoming, Colorado, and New Mexico. Looking eastward from the base of the Rocky Mountains, a treeless, grass-covered prairie stretched as far as the eye could see. Hundreds of buffalo blackened the flat sea of grass. The plains also supported elk, deer, bear, coyote, and the Arapahoe and the Cheyenne Indians. Looking westward from the plains, the snow-capped Rockies towered and gleamed in the sun. From the site of Denver, they appeared to rise straight out of the prairie.[3]

On this wilderness, a group of men, surprisingly few in number, erected an industrial society, first in imagination, and then on the traditional lands of the Plains Indians. The concrete steps of this transformation, a microcosm of its occurrence elsewhere, is the subject of this chapter.

The western states are enormous by eastern standards: Some counties are as large as the state of Connecticut. In Wyoming, Colorado, and New

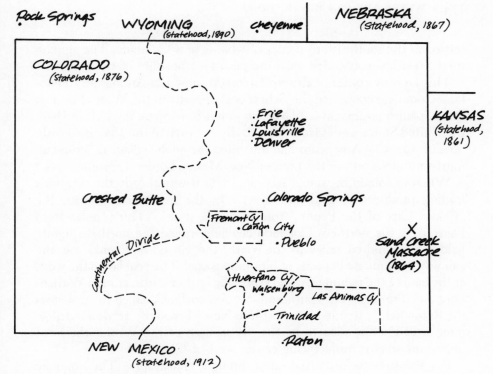

Rocky Mountain coal fields

Mexico, the plains rise a mile above sea level to meet the foothills of the Rockies. The Continental Divide, the line separating waters flowing east and south from those flowing west, bisects the region. East of the mountains, coal underlies sixty percent of the plains extending from Canada to Mexico—these are the Rocky Mountain coalfields. The coal is mostly lignite, the soft brown coal with visible plant components, and subbituminous. The exception is the rich bituminous coalfield extending from southern Colorado into northern New Mexico. In addition, one of the country's few anthracite seams outside of Pennsylvania lies beneath the mountains of central Colorado, near Crested Butte.[4]

In the Colorado Territory, the door to industrialization opened abruptly with the gold rush of 1859. Before the "Pike's Peak or Bust" excitement, the principal inhabitants were the Plains Indians and the Utes, a hunting people who lived in the mountains. The Plains Indians arrived in eastern Colorado in the late 1700s. Around the same time, Mexican sheepherders, the first carriers of European culture to the region, settled in the southeastern corner of what is now Colorado. The agricultural Mexicans contended with the persistent hostility of the Indians.[5]

This faraway corner of eastern Colorado (now Las Animas and Huerfano Counties) contained the richest coal deposits in the West. The area was a Spanish possession until Mexico won it from Spain in 1821; in 1848, the United States won it from Mexico. By the time of the 1859 gold rush, Spanish-speaking Americans had inhabited the adobe village of Trinidad, fourteen miles north of the present New Mexico border, for some years.[6]

When Colorado became a state in 1876, it was already the country's leading producer of gold and silver.[7] In the northeast, Denver, the "Queen City of the Plains," rose from the prairie. Thirty miles from Denver lay the northern Colorado coalfield, called the northern lignite field. One hundred miles to the south, the village of Pueblo was the gateway to immense deposits of bituminous coal. The coal lay to the west, at the base of the mountains, and especially farther south, around Walsenburg and Trinidad. The southern Colorado coalfield, called by geologists the Raton field, extended into northern New Mexico. By 1881, a red glow hung over Pueblo, soon to be the "Pittsburgh of the West," the most important steel manufacturing center west of Chicago.[8]

For the Indians, industrialization brought catastrophe. The nomadic Plains Indians followed herds of buffalo north and south as they migrated with the seasons. The Arapahoe and the Cheyenne lived in a symbiotic relationship with the animals, killing them for meat and using the skins for clothing and shelter. The two peoples, descended from different branches of the Algonquin family, did not speak the same language, but they communicated in sign language and a great friendship existed between them. They rode the horse, adopted from the Spaniards, and united in frequent wars against the Utes in the mountains, and against the Crows, Pawnees, and Sioux to the north.[9]

The Utes, or "blue sky people," had inhabited the Rockies for centuries. Their appearance, language, religion, and way of life differed radically from those of the Plains Indians. They were a short, dark-skinned people who hunted game on foot. The Utes were introduced to the expansionism of the United States, not by any Caucasian group of explorers, fur traders, gold seekers, or settlers, but by the arrival of the Plains Indians, who had been pushed out of their traditional lands.[10]

White encroachment came first in the form of fur trappers such as Kit Carson. The fur-trading period began in the 1810s and lasted until the

1850s. The Plains Indians became traders as well as hunters, bartering skins for beads, trinkets, coffee, guns, and whiskey. A portent arrived in the late 1820s, when fur traders transmitted smallpox to the Arapahoe, and half their number died.[11]

The fur business began with beaver but ended with the livelihood of the Indians—the buffalo. White men, unlike the Indians, did not kill the animals sparingly, taking only what was needed. Instead, they killed thousands of the huge bison, skinned them, and left the carcasses to rot.[12] During the 1850s, the intensification of Indian attacks (and fashion changes in the East) contributed to the decline of the fur trade. Gradually, the traders abandoned their forts, but by then the Indians were beginning to starve.[13]

In 1859, a decade after the California gold rush, a rumor of gold sent fifty thousand men streaming across the Great Plains toward Pike's Peak. Most rushed to Pike's Peak from the Midwest or Missouri. They were down on their luck, a human flood released to the West by the banking collapse of 1857.[14] The "fifty-niners" established a settlement near the present site of Denver, a favorite camping ground of the Arapahoe, where they found that the Indians "persistently haunted" the place. An Arapahoe chief told the whites:

> Our country for game has become very small. We see the white man everywhere. Their rifles kill some of the game, and the smoke of their camp fires scares the rest away. . . . It is but a few years ago when we encamped here . . . and remained many moons, for the buffalo were plenty, and made the prairie look black all around us. Now none are to be seen. . . . Our old people and little children are hungry for many days and some die for our hunters can get no meat. Our sufferings are increasing every winter.[15]

In the years following the gold rush, the small-scale enterprise dominated the white man's economy. The miner with his burro who disappeared into the Rockies to pan for gold was a budding entrepreneur with little information, even less capital, and quite frequently, no luck at all. Blacksmiths followed him West, perceiving that lucky or not, the gold miner had to purchase tools and other iron supplies. They established their works on the plains north of Denver near outcroppings of coal.[16]

At first, the gold miners panning the mountain streams built their campfires with timber cut from the lower slopes. This created "a general barrenness of landscape" around the mining camps. Thus began the turn from wood to coal in Colorado. As the price of firewood rose, enterprising individuals began transporting coal from the northern lignite field to Denver in ox-drawn wagons.[17]

Development proceeded hand in hand with the destruction of those specified in early treaties as the rightful occupants of the land.[18] "[T]his is a life and death struggle," proclaimed a Colorado guidebook, "to be ended by the death of the savage—and the sooner it can be finished the better."[19] In 1862, the Arapahoe and Cheyenne began collecting horses and guns for a long, sporadic, and ultimately hopeless war of resistance. The Sand Creek Massacre ended all hope of peace. On 29 November 1864, a unit of the Colorado militia entered an encampment of Arapahoe and Cheyenne at dawn and slaughtered more than one hundred Indians as they woke from their sleep, mostly women, children, and old people.[20] The Indian wars, fought with the Plains Indians on the prairie and with the Utes in the mountains, continued another twenty years.[21]

In 1870, coal gas lamps began to illuminate the Denver nights.[22] In June, Denver got its first railroad connection—the Cheyenne-Denver branch of the Union Pacific. A short time later, a locomotive of the Kansas Pacific Railroad came chugging and belching into town. Until then, horses, oxen, mules, and burros had provided transportation. The land-locked Rocky Mountain West required the railroad for its industrial revolution. In the 1870s, three railroads built parallel tracks into the Colorado Territory from the Missouri River: the Union Pacific; the Kansas Pacific; and the Atchison, Topeka, & Santa Fe. All three laid their tracks over the Great Plains, "a vast region wholly destitute of timber."[23] Wood fuel was not cheap. The matter of coal was urgent.

The Union Pacific, the country's first transcontinental railroad, took the northern route through Wyoming, a sagebrush landscape swept by sandstorms. In Wyoming, the railroad opened coal mines at Rock Springs, Almy, and later at Hanna.[24] Then, hungry for coal to fuel its locomotives, it built the Cheyenne-Denver branch through the northern Colorado lignite field. By the late 1870s, a subsidiary operated four large mines in the Colorado lignite field, a mile-wide horseshoe-shaped area extending twelve miles through highly cultivated agricultural land. Little coal towns—Erie, Louisville, Lafayette, Marshall, Como, Northrup, and Superior—grew up around these mines. By 1884, the Union Pacific was producing nearly a tenth of Colorado's coal output, and virtually every ton of Wyoming coal.[25]

The second parallel road, the Kansas Pacific, built from Kansas City to

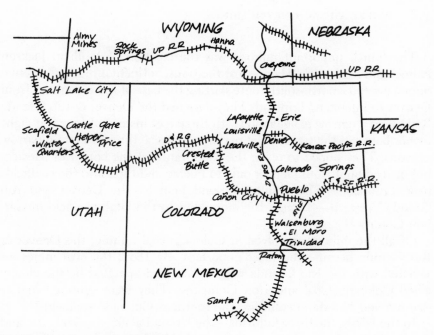

Railroad-coal nexus in the West

Denver, reaching the Queen City in the summer of 1870. This train carried a group of young men, managers of the railroad, who had been thinking about the Colorado wilderness for a long time. William Jackson Palmer and his associates had surveyed the region for the Kansas Pacific with the idea of building the first transcontinental railroad through the Colorado Territory.[26] When, much to their distress, the Union Pacific obtained from Congress the necessary land grants for a more northern route, they turned to other plans. In 1880, the Union Pacific, by then controlled by Jay Gould, obtained control of the Kansas Pacific.[27]

The Atchison, Topeka, & Santa Fe built the southernmost route through Kansas, reaching the Colorado Territory in 1872. The depression of the 1870s slowed construction, but in 1876 the yellow boxcars of the Santa Fe rolled into Pueblo. From there, the railroad built to the base of the mountains, where it opened a coal mine near Cañon City in Fremont County. The Santa Fe also inched southward toward the rich coalfields surrounding Trinidad. It soon became the first large-scale coal and coke operator in the southern Colorado coalfield. It continued into New Mexico, opening coal mines all along the way.[28] By 1885, the Santa Fe produced one-third of all Colorado coal.[29]

The fourth important railroad was the creation of William Jackson Palmer and his friends. Having lost the transcontinental route, they envisioned instead a north-south route skirting the base of the mountains from Denver to Pueblo to Trinidad. They designed the Denver & Rio Grande Railway as a narrow gauge road (with three feet instead of four feet eight inches between the tracks), the better to squeeze through certain narrow canyons. They hoped to reach the Rio Grande at the border of Mexico, but instead built from Denver only to Pueblo and then into the coalfields near Trinidad.[30] They built a second branch—the Denver and Rio Grande Western—through the mountains to the Utah coalfield in Carbon County.[31]

Of all the railroads involved in Colorado coal mining, the Denver & Rio Grande became the most important. By 1882, the coal mines associated with the Rio Grande were owned and operated by the closely allied Colorado Coal and Iron Company. They were centered in Las Animas and Huerfano counties, the southern Colorado coalfield.[32]

In the 1870s, the three railroads—the Union Pacific, the Santa Fe, and the Rio Grande—competed over Colorado like dogs snarling over a bone. Competitive capitalism reached high drama in the contest between the Rio Grande and the Santa Fe for the route from Pueblo to the silver-booming town of Leadville. The route west was no flat prairie. It squeezed through the Royal Gorge, a high-walled canyon susceptible of threading by one road's track. In April 1878, the opposing parties dropped their two-year-old pooling arrangement like a hotcake and raced to the gorge to furiously begin grading the roadbed at different points. Each side built fortifications and engaged armed guards to overlook the proceedings. From time to time, construction gave way to fist fights. Finally, the courts enjoined the combatants, and the battle shifted to the legal arena, raging on for two more years before the Rio Grande emerged victorious.[33]

Relations between the Rio Grande and the Union Pacific were less theatrical but no less acrimonious. "The conduct of the Rio Grande toward our company," reported a Union Pacific manager to the company president in 1885, "has been aggressive, unfair and at times scandalously abusive in its advertising matter and newspaper organs. . . ." Not only the company but "our officers and agents . . . were lied about and ridiculed by name in published matter emanating from the headquarters of the company."[34]

Wrangling among themselves did not prevent these three corporations from quickly superseding the small wagon mines to become the largest coal operators in the West. In this region, the era of small-scale enterprise was quite short-lived. By 1886, mines operated in the interest of the railroads produced 1.1 million out of 1.4 million tons of Colorado coal. By the end of the decade, the railroads had developed a "community of interest" overshadowed by Jay Gould and the Union Pacific Railroad.[35]

The leading figure in Colorado's early railroad-coal industry was the Rio Grande founder, William Jackson Palmer (1836–1909). Although he was an exceptional entrepreneur, Palmer's biography also illustrates the life pattern of the vast majority of American businessmen. He made his contributions in the context of a family that offered him every advantage. In this, he was not exceptional: Most American entrepreneurs did not proceed from rags to riches. The Horatio Alger myth has been laid to rest by scholars writing on the career paths of other American businessmen.[36] Palmer, like most of his peers, was born in the United States to parents who were not immigrants. The son of a Quaker merchant, he grew up in Philadelphia, beginning his working life (as a commercial clerk) at age seventeen. By today's standards, he was young for full-time work, but for his own day he was older, with far more education behind him, than was possible for working-class boys. The well-connected social world in which he grew up and into which he married was crucial to his early training and opportunities, and later to his ability to attract capital into his business enterprises.[37]

Palmer soon left his first job to work as a surveyor for a local railroad. In 1855, the railroad's chief engineer helped him finance a trip to Britain, where he studied coal mining and especially the British practice of burning coal instead of wood in locomotives. He wrote up his observations and sent them in to that newspaper central to the developing coal industry—Benjamin Bannan's *Miners' Journal*, published in Pottsville, Pennsylvania.

Young Palmer's successful future was incubated and nurtured by his parents and also by an uncle who was both wealthy and quite fond of his nephew. Over the years, F. H. Jackson, a Quaker and an officer of the Westmoreland Coal Company, acted as a mentor knowledgeable in the coal business but, more importantly, knowledgeable in the business of getting ahead. "Study the [coal] business carefully," he wrote to William during his British sojourn, "and keep copious notes of thy information so that when thee returns thee may be looked upon as an authority on the subject of burning coal in locomotives. I anticipate a very conspicuous future for thee. . . ."[38]

Upon his return from England, Palmer spent a year as secretary-treasurer of the Westmoreland Coal Company, of which his "affectionate uncle Frank" was a director, and then became private secretary to J. Edgar Thomson, the president of the Pennsylvania Railroad. He was now twenty-one years old and earned nine hundred dollars per year. In his new position, he conducted extensive experiments on burning coal in locomotives; to him belongs much of the credit for the transfer of this technology to the United States. His uncle continued as a supporter. "Last evening I took up thy report of the experiments of coal!" he wrote Palmer in 1860, following this declaration with four pages of detailed critical praise.[39]

During the Civil War, Palmer fought on the Union side, emerging at age twenty-nine as a brigadier general. At the close of the war, he was chosen by Thomson and other managers of the Pennsylvania Railroad to look after their interests in their projected transcontinental railroad—the Kansas Pacific. Palmer moved to St. Louis to act as treasurer of the road at a yearly salary of five thousand dollars; he also became president of the construction company organized to build it. Again, his uncle's support acted as a leavening on his career. Jackson helped his nephew to plan strategy toward the stated career goal of winning a fortune:

> . . . I think your material interests will be much better advanced in St. Louis. 1st because . . . it is much better to grow up with a place than to start in it after it is already grown. 2. You will have much more unlimited connections with Eastern as well as Western capitalists in St. Louis than in New York. 3. A railroad being completed is a big thing. 4. Any capital that you may accumulate or control will pay much more largely in the West than in the East. 5. An influential position in a portion of the Great Pacific railway and an intimate connection with the probable controlling interest of the whole railway is "very large."[40]

Uncle and nephew both understood the importance of connections with substantial capitalists. As Palmer philosophized to his uncle, "Young men without money can only make a fortune by connecting themselves with capitalists." Jackson began to contribute not only advice and encouragement to Palmer, but capital to his enterprises.[41]

When Palmer left the Kansas Pacific in 1870 to found the Denver & Rio Grande, he continued his close association with the Pennsylvania Railroad, the president and vice president of that corporation appearing at various times as trustees of the mortgage of the Rio Grande. About this time, he married Mary Lincoln Mellen, the daughter of William Procter

Mellen, a prominent New York attorney with many social and financial connections. Palmer's father-in-law became a director of the railroad.

In the mid-1880s, Palmer lost control of his Colorado enterprises (although he kept control of the Denver & Rio Grande Western, the branch running from Pueblo through Leadville to Utah), and shifted his business interests to Utah and to Mexico. He lived to be an old man, retiring in 1901 to his Colorado Springs home. He died in 1909, a capitalist of the second rank, no Jay Gould or John D. Rockefeller but nevertheless a wealthy man. "General Palmer," his *New York Times* obituary explained, "was well known in this city both socially and financially. He was a member of the Metropolitan Club . . . and also a charter member of the City Midday Club. . . . [He was] the foremost citizen in Colorado. He leaves an estate valued at 15 million."[42]

Palmer's key associates in his Colorado enterprises also came from privileged family backgrounds. His close friend and business partner, Dr. William Abraham Bell (1841–1921), was the son of an affluent London physician. Bell was himself trained as a medical doctor, but after practicing for two years he left his profession to join General Palmer in the survey of the Colorado Territory for the Kansas Pacific. In 1870, he returned to London, calling upon friends and his father's well-to-do clients to raise capital for the Denver & Rio Grande. According to one historian, Bell "frequently returned to Great Britain to find investors for the numerous business enterprises in which he engaged with General Palmer."[43]

A third prominent associate, Alexander Hunt (1825–1894), was also the son of a physician. Hunt made his first fortune in the California gold rush by selling merchandise to the gold seekers. At the age of twenty-five, he was a wealthy man, but he lost his fortune in the banking collapse of 1857. In 1859, he joined the rush to the Colorado Territory, where he engaged in various business enterprises. In 1867, United States President Andrew Johnson appointed Hunt Governor of the Territory, as well as ex officio Superintendent of Indian Affairs. In this capacity, in 1868, he negotiated a treaty with the Utes by which they ceded all of eastern Colorado. This put Hunt in an excellent position to become a land developer: His chief function in the Denver & Rio Grande companies was acquiring real estate.[44]

Palmer and his business associates came to their projects buoyed by elite family backgrounds and social connections. Palmer, the prime mover of the group, became one of the key industrial developers of the Rocky Mountain West. His family background helped him in every way, but there was more than that to his success as a capitalist. He saw the world through a developer's eyes, a habit nurtured by his early association with the railroad. He received his training in what Alfred Chandler, Jr., has called "the largest business corporations the world had ever seen."[45]

The railroad functioned to connect many localities; it was a force that worked against the localism through which most Americans still saw the world. Further, railroad men were among the best-traveled Americans (and travel must have acted as an antidote to provincialism). For Palmer and other railroad managers, travel was both necessary and free. As a young man working for the Pennsylvania Railroad, he wrote to his parents, "I find the name of J. Edgar Thomson [president of the Railroad] a passport wherever I go—and believe, with his letter of credit, I could travel from Maine to Texas without the unpleasant necessity of putting my hand in my pocket for the pewter."[46]

When Palmer and his group arrived in Colorado in 1870, the territory was sparsely settled, with only forty thousand citizens, about one for every three square miles of land.[47] The young entrepreneurs looked at this wilderness as clay waiting to be shaped. Palmer revealed how the shaping would be done in the Denver & Rio Grande's *First Annual Report*, which he wrote in 1873 for stockholders and potential investors. This revealing document shows Palmer thinking across several industries and across time into the future. The narrow gauge road was to be built into the wilderness, and Palmer pictured there hotels for his passengers to stay in and towns for them to live in. To build the towns required lumber and the development of agriculture. To heat the houses and run the railroad required coal.

Before the railroad, Palmer wrote, the region was cut off from easy access "by the great plains and by wandering tribes of Indians." But the railroad itself would settle the country; its first business would be to transport emigrants. They and "all of their household and business plant, their first necessary food and other supplies, would alone furnish a considerable immediate traffic, while their subsequent trade, the building up of large towns, and the growth of the mines and manufactures, would yield a permanent and lucrative transportation."[48]

Palmer enthused over the region's geographical advantages. To the east lay four hundred miles of an arid, treeless plain, unsuitable in its natural state for anything but nomadic stock raising. This plain and the rugged mountains to the west shielded the projected north-south route from potential competition from a parallel road. Streams fed by melting snows poured water down from the Rockies, providing waterpower to drive machinery. The streams watered a narrow belt along the base of the mountains, potentially a fertile agricultural region that could be extended by irrigation. The Colorado Rockies contained a wealth of gold, silver, lead, copper, iron, and other metals. Pine trees flourished on the lower slopes, and in them Palmer saw a lumber business and the raw material for the wooden structures that would constitute the towns and cities of the high plains and foothills. Finally, there was an abundant supply of iron ore, and "at numerous points along the whole belt named . . . extensive deposits of good coal."[49] Of the Trinidad coalfield in the southeastern part of the territory, Palmer stated, "[T]he extent of this coal field is so great, and the seams so numerous that the coal may be considered as practically inexhaustible."[50] That these coal lands legally belonged to the Ute Indians Palmer considered of no importance, and he did not mention it.

During the next decade, Palmer and his group formed a cluster of companies, each with a particular function in the overall grand development plan. To his fiancée Palmer wrote in December 1869: "I have spent most of the day in answering half a dozen long and most fertile business letters of Dr. Bell, covering about eleven different schemes and projects in which we are jointly interested and out of which we expect to make our fortunes."[51] They formed one company as the railroad itself; they organized another to acquire thousands of acres of land between Denver and the New Mexico border. They created town companies and construction companies. They organized a firm to mine coal and iron ore, and to manufacture it into coke and into steel.[52]

They acquired the land at little or no cost. To begin with, the Rio Grande obtained the passage of two laws in the United States Congress which granted to it "the right of way over the public domain, one hundred feet in width on each side of the track, together with such public lands adjacent thereto as may be needed . . . not exceeding twenty acres at any one station, and not more than one station in every ten miles; and the right to take from the public lands adjacent thereto, stone, timber, earth, water, and other material required for the repair of its railway and telegraph line."[53]

In the second place, they stole vast tracts of land, through "bold, reckless, and gigantic schemes to rob the government," as the U. S. Secretary of the Interior put it. The Homestead Act of 1862 had offered fifty million acres at $1.25 an acre to anyone who would cultivate their land for five years. In Colorado as elsewhere, speculators and railroad corporations siphoned off much of this public domain. "[T]he southern portion of Colorado has furnished a prolific field for the perpetuation of frauds of the most audacious and flagrant description," reported the Secretary of the Interior. "[T]his section of the country was covetously scanned by unscrupulous parties and, as the prospective importance of the coal and agricultural lands became apparent, a scheme was matured to obtain possession. . . ." In 1874, a grand jury found that "great frauds" had been committed by persons claiming to be entering agricultural claims at the land office. Actually, they were buying land for a Rio Grande firm at rock-bottom prices. Many people sued, but before the cases came to trial, the accused slipped into a new suit of clothes and calmly walked away. The company in question merged with two other Rio Grande firms. Although the directors were virtually identical, the court held that the new company (which had a new name) was a "different entity" from the one named in the law suits. The plaintiffs lost their cases.[54]

The Palmer group was not uniquely corrupt; for these and other developers, bribing the land office was an accepted cost of business. Toward the end of the century, for instance, a superintendent of the Union Pacific Coal Company wrote to his superior of the need to acquire new coal lands, but, as it happened, under the law the company could not get title to the desired lands. "[W]e must have the land," the manager wrote, "and a little stretch of some man's conscience will enable us to get it, and three hundred dollars per entry of 160 acres or less is sufficient to cover the stretching process. Of course it is not absolutely safe, but we have the U. S. Land Department officials in Wyoming with us."[55]

The extent of the land that slipped into Denver & Rio Grande control in the 1870s was vast. One Rio Grande company, the Mountain Base Investment Fund, acquired 83,000 acres south of Denver, mostly from the public domain, at prices that William Bell asserted "will never again prevail in Colorado."[56] In addition, Alexander Hunt bought immense tracts near the site of Colorado Springs at eighty cents an acre and then sold it for fifteen dollars an acre to one of the Rio Grande companies of which he was a director. In turn, the company sold the land to affluent settlers (mostly British) who came to live near the health-giving Colorado Springs.[57]

Colorado's new capitalists also came into possession of large tracts known as the Mexican land grants. These were thousands of acres extending north from New Mexico into southern and central Colorado. The largest, the Maxwell Grant, consisted of more than a million acres. The

1848 treaty between Mexico and the United States stipulated that the lands would remain in the possession of the previous Mexican owners. In later years, much litigation ensued, but the heirs eventually found themselves in possession of an immense landed wealth. In 1870, they sold it at less than a dollar an acre to a land-holding firm, the Maxwell Land Grant and Railroad Company. The chief investors in the new company were British. Its president was General William J. Palmer.[58]

For the Colorado capitalists, acquiring land was relatively simple; acquiring construction funds to build the railroad presented more of a challenge. In 1870, Palmer and Bell traveled east hunting for investors. Palmer went to Philadelphia, where he enlisted the support of acquaintances, including the president of the Pennsylvania Railroad. Bell returned to England to peddle the stock of the various Rio Grande companies to his father's wealthy patients and others. In addition, a Dutch bank purchased nearly a million dollars worth of railroad bonds at two-thirds of their face value.[59]

British capital fertilized the Colorado enterprise in another way. Lacking sufficient funds to buy rails, the Palmer group bought them on credit from British iron manufacturers. In doing so, they were following a common method of transferring British capital. In 1872, forty-two percent of all railroad iron consumed in the United States was manufactured in Great Britain, and much of it was supplied on credit.[60] The credit extended was a form of capital, and it made up for the horrendous inconvenience and expense of transporting iron rails over such a distance.

The rails arrived four months late, but in October 1871 a Rio Grande locomotive chugged south from Denver to Colorado Springs at fifteen miles per hour. The first seventy-six miles of the picturesque narrow gauge road was in place.[61] This efficient transportation system now monopolized what instantly became an essential service. Within a year the town grew from a scenic hamlet to a community of fifteen hundred inhabitants. Before, a stage coach carrying five passengers had traveled the route three times weekly. In its first year, the Rio Grande carried five hundred passengers weekly over the same ground. It also transported more than fourteen thousand tons of lumber (at first the most important freight), various kinds of merchandise, and wood fuel cut from the pineries of the lower slopes. Coal as freight ranked fourth in importance; within the decade, it would be first.[62]

The mere prospect of the railroad caused both the population and the property values of a place to increase. As the company began grading the route south from Colorado Springs toward Pueblo, this town, and even Trinidad, one hundred miles farther south, began to prosper in expectation. This gave the entrepreneurs great power over the townspeople, because to divert the railroad by even a few miles could ruin those bypassed.

In 1872, 363 voting citizens inhabited the village of Pueblo. At Cañon City, a few miles to the West, a Rio Grande company had already opened a coal mine. The directors knew they were going to build the railroad, first to Pueblo and then to the Cañon City mines at the base of the mountains. Nonetheless, Alexander Hunt informed the citizens that they were not on the projected route. If they wanted the railroad, they would have to help pay for it. Responding to Hunt's threat, in June 1872 Pueblo voted a bond issue for the Rio Grande in the amount of $100,000. The company, playing coy, did not respond. As the summer wore on, the farmers and merchants of Pueblo became apprehensive and then agitated. Finally, in late September, they retaliated by inviting two railroads to build to Pueblo—the Kansas Pacific from the east and the Rio Grande from the north. The thought of competition sent Hunt scurrying to Pueblo to inform the people that the company did indeed intend to build to that point, but that to do so required an additional bond issue of $50,000. The townspeople complied.[63]

When the Rio Grande finally laid the track from Colorado Springs to Pueblo, it did not build the depot in the heart of town after all, instead terminating the road on company-owned land across the Arkansas River from the town. This infuriated the citizens. Although the railroad was close enough for Pueblo to continue developing anyway, they rescinded the bond issue. For its part, the railroad, through the sale and commercial development of its own real estate, made itself the principal beneficiary of the coming of the railroad.[64]

Policies deriving from the company's drive for profits aroused the loathing of many people. The railroad charged the highest passenger and freight rates the traffic would bear. The fare between Pueblo and Cañon City was so high that two people could hire a private cab for two-thirds the train fare. A ticket from Cañon City to Denver, a distance of under 150 miles, cost seventeen dollars at a time when the 400-mile train ride from Cincinnati to Washington, D. C., came to fourteen dollars. Freight rates were so high that numerous merchants reverted to mule- or ox-drawn wagons. In 1873, the region's largest circulating newspaper, the *Rocky Mountain News*, reported that the railroad "shares the hate of all of southern Colorado." In the early 1880s, a settler from England wrote that the freight rates from Colorado Springs were so exorbitant that no particular advantage accrued to a farmer owning land in proximity to the railroad. Another writer asserted that the narrow gauge railroad was known all over Colorado as the "Narrow Gouge." These unkind remarks filled the newspapers alongside other articles praising the railroad for bringing prosperity to the state. In a sense, both sides reflected the truth.[65]

Along with high prices, the Rio Grande capitalists dreamed of low wages. Low wages had originally attracted them to the idea of building to Mexico. The directors had enthused to stockholders that the cost of labor in Mexico averaged less than one-fifth of what it was in the United States.[66] However, the Santa Fe blocked the Rio Grande by placing its own tracks on the route to Mexico first.

Actually, in the sparsely settled West, wages were high. Just as the railroad had to import rails and capital from a great distance, so it had to import labor. In 1880, the Rio Grande construction manager complained that the cost of labor and the problem of obtaining a sufficient quantity of good labor were major impediments to progress. In 1882, a consulting engineer noted that in mining regions, railroad wages were forced up because miners could earn better wages.[67] Why shovel dirt when you could shovel coal for more?

The railroad played a major if unintentional role in supplying labor to the entire region. In 1880, the company employed 3,000 men, who were, however, continually leaving for greener pastures. To attract labor, the construction manager advanced the train fares to 250 men from St. Louis, 300 from Chicago, 1,000 from Kansas, and 200 from Canada. In each case, the manager reported, the laborers agreed "to refund the amount when earned in our service," but instead nearly all of them deserted. "[M]any went to the mines, a few returned to their homes, and the Lord probably knows where the rest are."[68]

Concluding that "any number of worthless fellows were anxious to get a free ride to Colorado, the manager stopped prepaying fares and turned instead to new strategies for obtaining labor. He made efforts to import blacks from the south but without success. So he turned to Italy. "I am now in correspondence with a gentleman in Milan," he reported, "with a view to securing laborers."[69] In the 1880s, Colorado's first Italians arrived. Called "new immigrants," they preceded many native-born Americans to the still rather vacant Centennial state. They began working on the railroad, but many ended up in the mines, joining the Britons or their sons, who predominated in the coal mine work force during these early years.

In 1880, the Denver & Rio Grande capitalists merged three of their firms to form the Colorado Coal and Iron Company. The new firm immediately began constructing a steel mill at South Pueblo. It soon became the largest coal mining, iron ore mining, and steel manufacturing enterprise in the West.[70]

The new firm's coal mine superintendents felt the shortage of labor acutely. As early as 1881, George Engle, the superintendent at El Moro, had turned to the Lechner undermining machine to make up for the lack of men. But even the machine cost more in the West. In May, Engle sent his superior a series of articles he had clipped on the Lechner machine with the comment, "Please note that estimate of wages [for machine operators] is far below what we will have to pay here."[71]

In the spring of 1881, Engle had to deal with the continual departures of his best men to the silver mines. He also contended with the absenteeism of his Mexican employees on the Easter holiday and on paydays. He blamed his labor problem on "the difficulty of getting men to work behind the machines at the present price per ton."[72]

For George Engle, the mining machine did not automatically solve the labor shortage. For one thing, certain key workmen resented the machine. The blacksmith, whose job included repairing the machines, expressed his firm opinion by charging exorbitant rates to fix or even to look at one. The machines had been installed for a trial period, their success depending upon how far they reduced costs. "You can depend these blacksmith bills put in against us are a concerted plan," wrote the manufacturer to the coal company manager, " . . . to kill us by the expense and at the same time show you what 'repairs' were costing and thus have a double effect in getting rid of us and our machines."[73]

For another thing, Engle could not obtain competent labor to man the machines. He had two mining machines but only one good operator, a miner named Dave Jones. Jones's work had to withstand constant interruption from the operators of the other machine, who were obliged to turn to him for assistance. Engle regularly reported that the cause of poor production by the second machine was "green men."[74] "In order to get better results we must have better men," he fumed to his superior, "and we must pay them better wages—I do not think that anything less than $3./day will secure the men we want. . . ."[75] Dave Jones received eighty dollars per month. Here was one of those well-paid miners to whom the operators frequently referred. Jones's skill was indispensable: Even the superintendent, it seems, lacked the knowledge to supervise the operation of the second machine.

Despite such troubles, company records reveal that the coal department of the Colorado Coal and Iron Company was highly profitable. The company's *Second Annual Report* printed in December 1880 emphasized that the *net* earnings from the coal and coke department alone paid the entire interest on the mortgage bonds, as well as the interest on capital necessary to build the steel plant at Pueblo. The following year, the net surplus, after interest payments, leaped from some 43,000 dollars to 239,000 dollars.[76] High prices boosted profits but so did the steady reduction of wages. The evidence on wages is contradictory, undoubtedly because the scarcity of labor and its high cost were solved in a sporadic rhythm influenced by the general state of the economy. But the overall trend was downward. In the mid-1880s, a mining engineer reporting on the Colorado Coal and Iron Company's coal properties asserted: "The wages in Colorado are not higher than in the East, notwithstanding all assertions to the contrary. Unskilled labor is even cheaper."[77]

The company made its mines pay; the problem was the steel mill at Pueblo. The furnaces consumed capital readily enough but failed to refine it adequately into profit.[78] It may have been a severe disadvantage to be so closely associated with the railroad. The original plan was for the different Rio Grande companies to offer advantages to each other, but in the mid-1880s, according to one engineer, the Coal and Iron company seemed to be working "more for the benefit of the railroad companies." Freight rates were high, and the mills at Pueblo were located far from the company's iron mines. "Every raw material only reaches its destiny for use after an expense of 50 to 100 percent for transportation have been added." He called attention to glaring inefficiencies at the steel mill, suggesting that "the Company has too many irons in the fire. This without fun."[79]

Engineers reporting to the company later in the decade revealed a coal department suffering from gross negligence. Necessary improvements had been neglected entirely or delayed. For instance, extremely long hauls got longer each working day, as the rooms receded farther from the mine opening. These hauls steadily increased the fixed expenses of mining coal, but no funds were being expended to construct new openings or to improve the haulage ways.[80]

In 1890, George Ramsay, the new general superintendent of the coal-mining department, found among his many problems a tipple at El Moro that "was in very bad shape, the timbers being rotten and considered unsafe." At the Walsen mine (near Walsenburg), "everything was in the worst condition when I took charge that I ever saw in a mine, the entire machinery and everything connected with it was worn out and has been continually repaired and patched up to make it run. . . ." The following year, Ramsay instituted longwall mining at the Cameron mine (also near Walsenburg), where he found the greatest trouble to be "the scarcity of miners. . . . [T]he men we were able to get were mostly not practical miners and unless they had a great big vein in which they could work standing straight up on their feet they were at a disadvantage and could do but little."[81]

These difficulties occurred during a decade (the 1880s) in which the region's coal industry had expanded with breathtaking rapidity. In one decade, production increased five times, from .5 to 2.5 million tons.[82] Profitability drove the growth of firms until, inevitably, they invaded one another's markets, whereupon the increased competition caused the rate of profit to decline.

In particular, the Colorado Coal and Iron Company ran up against the Colorado Fuel Company. This new firm was the creation of an Iowa coal operator named John C. Osgood (1851–1926). Osgood had come to Colorado in the 1880s as a coal dealer. Encouraged by his good fortune in an expanding industry, he had acquired mining properties and by the decade's end had emerged as an aggressive, successful operator. In 1890, however, he was obliged to report to his stockholders, "Competition has reduced the profit on a ton of coal or coke to so low a figure that no considerable reduction can be made in the future. In view of these facts, the present and future value of your properties cannot be estimated."[83]

About the same time, an engineer explained to the Colorado Coal and Iron company that the cause of high fixed expenses in coal mining was "the failure of our Selling department to furnish orders at the several mines in sufficient amount and regularity to run the works on approximately full time during the season."[84]

Both companies resolved the problem of competition in 1892, when they merged the two enterprises. The result was the Colorado Fuel and Iron Company. Before the end of the decade, it had become one of the hundred largest corporations in the United States.[85]

In 1880, the Centennial State was home to fifteen hundred coal mine workers. A decade later, their ranks had grown to five thousand men and boys. Their numbers included Italians, eastern Europeans, and blacks, but for the most part they constituted the western outpost of the industry's established, British-dominated work force. Something of who they were was thrown into the light of recorded history by Colorado's first major coal mine disaster.

NO TIME TO IMPROVE HIS INTELLECT: MINERS IN THE ROCKY MOUNTAIN WEST, 1880-1903

Eight hours is enough for any man employed underground to work in one day. He has no time to improve his intellect if he works more. . . .

Colorado coal miner, 1887[1]

Coal mining in Colorado . . . has usually been carried out by large corporations. . . . The workmen employed have been looked upon as so many machines, useful for the purpose of grinding out profits, but worthless from any other point of view.

Colorado Commissioner of Labor Statistics, 1902[2]

The Crested Butte mine blew up early in the morning of 24 January 1884, "shortly after the unfortunate men had commenced their labors." The explosion hurled fifty-nine souls into eternity. Ten survivors battled hard for dear life through deadly gases up the main hoisting slope, the strong half-carrying the weak. They were languishing badly when they came upon one of their fellows lying unconscious on the slope, but they managed to drag him to safety. After breathing the fresh air to regain their strength, they "peered with straining eyes through the portals of death with throbbing hearts, anticipating that there might be others who would make their escape; but after a few minutes had passed away, and not one anxiously expected form appearing astride the gloom, their fiercely cherished hopes were changing to feelings of wild despair. A father was looking for a son and a son for a father."[3]

Then spread the "horrible alarm that fifty-nine souls were entombed" in the mine.

The coal mine inspector, arriving at the little village some time later, read the doom of the trapped men in the "sadness and gloom visible on the faces" of the rescuers.[4] When, days later, the rescue party found the dead, it was evident how they had fought against their fate. Two young men had erected a barricade against the spreading fumes and had lain down next to the face, covering their mouths with flannel. They obviously had lived for some time. Another group had rushed for the air shaft. Rescuers found their bodies "huddled and piled in little groups in indiscriminate confusion."[5]

The fortunate were thrown to death by the explosion itself. The inspector described the scene: "I looked in silent awe on the dead, lying in solemn rest amidst the ruins of the ruthless blast. Mules lay swollen on the track, and mine-cars, stoppings, and air bridges were dashed to pieces, and it was clearly discernible, from the appearance of the bodies how far the fire-damp had wrought the work of death, and to what extent the after-damp had finished the sad havoc."[6]

Who were the dead? Their names—David Thomas, John McGregor, James Driscoll, John Price, Thomas Lyle, and so on—reveal their British origin.[7] The record of Colorado's first coal mine disaster establishes that most had been working far from home; most had journeyed West alone or with a son. Only eight of the fifty-nine dead had families living in Colorado. Of thirteen bodies claimed immediately after the explosion, only three were buried nearby. Ten others, migrants even in death, were shipped at company expense to coal towns in Ohio, Illinois, and the anthracite region of Pennsylvania.[8]

The extreme mobility of the coal-mining population offset the isolation of the coal camps. "We have a delegation of miners here this morning from three striking districts in Illinois," a miner wrote in 1889 from Rock Springs, Wyoming. The ninety-four Illinois miners had not decided whether or not to stay, as "the work and wages do not agree with their contract." Yet no one considered it the least bit unusual that they had traveled halfway across the continent to investigate.[9] To a large extent, reported the U. S. Industrial Commission, miners were a floating population. The commission examined the 1890 records of one Illinois firm and discovered that out of 210 employees, only 45 had worked there continuously throughout the year.[10]

The circulation of population provided a channel for the circulation of ideas and information. One piece of information (or misinformation) had to do with where work was available. Partly because such data flowed about the country easily, the demographic changes in the western coal industry mirrored those in the East. Brother followed uncle or friend, and each group drew more of its own to the region. Operators frequently advertised for labor in distant cities. "Never mind advertisements, stay away, stay away!" exclaimed a Rock Springs miner in the late 1880s. "Though you are out of work you can't mend matters by coming here."[11] In the West, it had not taken long for a labor surplus to assist the operators in reducing wages.

The free flow of miners from east to west and back resulted in a western union movement that (in many ways) paralleled that of the East. Colorado's first coal strike probably occurred in 1871, when Erie miners, most of them Scottish and Welsh, protested a new company rule requiring that coal be screened before it was weighed. (Any that dropped through the screen was not paid for: This was a wage cut.)[12] Despite losing this strike, the miners of the Rocky Mountain coalfields won many local victories during the next thirty years. It was a victory for labor when, in March 1883, Colorado enacted a coal mine inspection law. The state's first inspector, a Scottish-born miner named John McNeil, berated the companies for working their mines "in a rude, miserable, and even reckless manner."[13]

THE ERA OF THE KNIGHTS OF LABOR

From 1881 to 1886, Colorado coal miners embarked on sixteen strikes and won nine. Many of the militants were Knights of Labor. In 1882, miners at Union Coal's Louisville and Erie mines forced the company to rescind a wage cut. In 1883, miners at the Colorado Coal and Iron Company's Coal Creek mine (near Cañon City) won higher pay for nut coal. In 1884, workers on the Union Pacific Railroad blocked a wage cut. Like a strong magnet, this railroad victory attracted coal miners and others into the Knights of Labor.[14]

The first big strike began in July 1884. Miners at Rockvale walked out of the Santa Fe's Cañon City mine to protest a fifteen-cent reduction in the tonnage rate. Then the Colorado Coal and Iron Company dropped

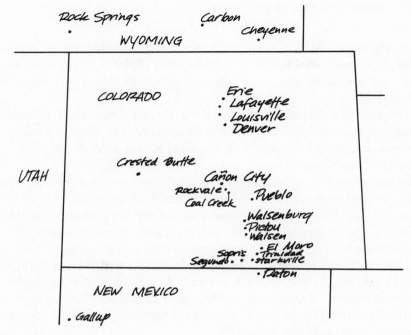

Western coal towns

the rate at nearby Coal Creek and Oak Creek. The workers quit and on second thought added to the list of demands: a reduction in the price of illuminating oil and of blasting powder; an end to the obligatory one dollar per month medical fee; and the abolition of the employment contract that forced miners to forswear union membership.[15]

The 1884 strike began in an atmosphere of festivity. A Rockvale striker declared that the men were enjoying themselves splendidly. Some of the young men played baseball, while others arranged "pleasure parties." However, he would not say how long peace and tranquility would reign, considering the "soulless treachery and despotism of a modern Nero," namely the manager at Rockvale, "a heartless, God-forsaken piece of human nature—yes sir, forsaken by God, left to the hardness of his own heart—if he has one at all. . . ."[16]

To fight the strike, the Colorado Coal and Iron Company used gangs of Italians (a miner suggested) "as screws to squeeze practical miners down to the lowest living wage."[17] Resistance, as the miners saw it, was synonymous with honorable manhood. "Fellowmen, manfully respond to the call!" cried one organizer. For him, the soul of the struggle was brotherly love: "As the storm increases we realize that we are going through an ordeal that will cement us together in the bonds of unity and love, that is making upward of 800 coal miners all told, as the heart of one man."[18]

Although the companies used Italians as strikebreakers, it is highly significant that, as early as this, a few Italians announced their readiness to stand shoulder to shoulder with the other nationalities in the effort to maintain their rights.[19]

In late August, a committee of Fremont County strikers boarded the train for Denver to seek the advice of James Buchanan, publisher of the Denver *Labor Enquirer* and regional leader of the Knights of Labor. The miners wanted a statewide strike, and Buchanan, a furious opponent of his organization's national policy of condemning strikes, supported them. But he cautioned against striking foolishly and advised the purveyors of heat to wait for the first snowfall before calling a convention to consider a general walkout. The miners saw the sense of this and decided to wait for the designated signal. On 25 October, it snowed. Within a week, coal mining in Colorado had ceased.[20]

The miners of the northern Colorado lignite field joined the strike despite the fact that their wages had not been cut. Theirs was essentially a sympathy strike, but for good measure they demanded a raise. Indignant Union Pacific managers could discover no grievances but found the men to be "silent and sulky."[21]

The lignite miners achieved their goals rather quickly. Some of them worked for independent operators—independent, that is, of the Union Pacific's subsidiary, Union Coal. The independents agreed to settle through a Board of Arbitration, and their settlement became the weak link in the Union Pacific's armor. Most of the miners going back to work for the independents had been fired by Union Coal the year before for union activism. These men felt able and quite willing to aid the miners striking their former employer.[22]

The Union Pacific's strength lay in Wyoming. At Carbon, miners were loading two hundred cars a day to make up for the dearth in northern Colorado. It was therefore a severe blow to the company when these miners joined the strike on 13 January. "During my railroad career I have had a good many hard nuts to crack," wrote one manager to the company president, "but the labor question here beats anything I have ever tackled." He reported that the Carbon miners struck because the company would not discharge the foreman and "everyone else who does not suit them. They claim that before they will go to work they are going to compel us to discharge all Findlanders and Chinese."[23]

The Union Pacific capitulated. The company joined the Board of Arbitration, and within a week the strike was settled; northern Colorado miners returned to work with increases ranging from 12.5 to 25 percent.[24] At Carbon, the miners agreed to drop the issue of foreign workers, while the company agreed to submit other differences to arbitration.[25] The northern lignite miners began to contribute the entire amount of their increase to strikers in the south.[26] But their heartfelt efforts were inadequate to the task. The southern Colorado companies introduced armed guards and a stream of strikebreakers to defeat their former employees.

The northern Colorado victory and the southern Colorado defeat established a wage differential between the two fields that remained in effect for thirty years. Also, because the lignite miners had more clout, they succeeded in blocking the introduction of the newer ethnic groups. The northern Colorado lignite field became a bastion of the old ways. The southern bituminous field came to symbolize every hateful thing about the new industrial capitalism.[27]

The 1884 coal strike set another pattern. The victorious firms paid a high price for victory. The U. S. Geological Survey noted that repeated and extended strikes had severely curtailed Colorado coal production in 1884, with the Colorado Coal and Iron Company suffering the heaviest loss.[28] Even after the strike was over, bringing the mines back up to production was an expensive proposition. The superintendent at Coal Creek deplored the "large amount of repairs necessary after the strike."[29] At the Cameron mine, "the expense of cleaning up falls of roof and retimbering after being so long shut down was quite heavy."[30] Responding directly to labor activism, the superintendent at El Moro (Las Animas County) added two mining machines, in part to "lessen the liability of strikes and labor troubles."[31]

Yet the Colorado Coal and Iron Company recovered, reporting to stockholders early in 1885 that despite the "many vicissitudes and trials" of the previous year "coal sales are in excess of any year since the organization of your company."[32]

The Union Pacific had a different problem on its hands: a victorious work force. In response, the company began agonizing over a proposal to lease its northern Colorado mines to an outside operator. Yet even with the labor troubles, the profit on the Colorado coal properties for 1884 had been more than 88,000 dollars. "I join with you," wrote the general manager to the company president, "in your anxiety to get rid of the labor difficulties connected with the operation of these mines. . . ." The problem he saw with the proposition currently on the table for leasing the mines to an outside party was that it would "virtually deprive us of this profit." Nevertheless, in the autumn of 1885 the Union Pacific abandoned its mine at Northrup and leased the mines at Louisville and Erie. Without relinquishing control over the careful maintenance of the mines, or over the use of its northern Colorado coal to fuel its locomotives, the Union Pacific handed the "hard nut"—the labor problem—to someone else to crack.[33]

Of all the companies, the Union Pacific adopted a flexible strategy for controlling its miners, a strategy including a hard approach, as in firing union activists, and a soft approach, as in negotiating with the union. In March 1884, the coal-mining superintendent at Grass Creek, Utah, reported to his superior that he had discharged 12 men: "In making the selection I was guided by the following objects, viz. to clear out a group who have been striking, breaking property . . . a few of the last comers; a few of the members of the Church [Mormons] because there is too many of them here; and a few who entirely ignore the store here, and take all their money to Coalville."[34]

A conscious policy of racial and ethnic mixing formed an important element of the strategy. In Wyoming, the company added Chinese mine workers to the mix; in the face of the implacable anti-Asian racism of the white workers, this guaranteed a divided work force. At the same time, it introduced tension into the coal towns that was nothing less than explosive.

Jay Gould, in 1874 the new controlling director of the Union Pacific Railroad and its subsidiary coal operations, first suggested the idea of introducing the Chinese.[35] Under Gould's direction, the company hired Chinese mine workers at Almy, Wyoming, at lower wages. The miners struck but returned to work with the issue unresolved, whereupon the company cut their wages. "With Chinese at Almy and native miners at the other point [Rock Springs]," Gould wrote to his general manager, "you can play one against the other and thus keep master of the situation."[36]

The company did not quite keep master of the situation, but it did force wages down at Rock Springs the next year by hiring Chinese workers there. The company had asked the five hundred Caucasian mine workers

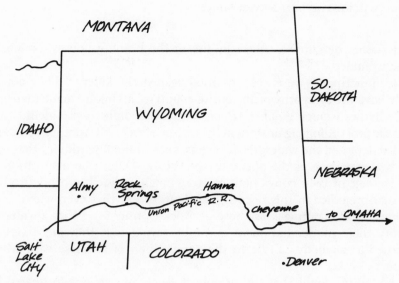

Rock Springs, Wyo.

to increase production, but they refused, accustomed to setting their own production levels. The company brought in Chinese laborers and "only by the exercise of a great deal of strategy and policy," recalled a former manager, "was a serious riot avoided on that occasion."[37]

For a decade, friction between Caucasians and Asians kept Rock Springs in a state of tension. Miners there did not join the strike of 1884, a mark of the success of the policy. By 1885, the company had increased the number of Chinese mine workers to 331, compared with 150 white miners. In 1885, a Knights official complained to the company, "[The] Chinese are having all the work they can do, whilst our men are left out in the cold. . . . [This] makes the situation terribly aggravating, and in spite of my efforts will undoubtedly result in a severe struggle. . . ."[38] The thought of inviting the Asians to join the Knights of Labor occurred to no one.

On 2 September 1885 at Rock Springs, a quarrel broke out between two white and two Chinese mine workers over the mine boss's distribution of underground rooms. The white miners attacked; a few hours later, one of the Chinese died of his injuries. Meanwhile, white miners, mostly English, Scottish, Irish, Welsh, and Swedish immigrants, rampaged through the Chinese section of Rock Springs shooting and setting houses on fire. In the Rock Springs Massacre, twenty-eight Chinese persons died—some by gunfire, others by fire in the burning houses. Nineteen were wounded.[39]

No indictments were ever returned against the killers. "They apparently have the sympathy of the entire country," a Union Pacific manager told Charles Francis Adams, Jr., currently the company president, "and thus far proceedings against them have been a farce."[40] Another company official deplored the widespread, lawless, anti-Chinese sentiment that had the better portion of the public by the throat: "Men who ought to have the courage of their convictions are very much inclined to pander to the popular prejudice in this regard to such an extent that even where they deprecate such outrages as those at Rock Springs . . . they invariably qualify their condemnation of crime by saying that they are as much opposed as anybody can be to the competition of Chinese with white labor. . . ."[41]

"I have no doubt that the trouble at Rock Springs was instigated by the Knights of Labor," wrote the Union Pacific General Superintendent.[42] The Knights committee in Denver denied the charge, but in such a way as to suggest that in any case they did not disapprove:

> Permit us to state that we abhor the action taken by these outraged miners as much as anyone. . . . As to the cause that led to the introduction of Chinese labor on the Union Pacific system, we have no desire to discuss. There is one fact we do wish to state. That is this: the company can get all the labor they require from citizens of this country. Whatever the excuse the company had for introducing this element into civilization there is no excuse for retention of the same.[43]

Not a month had passed since the massacre. These Knights hardly paused for breath before denouncing the victims. At the national level, the leader Terence V. Powderly issued the ludicrous statement that no blame could be attached to organized labor for the Rock Springs Massacre.[44]

For the company, the massacre presented an opening through which to accelerate certain changes. "Such an opportunity as the Rock Springs Massacre is not likely to offer itself again," President Adams declared. Earlier he had stressed, "It is in our interest to have Rock Springs worked entirely by Chinese and machinery."[45] That autumn, the management introduced undercutting machines. One machine, the superintendent reported, would eliminate forty miners. In their places would go "1 machine man, 1 helper, probably Chinese, 2 miners for blasting, timbering, etc., and 15 Chinese for breaking and loading coal."[46]

The Union Pacific fought the union by hiring Chinese, by firing activists, and by conscious racial mixing of the work force. Yet at other times, the firm's pragmatic style of coping with its labor problem led it to accommodate the Knights. In 1884, the company had settled with strikers rather than fight to the finish at any cost. Again in 1889, the company and its employees signed a "Memorandum of Understanding" that regulated wages and other conditions. Two years later, the general manager received a letter from the Employees' Association (members of the Knights) expressing their appreciation of "the kindly spirit you have met us in." The manager sent this to his superior with the note, "Believe the feeling is now better among the employes than it has been for years and the company will receive good results therefrom."[47]

About this time, a Colorado Coal and Iron Company manager made an unauthorized tour of the Union Pacific mines in Wyoming. After surveying the competing firm as best he could under the circumstances (he had not requested permission and was not invited), he reported to his superior that the miners at Rock Springs were paid seventy-five cents per ton, "which I think is very high wages and you can see at once that it is very hard for the coal companies operating in the southern part of the state to keep miners when the Union Pacific Mines are paying such high wages to their men in Wyoming."[48]

THE NEW ERA IN THE WEST

In the 1880s, the companies may have dominated labor, but they were not comfortably in control. In the 1890s, the firms enhanced their collective strength through the concentration of ownership. The 1892 formation of the Colorado Fuel and Iron Company through the merger of competing firms was especially significant.[49] The new company, reported its geologist, occupied "a commanding and unassailable position in the coal trade of the West."[50] In 1896, it added to its power by leasing the coal mines of its remaining competitor in central and southern Colorado—the Atchison, Topeka, & Santa Fe Railroad.[51]

The trend toward concentration was not of course uniquely western. Neither was the inflow of eastern and southern Europeans. As time passed, it became difficult to distinguish El Moro, Colorado, from Connellsville, Pennsylvania. In the early 1890s, new immigrants in the West began fighting to better their conditions in tandem with their counterparts in the East. Before the merger, on 4 March 1891, they struck the Colorado Coal and Iron Company's Crested Butte mine to protest the "accumulation of gas and dust in the old rooms, entries, and air courses of the mine." Within days, the strike disintegrated, when the Austrians began to return to work.[52]

Later that year, workers at Crested Butte were tested again when the Colorado Coal and Iron Company announced a reduction. They responded by walking out on 1 December 1891, although they had no formal connection with any union. The strikers included 70 Italians, 130 Austrians, and 30 English-speaking miners.[53] "Since the men went out," reported the *United Mine Workers Journal*, "they have been very ugly and have paraded the streets heavily armed, threatening death to anyone who should assist the company."[54] But only strikers were hurt, when a sheriff's posse fired into a crowd and several were wounded.[55] As before, Austrians ended this strike by returning to work. The company blamed the Italians for initiating the walkout and declined to rehire this nationality.[56]

For coal operators, the new immigrants were no panacea. Even the Chinese at Rock Springs, who endured the daily hostility of white union miners, were themselves less than perfectly subservient. Welcomed into the mines to solve the labor problem, they soon became the new labor problem. The unskilled Chinese laborer shared one characteristic with the skilled craft miner. Coming as he did from a preindustrial environment, he, too, was accustomed to setting his own work pace. In February 1886, production at Rock Springs ground to a halt due to "Chinese still celebrating" the New Year. The following year, the manager reported, "Most of the Chinamen idle; would not give reason except that 'they were tired.' " Their collective fatigue continued a full two weeks.[57] In general, the Chinese laborer did not like being driven to work longer or harder any more than the Briton did. A journalist visiting Rock Springs reported: "The Chinaman . . . can do as much work in the mines as his white competitor but he works spasmodically. At Rock Springs the local manager of the mines told me that no Chinaman pretends to work more than four days in the week—the rest of the time he spends in lolling around, gambling, smoking opium, and worshipping superstitions. . . ."[58]

THE ERA OF THE UNITED MINE WORKERS OF AMERICA

With the founding of the United Mine Workers of America in 1890, the Erie Assembly of the Knights of Labor promptly went over to the new national coal miners' union. In 1892, the UMWA designated Colorado, Wyoming, Utah, and New Mexico as District 15. Miners elected William Howells, a union activist from Coal Creek, as their district president. For years thereafter, Howells went around trying to organize the coal miners. But despite the several strikes of the decade, he could not consolidate his forces. By 1900, District 15 was only a faint shadow of its parent in the East.[59]

Western miners had an alternative to the UMWA, and this created problems for Howells. The Western Federation of Miners (WFM), founded in 1893 by metal miners in Butte, Montana, welcomed into its ranks all western miners, including the coal diggers. The organization's influence spread rapidly through Colorado, with the majority of lignite miners and even many southern bituminous miners joining up. In June 1896, Louisville coal miners negotiated a contract with their employers. The Western Federation of Miners signed it on their behalf.[60]

The Western Federation of Miners was no United Mine Workers of America. Although the western union struggled mainly to reduce hours and increase wages, it gradually came to espouse an anticapitalist philosophy. In 1897, it withdrew from the American Federation of Labor to form a new umbrella organization, the Western Labor Union, a rival center of trade unionism. By the late 1890s, many of the more radical activists in the labor movement were turning away from Populism, with its emphasis on a return to the small-scale enterprise, toward socialism, an eclectic philosophy that saw working people as having class interests. In 1900, the WFM endorsed the popular Socialist Party candidate, Eugene V. Debs. Two years later, the outgoing WFM president proclaimed, "There are only two classes of people in the world. One is composed of the men and women who produce all; the other is composed of men and women who produce nothing, but live in luxury upon the wealth produced by others." His statement reflected a new awareness of the opposed interests of capital and labor. But then he added a thought that could have come straight from the mouth of any Knights of Labor official speaking in the 1880s: The answer to labor's problems, he said, was to abolish the wage system, "which is more destructive of human rights and liberty than any other slave system devised."[61]

Philosophical differences divided the two coal miners' unions. "For a long time these two organizations have not been on friendly terms," explained the *United Mine Workers' Journal* in 1903, "the Western Federation of Miners going for socialism, while the United Mine Workers held aloof from entering political lines whatever."[62] At second glance however, the political distinctions between the two unions become less clear. Despite the UMWA's official position regarding politics, many members and a significant segment of the leadership embraced socialism. Even as the *United Mine Workers Journal* emphasized the union's abstinence from politics, it also ran a weekly column under the rubric "Why I am a Socialist." Some districts, such as the powerful District 12 (Illinois), were dominated by socialists of a rather moderate variety.[63]

Jurisdictional disagreements caused as much friction as politics. The WFM welcomed western coal miners, while the UMWA fought for jurisdiction over all American coal miners. For many miners in the Rocky Mountain West, the dispute became entangled in their practice of going back and forth between metal and coal mines; these miners often belonged to both unions. They required a transfer card between the two unions to avoid having to pay two initiation fees and two sets of union dues. For years, hostility between their two organizations at the national level prevented the resolution of this difficulty.[64]

During the 1890s, the minimal presence of the UMWA in the West minimized the conflict. But in 1900 UMWA organizers again tackled the southern Colorado coal camps, on 1 August organizing a new local at the Colorado Fuel and Iron Company's Pictou mine. From Pictou, two new unionists set out to proselytize in other southern camps. Before long, fifteen locals were meeting to discuss grievances.[65] In October 1900, unionists from north and south convened at Pueblo to reorganize District 15, electing a Rockvale miner, John Gehr, as district president. (William Howells continued as an organizer.) Immediately, a number of coal miners' locals transferred their charters from the WFM to the UMWA, but other locals and individuals chose to stay with the western union.[66]

Organizers faced the unremitting opposition of the companies. The state labor commissioner observed that the campaign in the south was "attended with difficulty and resulted in failure." Company detectives mingled among the miners and freely attended union meetings. Unionists simplified matters for informers by campaigning openly, in the belief that a state law enacted in 1897 had secured their right to organize. The companies fought back with every weapon at their disposal: "They had not hesitated to use the sheriff . . . a pliant tool of those corporations, to imprison upon trumped up charges, beat, intimidate and run out of the county any person whom it was believed was there for the purpose of organizing the coal miners."

This state of affairs had gone on for years, resulting in "a condition of peonage . . . that was but little removed from downright slavery."[67]

By 1900, the Union Pacific had abandoned its moment of accommodation. In August, the company brought carefully selected German miners from Pennsylvania to Wyoming. Each one signed an agreement, the superintendent reported, "to withdraw from all labor organization and not join one while in our employ. While this might not be any good legally, it has a moral effect. Every possible care is taken to keep nationalities mixed, and not allow any nationality to predominate and no member of a labor organization is knowingly employed. If by accident we get one, he is dropped on the first indication."[68]

The firms also used political clout to insure that public servants served the company, not the public. Coal mine inspectors were high on the list of valued company assets. For instance, in 1900 one Union Pacific manager wrote to another that Mr. Thomas, "one of our old mine superintendents, [is] leaving us at our request to accept position of State Mine Superintendent."[69] A Colorado inspector, D. J. Griffiths, alternated his career as mine inspector with a second career as a Colorado Fuel and Iron Company coal mine superintendent, a position not open to critics of the company. Such relationships prompted miners to refer derisively to some inspectors as "tools of monopoly."[70]

The companies used political corruption, consciously, openly, and systematically, to consolidate and extend their power. "You know that Colorado has about 50% of foreign population, largely under the heel of labor union agitators and shyster politicians," confided a key Colorado Fuel and Iron Company manager to his superior in New York. "The shyster politicians are represented in our legislature, and they are appointed to responsible positions as mine inspectors, commissioners of labor, etc. . . . They introduce . . . laws which if enacted would hamper and cripple operations and reduce profits to zero. To overcome this corporations in this state follow one of two methods,—one is by graft and bribery undercover. . . ."[71]

THE 1901 STRIKE

But even with the variety of weapons at their disposal, in 1901 the corporations were still waging an uphill fight to defeat unionism. The year opened with "The Great Strike of the Coal Miners," a strike that culminated in a partial victory for the miners. By this time, some seventy percent of the northern Colorado lignite miners worked for the Northern Coal and Coke Company, a firm recently organized to lease the Union Pacific's northern Colorado mines. On New Year's Day 1901, a number of the Northern's employees, all members of the UMWA, met in an unofficial mass meeting in Louisville to discuss their grievances. They formulated demands—a ten-percent increase in the tonnage rate and a reduction in the charges for blasting powder—and presented them to the company. The company turned a deaf ear, in response to which the miners struck. "No men are working at Marshall, Lafayette, Erie, or Canfield," reported the Boulder *Labor Leader.* "Now is the time for every miner to join the union and stick to it. If he don't, his cake is all dough."[72]

With the morning temperature on New Years' Day hovering at nineteen below zero, coal production ceased. Within days, the miners expanded their demands to include the eight-hour day and recognition of the UMWA as their bargaining agent.[73] The strike spread south. "The whistles at the local coal fields," a Fremont County newspaper reported, "in accordance with an old custom, blew at 7 o'clock last evening as a signal that the mines would work today. But the miners were deaf to the call, and not a miner in the fields at Rockvale, Brookside, or Coal Creek appeared for work this morning."[74]

District President John Gehr, accompanied by both English-speaking and foreign-speaking miners, began to pay organizing calls on the Colorado Fuel and Iron Company's camps in Las Animas and Huerfano Counties, and in Gallup, New Mexico.[75] In Gallup, superintendents told reporters not to expect a strike, because the men had no grievances. The men informed them otherwise: The companies systematically cheated in weights and forced them to pay exorbitant prices at the company store.[76]

For the providers of heat, cold weather was good strike news. A greenhouse owner became hysterical, and he was probably typical: For businesses like his, the lack of coal to fuel boilers meant ruin. With the supply of the relatively smokeless lignite cut off, Denver turned to bituminous coal for heating buildings. On still days, "great clouds of black, sooty and vile smelling smoke roll down into the streets, almost stifling the pedestrians below." But later in January, the weather turned against the strikers, bringing unusual summerlike temperatures to the high plains east of the Rockies.[77]

The strike was not planned; rather, it spread and evolved. At the end of January, an organizer from the national office arrived to take charge. By this time, miners had added demands for check-weighmen and for a regular payday every two weeks.[78]

The press opened a debate on exactly how much miners did get paid. This was grist for anyone's propaganda mill because the question had no simple answer. Companies modified the tonnage rate with their determination of how many pounds constituted a ton, and modified it again by weighing and paying for only part of any given carload of coal. Added to the complexity of such mathematics was the questionable accuracy of the scales. All these matters affected the rate paid to the miner.

Then again, the tonnage rate and the actual amount paid were two different things. From the miner's pay envelope, companies deducted the amounts owed the company for supplies and necessities. Of course, the company controlled the prices of such items. One miner's pay statement for December showed the amount earned to be $40.40. Expenses deducted for groceries, blacksmithing, and so forth came to exactly $40.40, leaving the man with an empty pay envelope.[79] This sort of coincidence moved a miner's wife to quip that if they were going to starve to death anyway, they might as well die rested: They planned to stay on strike.[80]

But there was even more to confound a simple answer to the question of how much a miner got paid. Out of what remained after company deductions, the miner paid his laborer. Finally, so seldom did the miner work a full week that the tonnage rate was no measure at all of his standard of living. "This is the kind of thing," wrote one journalist, "that makes one disbelieve the old proverb that figures can't lie."[81]

Early in the strike, the president of the Northern Coal and Coke Company told reporters that miners earned eighty to one hundred dollars per month, an excellent wage for the skilled worker of the day. "People ought to know," the miners countered, "whether the miners as a whole are making living wages or not, or whether there are a few favored ones who for reasons best known to the company and its bosses, are [e]nabled to make, as Mr. Cannon has stated, $80 to $100."

The miners contended that the average digger lived with the wolf at the door. They were fighting for "the wherewithall to pay board" and "have a few dollars over . . . to pay for such needed articles of clothing as we must have." Less than full-time work was a major problem: a miner might earn sixty dollars one month but only twenty the next.[82]

A state Bureau of Labor Statistics survey conducted the year before revealed the average annual wages of coal miners to be 370 dollars. This contrasted unfavorably with the average wage of 772 dollars for thirty occupations; a clerk earned 482 dollars, a bricklayer 969. The question really was, as the Labor Commissioner put it to a coal miner, "How in the world do you and your family manage to live?" The man shrugged his shoulders in reply:

> I have often asked myself the same question. You see, we don't live, merely exist. We only stay during a portion of the year. We sometimes cultivate a small garden patch, our wives do a little washing or something of the kind, and earn small sums. We occasionally get a day's work outside of the coal mine. We occasionally have a little money saved from the winter season, and we pull through, I hardly know how, until work again becomes more plentiful. I sometimes wonder why the soul will be satisfied with a body that is treated so shabbily as is that of the average working man. We feel ourselves compelled to put our children to work . . . as soon as they are old enough to earn a dollar.[83]

Today, we tend to think of the demand for increased wages as a "bread and butter" issue, one that seeks not to change the system but to improve conditions within it. Yet among coal miners at the turn of the century, this simple wage demand often reflected a radical critique of class relations. Consider the class awareness exhibited on the editorial page of the *Pueblo Courier*, a labor paper. Advocating an increase for miners, the editorial proclaimed: "Everybody is aware of the fact that the coal companies are paying non-resident stockholders fat dividends every quarter, while the men who actually do the work barely maintain a miserable existence, and that by practicing the shrewdest economy." Similarly, a coal miner asserted that God put the coal in the earth "for the benefit of the whole human family, and . . . it takes quite a stretch of the imagination to see why anybody should be permitted to stand between God's blessing and mankind and compel us to freeze or starve to satisfy a lot of idle drones or to declare dividends for a lot of idle loafers. . . ."[84]

On the other hand, District 15 officials carefully disavowed explicitly anticapitalist opinions during the strike. Organizer William Howells insisted: "We do not wish to cripple the investment of capital in the coal production of the state."[85] Another organizer argued that if the Colorado coal operators talked to their counterparts in the Midwest, they would discover that the union, "instead of being a menace to capital, had become an anchor of safety for it."[86]

In 1901, miners attempted a statewide action, but their efforts were hampered by the fact that mine workers throughout the state shared neither the same conditions nor the same history. The one thousand northern lignite miners constituted a strongly British population (sixty percent foreign-born, mostly English), whose living and working conditions were a far cry from those of miners living farther south.

According to a Colorado legislative committee investigating the strike, the northern Colorado miners felt attached to their communities, because (unlike many mine workers across the country) they had really settled there. They lived in their own "pretty little cottages kept nicely painted." Brightly flowered vines tumbled over the cottages in the summer, and many householders kept gardens. The miners had purchased their homes with small monthly payments made over a long period of years when "conditions in the northern fields were more prosperous than at the present time."[87] Many had lived in the shadow of the Rockies for eight or more years; a number could recount firsthand the story of the 1884 strike.

The coal towns of the northern lignite field were unique, democratic places in which the union stood at the center of social and cultural life. A picture of the social fabric of these towns emerges in accounts of the Lafayette Labor Day celebration held in 1901. On that day, by ten in the morning the streets were full of people, "old and young, men with their wives, boys with their sweethearts," all strolling to the new union hall to the "sweet strains of music" played by the Mine Workers' Brass Band. The day's entertainment included speeches, singing by the Lafayette male quartet and by the ladies' quartet, a bicycle race, a foot race, a sack race, a wheelbarrow race, a ladies' race, a horse race, and, finally, a pony race. These events "passed off pleasantly without a hitch or quarrel as too often occurs."[88]

A grande ball concluded the day. Miners provided the music; their dance band included a clarinet, a piano, a concertina, and a slide trombone. The women, members of the "Ladies Lignite Temple No. 10, Rathbone Sisters," served supper to 150 people, who declared (a miner reported) "they never partook of a better repast and the ladies realized a neat little sum." The ball "was one of the sweetest affairs ever seen around here."[89]

The lignite miners enjoyed an independent community life, electing town and county officials out of their own ranks. Nevertheless, the element of company coercion was not entirely absent. At Erie "a feeling in the air" told the men that if they purchased items at the company store they would get more work to do in the mine. At the store, blasting powder, a necessary mining supply, cost $2.75 per keg, in contrast to $2.25 per keg at independent stores.[90] In northern Colorado, company stores ruined the independent stores by opening with extremely low prices. Having driven off the competition, they raised their prices. During the strike, they cut off credit to striking miners who had nowhere else to go for necessities.[91]

Conditions underground were also far from desirable. The Legislative Committee found the air in half the northern mines to be "extremely poor." At the Northern Coal and Coke Company's Hecla mine near Louisville, a fire had been burning in the gob (the waste thrown along the edge of the tunnel) for two years, creating a high risk of explosion and a suffocating atmosphere.[92]

At the Superior mine, the miners were obliged to lay their own track and to furnish their own timbers. "If there is water in the room," one worker joked, "the miner can bail it out or lie down in it just as he pleases."[93] Not one Northern Coal and Coke Company mine possessed a second means of escape. Air shafts would have served the purpose, but, inexcusably, they lacked ladders. A journalist made the point that in a fire the miners would be smothered like rats in a hole.[94]

This was deplorable, but the farther south the legislative committee traveled, the worse they found conditions to be. When they arrived at the Colorado Fuel and Iron Company's camps in Huerfano and Las Animas counties, the investigation ground to a halt. No workers would appear as witnesses. At Pictou, the miners, mostly Italians, Slavs, Mexicans, and blacks, "frankly stated that should they appear on the stand and tell the truth they, to use a miners' vernacular, would be compelled to 'hit the road.' "[95] Democracy in this corner of America was unknown.

In the view of Colorado Fuel and Iron Company president John C. Osgood, coal mine unions were "a curse to the men as well as to the employers." He denounced UMWA President John Mitchell as a tyrant worse than the czar of Russia: "No selfish and cold-blooded employer ever exacted the blind obedience, absolute surrender of independence, or contribution of hard-won earnings that he and his organization exacts from his dupes. No slavery can be worse than the slavery which his organization imposes on its members."[96]

Projection—the unconscious process of attributing one's own attitudes or feelings onto another—comes to mind as a possible influence on Osgood's opinions. For the tyranny imposed by the Colorado Fuel and Iron Company on its employees was no small matter, especially in the southern coal counties where the company controlled the law-enforcement apparatus. On one occasion, the pit bosses from Starkville (in Las Animas County) rode out on horses to break up an open-air union meeting, and drove the miners back to camp. In Huerfano County, Sheriff Jeff Farr and his deputies drove the attendants of a similar meeting into an arroyo and beat them with revolvers. The sheriff told the men, "I am the chief of this county and if you don't believe it I will blow some of you to hell, you sons of bitches." The union accused this man of grossly and brutally violating the laws to carry out the will of the coal-mining companies.[97]

The miners saw such antidemocratic methods as anti-American, for, in their view, America meant freedom. Consider the particular way a unionist expressed himself the week Gallup, New Mexico, joined the strike. The company, "in pursuance of its policy of hatred for all that is honest and just and American," discharged seventy union members the day after they joined.[98]

The outcome of the 1901 strike reflected original conditions. In southern Colorado, the strike never got off the ground; in New Mexico, it was lost. In Fremont County, however, the Colorado Fuel and Iron Company granted all demands except that of union recognition. A month later, the Northern Coal and Coke Company did the same. When the lignite miners returned to work, the UMWA claimed ninety percent of them as members.[99]

A SHIFTING LANDSCAPE, 1901–1903

Reacting to the strike, the Colorado Fuel and Iron Company entered the era of paternalism by establishing a "Sociological Department." Whether to ameliorate conflict, to polish the company image, or to actually better conditions, the new department presented gifts to the miners' children at Christmas and operated kindergartens in the coal camps. It did nothing to address the core of miners' grievances: poverty, and a lack of safety and freedom. Nevertheless, the company acquired a new look. In September 1902, an article in *Outlook* beamed, "While the relations of coal miners and their employers in the East remain so unsatisfactory . . . it is pleasant to note the good feeling that exists between employers and employed in a great mining industry in the West."[100] The Sociological Department's own *Camp and Plant* described the Pictou coal camp: "The Italian climate, the sunshine and purified air, cool breezes in summer with warm winds for winter; . . . six feet of the finest domestic coal overlaid with self supporting sand rock for roof, thereby affording protection for the sturdy miner—all aid to make the life of this community one of peace, joy and happiness, and a comfort to those who dig for coal."[101]

The reality was otherwise. Consider an explosion at Pictou in 1902 that killed two miners. Gas had troubled the miners for more than a week; days before, small explosions had severely burned three men. Nevertheless, on the day of the fatal explosion, the fireboss reported no gas, and no equipment was on hand for the rescue effort. Although the company-controlled coroner's jury exonerated the company, a union committee announced that the Colorado Fuel and Iron Company "is to blame and is guilty of murder."[102]

The lack of freedom at Pictou was palpable. On the occasion of the accident, a local union official excoriated the company:

> The union men at this place sneak around most of the time like they were stealing something when they come to a union meeting; yet the C.F. & I. Co. is going around advertising their camps and plants as a Garden of Eden and getting men to come to their black, dirty holes, where there is not a living to be made, and where a man can not have any enjoyment of life at all, and where the tyrannical superintendents . . . give [the men] to understand that he will lose his job if he joins any labor union.[103]

Fear stalked the union organizer. "We men dare not go through the streets of our towns, or out on the trails alone," one organizer revealed. "We must go in twos and threes."[104] The company had "inaugurated and maintained a reign of terror," reported the Commissioner of Labor Statistics in 1902, "so completely subjugating the miners in these fields with the fear of arrest, imprisonment, and discharge, that, while anxious to organize, [the miners] feared to do so."[105] Indeed, from 1900 to 1903, according to union estimates, Colorado coal companies discharged and blacklisted six thousand of the eight thousand new union members.[106]

By 1900, the Colorado Fuel and Iron Company had shouldered its way into first place among Colorado corporations; it was one of the hundred largest firms in the country.[107] To its employees, it represented absolute power. But from management's perspective things were slipping out of control. Beginning in 1901, John C. Osgood, financially overextended due to an ambitious expansion of the steel plant, and sucked dry by unanticipated labor troubles, began a struggle to retain control of the company. The vultures circling were financial and railroad capitalists of national stature, men like J. P. Morgan and John D. Rockefeller. The silent exchanges of stock that constitute the plot of this story would be too tedious to recount here. Suffice it to say that, most significantly, in 1901 John D. Rockefeller purchased six million dollars of Colorado Fuel and Iron Company stock.[108]

In 1902, the fight went public with Osgood and his associates lined up against railroad capitalists Edwin Hawley, E. H. Harriman, and George Gould, who were allied with the Chicago financier John W. Gates. These railroad capitalists were minority stockholders in the Colorado Fuel and Iron Company. They held the controlling interest in the Union Pacific Railroad and in the Denver & Rio Grande, both large purchasers of Colorado Fuel and Iron coal. They also controlled the Utah Fuel Company, a subsidiary of the Denver & Rio Grande Western and by now a competitor of the Colorado Fuel and Iron Company. Under the circumstances, it was easy for Osgood to hold up to stockholders the specter of the company being purchased and gutted in the interest of competing corporations.[109] On their part, the big capitalists accused Osgood of operating the company (through his control of a galaxy of subsidiaries), for the benefit of himself and his friends, leaving the stockholders out in the cold.[110]

Osgood clung to the cliff of bankruptcy until 1903, and then let go. In July 1903, amid daily speculation in the Denver papers, control of the Colorado Fuel and Iron Company passed into the hands of George Gould and John D. Rockefeller.[111] (The closely allied Harriman of the Union Pacific remained a minority stockholder.) Osgood was forced to part with his company, but he remained a formidable anti-union force in Colorado through his ownership of the second and third largest coal firms in the state, the Victor Fuel Company and the American Fuel Company.[112]

Unionists detested John C. Osgood. Reacting to a comment in the Denver *Post* that Osgood "has been looked at as a sort of god by the 20,000 employees of the company by reason of his sociological work on their behalf," a district UMWA officer retorted: "How could men learn to respect a tyrannical villain that has said thou shalt work eleven or twelve hours a day, thou shalt trade at my store, pay my prices and receive just what I want to pay you, thou shalt read and belong to the kind of societies that is my pleasure, thou shalt not even speak to thy friend if he is distasteful to me, for he might be an agitator." The official conceded that Osgood had once elevated the tonnage rate without solicitation, but claimed that he got it back by cheating in weighing. He concluded that if Rockefeller and Gould turned out to be worse than Osgood, they would be the devil incarnate.[113]

The takeover was consummated at a board meeting in New York. Soon John D. Rockefeller, Jr., and Frederick Gates, a key Rockefeller executive, were in Colorado touring the properties. The *Rocky Mountain News* spread the trappings of the wealth they represented before the public eye.

The little inspection party, including Osgood and his associates, traveled from Pueblo to Denver and into southern Wyoming in a three-car train, "one of the most sumptuous that has ever pulled into the union depot of this city." Later, the group was entertained at *Redstone,* Osgood's palatial forty-two room Tudor mansion.[114]

The company they surveyed was an impressive conglomerate of coal, iron, and steel.[115] Company assets included six hundred square miles of coal lands on which were operated thirty-nine large mines and 3,500 coke ovens. Annual coal output exceeded six million tons a year. "The profit on coal and coke alone," asserted one new director, "pays the fixed charges of the company." Moreover, the company's "legitimate territory" included the railroad mileage of the West belonging to the Rockefeller/ Gould interests, railroads whose freight bills to the Colorado Fuel and Iron Company amounted to more than a million dollars a month.[116]

Even without Osgood, in the reorganized Colorado Fuel and Iron Company, unionists faced a powerful foe. The concentration of capital, and of the power it represented, had increased dramatically. To make matters worse for labor, in 1903 businessmen began to coordinate their anti-union campaign. Across the nation, the formation of the Citizen's Alliance represented an increase in class consciousness among capitalists and their followers. In April 1903, local opponents of organized labor formed a state Citizen's Alliance to oppose the "arbitrary exactions and unjust regulations of labor unions in this community. . . ." They worked to correct and prevent "pernicious class legislation," such as the eight-hour bill.[117] By October, the Colorado organization was strong enough to establish a statewide organization of over thirty thousand members.[118]

As the owners drew closer together through mergers and alliances, unionists found themselves split apart by the gulf dividing the United Mine Workers of America and the Western Federation of Miners. In 1903, UMWA President Mitchell wrote to WFM President Moyer to request that the western union cede jurisdiction of western coal miners. Moyer's chilly reply was in the negative.[119]

Despite this setback, in the summer of 1903 UMWA District 15 organizers began a new recruitment campaign in the southern coal camps. Company spies "dogged" them and their recruits. "Whenever we organized any men they were discharged the next morning," an organizer reflected. "We soon found we were only sacrificing the good men of the district." Hundreds of men were beaten up and fired at Primero, Hastings, and other large camps because they were suspected of being union men.[120]

Even in the independent lignite towns, organizing was not going particularly well. A Lafayette miner reported early in 1903 that the district was sorely in need of finances. Most of District 15 was unorganized and "we have no system to work under." He requested more attention from the national officers, "so we may become men as our brothers in the east are and not pieces of machinery."[121]

At Rock Springs, a union activist found himself without a job "although I had not opened my mouth to anyone in regard to the union except to men I knew had cards." He had helped to organize a small local two miles outside of town one Sunday morning before breakfast. Shortly thereafter, every man in the new local was discharged.[122]

One activist declared rather cheerfully, "the firing and blacklisting goes merrily on. They have . . . spies on every corner. . . . I consider myself the luckiest man in Rock Springs. I have been fired three times and ordered out of town by the marshall yesterday morning." The false claims of a superintendent had lured him from Salt Lake City to Rock Springs. "Well, the long and short of it was I came out here. But oh! What a difference! When I got here I found they were paying 47 cents only and that it took 2240 pounds to make a ton and the cars run three to four to the man . . . and instead of shoveling your coal into the car you had to shovel it into the back end of the company's store to pay your board bill with."[123]

There is no evidence, however, that these miners recognized the real barrier to organizing Rock Springs: their own racism. In 1903, the UMWA confessed that, because of "Chinese and Japs employed," the organizers were not able to get full representations in the camps; consequently Wyoming as a whole was poorly organized. "It is expected however," the union concluded, "that all except the Chinese and Japs will be members of the movement before the time comes to strike."[124]

Hanna presented another problem, partly because of its location in a sparsely settled "desert." The town's isolation and its domination by one firm meant that once discharged a man had to leave immediately; his recruitment was thereby lost to the organizing effort. By firing the "most chronic ones," an activist declared, the company managed to "scare the suckers back to their hiding places."[125]

At Hanna, the slope dipped into the earth for a mile and a half; the underground workings extended for more than twenty miles. On 30 June 1903, an explosion ripped through the mine, snuffing out 169 lives. The Hanna disaster overshadowed the stirrings and preparations for the great

1903 coal strike. Denver newspapers were filled with the gruesome details: "For hours tonight the scene at the mouth . . . was heartmoving. With clothes and hair awry, mothers, wives, sweethearts and children huddled together waiting or walking to and fro wringing their hands in utter collapse. Many sat on shattered timbers . . . insensible to their surroundings. The most frantic pushed to the edge of the gap and tried to force a way into the slope. . . ."[126]

It was months before the last body was removed. By then, the greatest strike the region had ever known was under way.

10

WE HAVE GRIEVOUS WRONGS: THE COLORADO COAL STRIKE, 1903-1904

There are . . . places, especially in southern Colorado . . . where one has to set aside his imaginative powers to convince himself that he is still in free (?) America.

Duncan McDonald, UMWA organizer, October 1903 [1]

We have grievous wrongs. We are Americans.

Southern Colorado coal miner, October 1903 [2]

"The miners [are] destined to be exiled from all civilized privileges by being forced to live in isolated hollows and cañons," proclaimed a Manifesto issued by United Mine Workers District 15 in August 1903. The document recited a litany of grievances from lack of ventilation to forced buying in the company store. It blamed the "servile policy adopted by organized capital" for the discharge of all "who show an inclination to unite themselves with their fellow workers." Company policies were "strongly tinctured with the old time feudal slavery."[3]

To publish the Manifesto was to throw down the gauntlet, but UMWA officers also invited the coal operators to a conference.[4] The operators declined. Colorado Fuel and Iron Company Vice President Jesse Welborn stated that the company would meet at any time with its own employees, who in any case did not desire to strike, but not with the union. A Victor Fuel Company manager insisted that the firm's mine workers were busy and contented.[5]

The managers believed their own statements fervently, but they were dead wrong. In reality, union organizers had been struggling all summer to keep the lid on a strike-prone work force. By September, conditions had become so unsettled that district officials quickly reorganized a regional convention to take place two weeks earlier than planned. On 23 September 1903, delegates met in Pueblo to determine whether or not to declare a general strike involving the 22,000 miners working in District 15— Colorado, Wyoming, Utah, and New Mexico. A union officer told reporters that "the intention to strike is firmly fixed in the minds of the men."[6]

For the union, the question was, could the UMWA afford not to endorse a strike? Could a union that held on to an antistrike position hold on to a membership that was itching to strike? Yet some union leaders believed that a strike would be a disastrous mistake. At the convention, John Gehr, former district president, now an organizer appointed by the national office, chided the delegates, "I do not think it wise, in fact I believe that it will mean the death of our organization in the Western field. . . ."[7]

The UMWA found itself squeezed between a rock and a hard place. The rock was the militance of its own membership; the hard place was the radical alternative presented by the Western Federation of Miners. The WFM was at this moment leading its metal miners in a ferocious battle for the eight-hour day—the Cripple Creek strike. Indeed, discontent was boiling among workers throughout the state due to the events surrounding the struggle for eight-hour legislation. For years, labor had pushed for such a bill; in 1902, a state referendum overwhelmingly mandated the state legislature to draw one up. This done, the document was introduced to the legislature, which sent it on to a conference committee of the House and Senate. From there, it simply vanished, never to be referred to again.[8] As a labor paper described what happened, "the mine owners, by lobbying, bribing, and distributing other species of graft, managed to thwart public sentiment by subsidizing the legislature, and thus precipitated most of the trouble."[9] Advocates of the eight-hour day saw no obvious alternative to a strike. WFM officers believed that a statewide victory for miners required a walkout by coal miners led by the UMWA. WFM President Moyer argued for this point of view before the District 15 convention.[10]

Conservatives in the UMWA, a minority at the convention, waxed furious at the spectacle of the WFM president urging the coal miners to strike.[11] In their view, the western union was manipulating the coal miners purely in the interest of their own strike. Neither were conservatives pleased with the proposed resolution endorsing socialism. "It has already brought up much favorable comment," reported the *Rocky Mountain News*, "and many delegates state that it will pass." The following day, "a lively fight" preceded its passage, and though many delegates demanded that their negative vote be recorded, the resolution passed by a large majority.[12]

The convention also voted to strike. This did not settle the matter however, because the final decision had to be made by the UMWA national executive board in Indianapolis.[13] It was perfectly legal for District 15 to strike on its own, but also obvious, given the scale of the trouble brewing in Colorado, that the district needed the backing of the national union. The question on the table at the board meeting was whether a strike could succeed even if sanctioned and subsidized by the national office.

District 15 officers favored a strike. They argued that thousands of Colorado mine workers were eager to join the union but dared not. Four-fifths of the work force would strike, but only if they were called out by President Mitchell. "It all hangs on that," one speaker emphasized. Then he added an important point. "The men of the northern field," he said, "are desirous of forcing the issue at this time."[14] Yet it was the response of the south that would be critical. Union strength centered in the northern lignite field, while corporate strength centered in the south.

John Gehr agreed that a strike could be won if four-fifths of the miners joined it, but "if they do," he said, "you can have my head. I have tried them and I know." He questioned the loyalty of the southern coal miners, particularly the Italians. "If four-fifths of the men will strike, why do they refuse to meet our organizers? We've had a number of strikes that were illegal. There is no discipline among the men. They have no respect for the constitution of the UMWA. They have taken the bit in their teeth. Some have threatened that if a general strike is not called, they will withdraw and join the Western Federation of Miners."[15]

Gehr berated the open organizing methods still being pursued in the south. "Nothing will be accomplished," he bristled, "by public meetings, parades and big hurrahs. The work will have to be done secretly. The operators have the support of the Governor, the citizens Alliance and the National Guards. I believe the CF&I wants a strike to be able to crush the union."[16]

The United Mine Workers of America, the largest American trade union of the day, was weak at the center. The discussion on the table illustrates the point. Men committed to the district argued for district interests, never considering how a regional strategy might fit into the national picture. Indeed, executive board minutes reveal no evidence of a national strategy.

Even within the context of regional strategy, the debaters ignored a crucial factor. While arguing the merits of a strike in terms of its potential support among miners, both sides focused on Colorado only, never engaging the question of whether District 15 miners in Wyoming, Utah, and New Mexico could be counted upon.

The board voted to leave the strike question up to a referendum of District 15 union members, a nondecision that, given the prevailing mood, amounted to authorizing the strike. The referendum was held. Colorado union miners endorsed a strike scheduled to begin on 9 November 1903 unless the operators came to a conference first. It would be directed from Trinidad (the county seat of Las Animas County) and was planned to include all coal miners in Colorado, Wyoming, Utah, and New Mexico.[17]

The strikers demanded the eight-hour day, an increase in the tonnage rate, a fair system of weighing, a biweekly payday with payment in legal United States currency, improvements in ventilation, and the definition of a ton as 2,000 rather than 2,400 pounds. A state law ignored by the operators already mandated the latter four demands.[18] The demand for union recognition was conspicuously absent.

The dry list of demands registered anguish and outrage. "We have grievous wrongs. We are American!" protested a southern Colorado miner in late October. "We are often robbed of 50 per cent of the coal mined. To complain of the injustice means instant dismissal." The company stores robbed the miners "disgracefully," and in the coal camps freedom of speech was unknown: "A man . . . dare not call his soul his own. If two men are talking together, a deputy sheriff or paid tool of the company will edge up alongside of him to find out if he is talking about his work or labor unions."[19]

Under these conditions, known organizers could make no headway: Any miner observed speaking to such a person would lose his job immediately. Therefore, the flurry of press reports on Mother Jones in mid-October prognosticated the beginning of an open fight. "Word was received," reported the Denver *Republican*, "that . . . the 'Joan of Arc' of the Pennsylvania coal miners would come to instill courage among the men in the West."[20] The Denver *Post* hyperbolized that Mother Jones could "sway thousands to a spirit of frenzy or with a shake of her head and a few soft-spoken words check the mob seeking to burn and slay."[21]

When the real Mother Jones, an elderly, white-haired woman wearing a plain dark gown, stepped off the train at Trinidad, the news quickly spread about town and many people "nearly broke the rubber in their neck in an effort to get a good look."[22] The labor press beat the welcome drum, particularly extolling her maternal qualities. A widow whose husband and children had died in an epidemic forty years before, Mother Jones had made the coalfields her home and the miners "her boys."[23] She personified the militant mother—the ideal of womanhood as it was conceived in the coalfields. "This brave little woman," gushed one labor paper, "has won for herself a simple appellation, the most loving and endearing that the human tongue or pen of man ever couched in the language of a word, 'Mother.' There is not a word in all the dialects of nations that recalls such sacred memories or appeals so strongly to all the good that lies in the human breast."[24]

With Mother Jones and other organizers in place, the strike was on. On Monday morning, 9 November, ten thousand miners—ninety-five percent of the Colorado work force—refrained in a body from going to work.[25] Even the union was stunned. It was the "biggest surprise party in the history of the West," declared the *United Mine Workers Journal.* "Union and non-union, white and Mexican, all obeyed the strike order of President Mitchell and the tie up is complete."[26]

Organizers watched hundreds of families streaming out of the Colorado Fuel and Iron Company's Berwind camp. "There were many wagons loaded high with furniture," reported the *Rocky Mountain News*. "There were men afoot with trunks on their shoulders, and women and children walking. . . . Nearly all of the miners recognized Mother Jones and took off their hats to her, saluting her with vive Mother Jones." Companies had begun evictions the moment the strike began. This was more than a workplace action: The entire community was immediately affected. The mining people clogged the streets of Trinidad. Cheap furniture glutted the secondhand stores as half the population put up their things for sale. Italian miners, "fairly well dressed and all decent and quiet," were seen walking toward the depot in threes and fours. Some planned to return to Italy. Others left wives and children in the union's care while they headed East to look for work.[27]

A few union miners "went back on us," as the president of the Rugby local put it. "One was our late Trustee, and as soon as the strike came on he flopped and said, 'I am with the company.' He picked up his rifle and started right up as a guard."[28]

But the real problems were in Wyoming and New Mexico. The UMWA needed these miners.[29] If the operators could supply their customers with coal from nearby coalfields (avoiding otherwise high transportation costs), victory would be theirs.

By this time, operators in the four-state region could present strikers with a unified front; the boards of directors of the region's major coal and railroad corporations overlapped significantly.[30] Given the unanimous anti-union views of the directors, the firms could handily agree on concerted opposition to the union. Only the Northern Coal and Coke Company, a relatively small firm but the largest in the northern Colorado lignite field, stood apart. Even its independence can be questioned, in that it leased its coal from the Union Pacific Coal Company.[31]

Considering the resources of the operators, the strike news from the outlying coalfields was considerably less than heartening. In southern Wyoming, where the Union Pacific had opportunely offered a ten-percent raise, miners continued working.[32] The strike in New Mexico barely got off the ground.[33]

In Utah, the only state besides Colorado where a real contest developed, the struggle began slowly. Two days before the scheduled walkout, the Utah Fuel Company, a subsidiary of the Denver & Rio Grande Western, offered its own ten-percent raise.[34] Nevertheless, striking miners succeeded in closing the mine at Castle Gate. Unlike their counterparts in Colorado, they included union recognition in their list of demands. This, announced a manager, "would be fought to the bitter end." Besides raising wages (and prices), the company rescinded the credit of strikers at the company store. The carrot/stick strategy worked. Of the four major Utah mines, Castle Gate alone remained closed until the Utah operators broke the strike.[35]

In Colorado, District 15 had its hands full managing a strike involving nineteen thousand men, women, and children.[36] Through its distribution of strike relief, it became, instantly and with little preparation, an enor-

mous social service agency. The District had years of experience in organizing but none in managing a large welfare operation. Although the
committee that oversaw the strike included experienced men from the
national office, it was beset with logistical and financial difficulties from
the outset.

Providing food and shelter to strikers pouring into Trinidad was the first
necessity. The union rented every available vacant dwelling, set up tent
villages on the prairies beyond company property, and established a soup
kitchen in Trinidad.[37] (In a weirdly callous decision, the strike committee
ruled that only men and boys could eat there.) The strike committee
distributed food and clothing and provided a doctor to any striker who
presented himself, but did not dispense money in a specific weekly
amount. To do so, the committee felt, would draw hundreds of "grafters,"
whose object in coming to the region would be to take advantage of strike
relief.[38]

Despite such precautions, the strike was in financial trouble from the
first week. The worm was the open-ended definition of who was eligible
for strike benefits. Of necessity, most southern miners joined the union
just as they joined the strike. Company repression had prevented the
union from keeping an accurate list of members eligible for relief in the
event of a strike. The union could have counted all new members at
the beginning of the strike, but this went against its policy of continually
accepting newcomers in the hope of recruiting strikebreakers. Men coming in to work (whether misinformed or not), and joining the strike
instead, swelled the relief list. In one typical instance, 140 men arrived
in Raton, New Mexico, only to discover that a strike was on. Their
response was to go to Trinidad "to be taken care of by the union."[39]

The union paid the train fare of strikers or strikebreakers leaving the
region in order to relieve the strike fund of their continuous support. But
the stream of new arrivals brought to the coalfields by company agents
made this particular expense open-ended and unending. (Ironically, the
policy became a bonanza for railroads with connections in Trinidad,
railroads which had close ties to the coal companies.)[40] Such policies
made it impossible to establish a budget and to stay within it. The strike
began to drain the national treasury of twelve to fifteen thousand dollars
a week.[41]

For the companies, fighting the strike was also a major logistical and
financial undertaking. At no time did operators consider negotiating. A

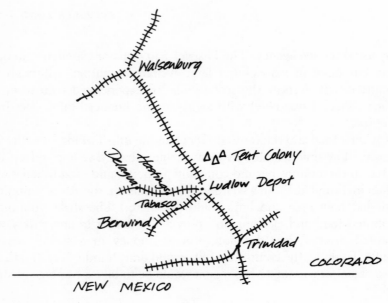

Walsenburg

Delagua Hastings △△△ Tent Colony

Ludlow Depot

Tabasco

Berwind

Trinidad

COLORADO

NEW MEXICO

So. Colorado coal towns, 1903

Colorado Fuel and Iron Company manager accurately summarized the position of every company in the region. "The strike will be fought out to a finish, whether it takes ten days, six months or ten years."[42]

Early in November, Colorado Fuel and Iron Company managers met in Trinidad to discuss tactics. Every superintendent and pit boss attended.[43] The following week, the Colorado Fuel and Iron Company and the Victor Fuel Company announced that they would act in conjunction.[44] They planned to resume production by concentrating all forces in one place—the Colorado Fuel and Iron Company's Walsen mine in Huerfano County.[45] Before long, strikebreakers escorted by armed guards could be seen going to work at the Walsen mine.

For the operators, the guards and their guns were the costliest aspect of the strike. In the first week, the Victor Fuel Company purchased the entire stock of rifles and revolvers at a Trinidad hardware store, while another coal company received a shipment of fifty rifles.[46]

The county sheriffs deputized the guards, but they were paid by the coal operators; no pretense of neutrality hampered their activities. To guarantee the guards' hostility to the strikers, the sheriff of Las Animas County dismissed fifty-eight of the sixty local undersheriffs because they could not be depended upon. He replaced them with ninety new gunmen recruited from outside the coalfields who "will not flinch if a test comes." Some were cowboys glad to earn five dollars a day.[47] And if introducing strikebreakers was one way to break a strike, introducing machinery was another. The Colorado Fuel and Iron Company used the lull in production to install undercutting machines in two mines, including the Walsen.[48]

Opposing the company goal of hiring strikebreakers was the union goal of getting them to quit. This and the task of keeping the ranks of strikers solidly behind the strike required unremitting organizing. Every night, organizers held meetings and rallies throughout the district. In this way, reported the *United Mine Workers Journal*, "the leaders will keep in constant touch with the men and cheer them on to the end."[49]

One reporter attended a meeting at Tobasco along with a crowd 2,500 strong, and wrote, "The side of the mountain was covered with people from Hastings, Delagua, Berwind, and Tobasco . . . miners and their families, superintendents and pit bosses . . . deputy sheriffs and a great many women." A chair placed above the crowd served as a platform. The first to mount it was organizer James Mooney, who spoke on the miners' conditions. Next a Welsh-born organizer named William Wardjon got up and sang "The Miners' Battle Cry." Then the crowd welcomed Mother Jones with tremendous applause. She spoke first to the strikers, next to the operators and superintendents, and finally to the "officers of the law." To these, she promised to assist in maintaining order, and she asked them to treat the miners with kindness and consideration. Concluding the meeting, Italian speakers reiterated what had been said in English. On this occasion, the union took in more than 1,500 new members, mine workers who had quit at the strike call but had not yet joined the union.[50]

Persuasion was the organizer's tool. The good organizers were good speakers, and some were great speakers. Crowds in this pretelevision era rewarded them by listening for two or even three hours at a stretch. One night in Pueblo, according to a reporter, William Wardjon spoke for nearly two hours, "and never once did he lose control of his large audience. With tender pathos and flashing wit he moved his hearers to tears or laughter at will, while all the time he delivered sledge-hammer blows for the cause to which he has consecrated his life."[51]

Later in the strike, Wardjon inspired a miner from Lafayette to write, "Mr. Editor, for nearly three hours he kept this large crowd of men spellbound. You could hear a pin drop at any time. I have been a miner for thirty years, and have heard hundreds of prominent speakers on the labor question, but I never heard a speech like Wardjon gave on this occasion."[52]

Mother Jones, as one scholar has noted, "lived by her tongue." Her ability to sway a crowd was legendary, and, of course, it depended in part upon the legend. At Rockvale, the announcement of her impending arrival "produced such universal excitement as is very seldom witnessed in these serious coal communities. From the early morning this day was given up entirely to this strange . . . excitement. Men and women, old and young, were all equally under the spell of expectancy." As she entered the Opera House, a full house rose and cheered. She spoke for nearly two hours, during which time "the audience was truly hers."[53]

Some of the union's best organizers worked in Italian, the language spoken by one-third of the district's miners.[54] One such was Joseph Poggiani, a "fluent and able talker, thinker, and organizer."[55] Another earned the appellation "Demolli the silver-tongued." Charles Demolli worked mainly in Utah and, according to the Salt Lake *Herald*, was "eloquent with tongue and pen in the Italian language. Not only this but he can talk in their native tongues with Finlanders, Slavs, French or representatives of other nationalities. With his level head, shrewd judgement, college education, suave manner and great magnetism, he is regarded as one of the strongest men affiliated with the United Mine Workers and he is idolized by his followers."[56]

The Italians became enthusiastic militants for the strike, but their activism originated in their own communities and flourished independently of the UMWA; their support for the union was conditional.[57] For them, ethnic and national loyalties complemented the desire to improve their condition as workers. For instance, the constitution of an Italian fraternal organization at Sunnyside, Utah, stipulated that any member who became a strikebreaker would be expelled.[58]

At Hastings, a large coal camp in Las Animas County, the numerous Italian residents resented the Victor Fuel Company's treatment of their fraternal organization. Prior to the strike, the society had made elaborate arrangements for its annual election, including the rental of a special train to bring members to Hastings for the day. After every detail was finalized, the superintendent refused the Italians permission to conduct the meeting. "This created so much feeling," reported the *Rocky Mountain News*, "that nearly all the men are now members of the miners' organization." The following morning, the paper added, "The Italians have quit almost to a man. They are greatly angered by the treatment of their countrymen by the Hastings management. They have not joined the union to any extent, giving as their reason that they were sold out by labor organizations on two or three occasions, and they prefer to go out on their own responsibility on this occasion." Two weeks later, company gunmen at Hastings arrested the vice president of the society and "forcibly ejected him from the camp." The Italians felt much enraged.[59]

Another center of Italian activism was Segundo, located a few miles from Hastings. The older section of Segundo had been settled prior to the arrival of the company and had remained independent. Hundreds of strikers from the surrounding coal camps congregated in "old Segundo." Seven or eight saloons catered to miners roaming the streets, and several dog fights furnished entertainment. At midday, the strikers pulled out an old wagon and decorated it with American flags "and from it an Italian speaker addressed the men in heated terms against the Fuel and Iron Co." He received demonstrations of approval.[60]

Among Italians, the strike was very much a family affair. Wives and daughters expressed open hostility toward the coal companies, which, after all, oppressed them directly in the company towns. The women's traditional female roles did not translate into passive behavior. Their treatment of a certain Father Berta, a procompany priest, was a case in point.[61] One day early in the strike, Father Berta came into Segundo to conduct services at his church. During his visit, a crowd of women attacked him and "compelled him to seek safety in flight." The men pursued him, "handled him roughly," and cursed him roundly. It was a group of UMWA organizers who rescued the priest, escorted him to the depot, and declared that the incident had "outraged decency and brought odium on the conduct of the strike."[62]

The women at Hastings echoed the aggressive spirit of those at Segundo. One day, a mine guard and his deputies began to demolish an empty shack in the Sicilian quarter of the camp known as Ragtown. They were forced to desist by one Marie Vanelle who assumed they were beginning evictions. As a deputy wrestled a revolver out of her hands, she raised a cry that produced ten other women. She then ran away, returning with a meat cleaver. With it, she struck one of the men, nearly severing his right ear. The other women also attacked, and the deputies restrained them only with difficulty.[63]

Married to such women, the miners' commitment to their union did not interfere with their devotion to their families. Neither did it interfere with their ethnic loyalties. An important resource available to Italian coal-mining families was the support of a community more than thirty years old that extended well beyond the coalfields. A newspaper reported that in a nearby farming region Italian strikers were "quartered on their countrymen," although they did their own cooking and supplemented the food supply by hunting rabbits and quail.[64] Here again, ethnic and national solidarity did not work against the class struggle but instead enhanced it.

THE SEPARATE SETTLEMENT IN THE NORTH

Measured by its support in the southern coal communities, the strike was supremely successful. The northern lignite miners had also quit work almost to the man, an expected outcome since it was they who initiated the strike. But within days tension between these two fields, one populated with the newer coal mine workers, the other with the traditional English-speaking work force, boiled up into a crisis that many believed turned the strike toward defeat.

Less than a week after the strike began, the Northern Coal and Coke Company proposed a settlement. Its managers had been talking with employees prior to the strike, and they continued talking after work had ceased. This contrasted noticeably with the lack of conversation between opposing sides in the south. Nevertheless, some unionists suspected the northern company of collusion with the overtly anti-union companies in the south. In their view, a settlement in the north would release a supply of cheap coal to the struck companies in the south, thereby increasing their ability to crush the union. Conservative unionists countered that the Northern Coal and Coke Company had extended the hand of genuine friendship to organized labor.[65]

In its proposed settlement, the Northern agreed to the eight-hour day (but only if the miners in the south won it), to increase the tonnage rate, and to abide by the state mining law. Miners had not demanded the recognition of the UMWA as a bargaining agent, and it was not an issue. The proposal was nothing to write home about. But in time the company improved it by dropping the condition on the eight hours.[66] The *Rocky Mountain News* editorialized that the lignite miners, "the most intelligent and best read coal miners in the world," should go back to work in order to narrow the contest down to companies refusing to make any concessions at all.[67]

Among unionists, controversy raged over whether or not the lignite miners should return to work. On 21 November, they met to consider their employer's proposal. On the one side, District President Howells, Mother Jones, and William Wardjon believed that if the miners returned to work they would kill the strike. Victory in the south required a coal famine. A lasting settlement in the north required victory in the south. On the other side, UMWA President John Mitchell and those loyal to him believed that a settlement in the north would enable the union to concentrate resources on the south and thus increase the chance of winning there. The lignite miners divided along the same lines; they passed the day prior to the vote debating the issue on the street corners of Lafayette and Louisville.[68]

On the day of the meeting, William Howells and Mother Jones hastened to Louisville. First, Howells took the platform. "What would your settlement be worth if we lose the fight in the south?" he asked. "If the operators in the southern field prove victorious, how long will you have your eight hours here?" His speech, reported the Bureau of Labor Statistics, had little effect. But then Mother Jones made one of her impassioned appeals. She carried the miners with her, and they voted against going to work.[69] Her talk was important, because it reflected her understanding of the fundamental task the union faced at the national as well as local level. She argued for solidarity between the established work force of the north and the newer immigrants of the south:

> Brothers, you English speaking miners of the northern fields promised your southern brothers, seventy percent of whom do not speak English, that you would support them to the end. Now you are asked to betray them, to make a separate settlement. You have a common enemy and your duty is to fight to the finish. . . . If you go back to work here and your brothers fall in the south, you will be responsible for their defeat. . . . I would say we will all go to glory together or we will all die and go down together.[70]

By now these English-speaking miners included Italians, many of whom had lived in Colorado for many years, if not all their lives. The Italians voted against the settlement in the north, almost to a man.[71]

The controversy brought out the extent to which the Western Federation of Miners was involved in the affairs of UMWA District 15. Before the vote, two WFM officers (President Moyer and Secretary William "Big Bill" Haywood) attended a daylong meeting with coal miners from Louisville and Lafayette. Afterward, one of them told the press that a settlement in the north would defeat the strike for both the coal miners and the metalliferous miners.[72]

Within the UMWA, supporters of a settlement regarded this as rank interference in the internal affairs of one union by the officers of another. Newspaper accounts at the time contained hints of their rancor, but it was later, after the strike was defeated, that full-blown recriminations exploded into public view.[73]

The struggle over the separate settlement in the north was a local skirmish on a national battlefield. Before the fight in Colorado, John Mitchell had come under fire from unionists who felt he was all too ready to compromise with the coal operators.[74] By now, his social circle included operators and excluded working miners. His outlook was reflected in his membership in the National Civic Federation, an organization begun in 1900 by Mark Hanna, an Ohio coal operator, and other men associated with large-scale enterprises. Influential in the movement toward what was later named corporate liberalism, the National Civic Federation included representatives of capital, "the public," and labor. Its mission was to reduce class conflict by opposing both socialism and the extreme anti-unionism of the National Association of Manufacturers.[75] However, many miners did not see it as benefiting labor, and they criticized John Mitchell's participation.

Mitchell represented one side of a growing split in the labor movement, while the opponents of the settlement in the northern lignite field represented the other. Organizers like William Howells, Mother Jones, and William Wardjon freely espoused socialist ideas, a fact that did not reduce their popularity. In a speech at Pueblo, Mother Jones "appealed to the unions to go into politics, for the reason that the capitalists are in it." She declared socialism "to be the remedy that would solve the labor problem of the world. . . ."[76]

These organizers openly denounced Mitchell's position on the northern settlement. Wardjon informed a large audience at Pueblo that a settlement would cause miners in both the north and the south to lose everything. Although he "had the greatest admiration for John Mitchell and believed him to be the world's greatest labor leader . . . he did not believe him infallible and he did not believe he could remain in Boston [where

Mitchell was attending a convention] and wage a successful strike in Colorado."[77] William Howells told another crowd that Mitchell could not understand the situation in Colorado from his office in Indianapolis.[78]

But Mitchell also had supporters in the West, and he had detractors who could be persuaded to change sides. Shortly after the vote against the settlement, efforts were made to arrange another meeting. At the second meeting, the northern lignite miners voted to return to work, 480 in favor, 130 opposed. The mines started up the following Monday. They began producing seven thousand tons a day, six thousand tons of which was the product of the Northern Coal and Coke Company. This, a newspaper reported, "went a long way in relieving the anticipated [coal] famine."[79]

Did the separate settlement break the strike? Given the virtually uninterrupted production in Wyoming, Utah, and New Mexico, and the frequently announced plans for bringing coal from the East, a total coal famine was impossible even if the lignite miners had stayed out.[80] The radicals' pronouncements about this were off the mark. Nevertheless, the settlement unquestionably assisted the large corporations in their fight to crush the union. Coal produced in the lignite field allowed them to keep their largest customers supplied more cheaply than by bringing coal from the East. It is conceivable, as one dissident unionist insisted, that the large operators encouraged the Northern Coal and Coke Company to settle. Anticipating the settlement, in late November a railroad subsidiary of the Colorado Fuel and Iron Company positioned a large number of cars in the northern field, prepared to carry coal south just as fast as it was produced.[81]

The separate settlement shattered the fragile unity holding together radical and conservative unionists. Local UMWA officers and members tended to feel more comfortable with the WFM than with the settlement. Organizers appointed by the UMWA national office supported the settlement (as well as other policies of John Mitchell). In November, the friction between the two reached the point that the Easterners (who controlled the strike) refused to use the Trinidad strike headquarters, instead working out of their hotel rooms. This prompted district officials to consider resigning, a step they felt would cause the district to fall to pieces.[82] On the side of the national office, a Board Member indignantly recalled, "When I went to Trinidad I was told by every national officer out there that they never went into the district office that they were not insulted by the district officers. . . . [M]y reception was in keeping with the reception of those who had preceded me."[83]

THE SOUTH AGAIN

Under the direction of a feuding leadership, the strike in the south staggered forward in the face of an incoming stream of strikebreakers. The strike itself created a new source of willing workers. In early December, lack of coal forced the Colorado Fuel and Iron Company to close its Pueblo steel plant, and this released hundreds of steel workers to the coalfields. Company managers escorted them south and introduced them to the art of mining coal.[84]

Armed guards swarmed the region, escalating the tension between the two sides. Although both strikers and company gunmen carried guns, the union worked hard to preserve peace in the well-founded belief that armed conflict would result in certain defeat. The UMWA issued the original strike call with the warning, "It is of course, unnecessary to remind you how important, how essential it is that the miners and their friends conduct themselves in a manner which will command respect, sympathy and support from the public. The men should be admonished to observe the law. . . ."[85] Union literature frequently instructed the miners to "[b]e patient, calm, and steadfast, and permit no provocation, no matter how severe, to goad you into lawlessness."[86]

Strikers kept the peace, but imperfectly. In early December, shooting broke out between deputies and Italian strikers near the coke ovens at Segundo, leaving two strikers dead. They were accused of "interfering with scabs between Segundo and Primero."[87]

Other incidents perpetrated by strikers were trivial. At Segundo, they stoned a superintendent's house. On a Trinidad street, an Italian striker walked up to a coal company manager and knocked the pipe out of his mouth.[88] At Sopris, three masked men entered a saloon, pointed their Winchesters at the patrons, and ordered them to put up their hands. They escorted a Mexican recruiter of strikebreakers to the outdoors, where they were joined by nine more armed men. The party marched the man down the street and into the darkness. Newspapers confidently reported him murdered, but he turned up later in good health.[89]

At Hastings, an explosion damaged the windows of the Victor Fuel Company's powerhouse. Strikers were blamed for the deed, but they ridiculed the accusation, pointing out that if a coal digger dynamited the powerhouse there would have been ruins to show for the attempt.[90]

Verbal fireworks flared daily. For the union, the propaganda war involved convincing the public of the righteousness of the cause and, even more difficult, convincing strikebreakers to quit work, a task of potentially immeasurable proportions. Company propaganda portrayed the union as an outside force. It portrayed organizers as "grafters," "tyrants," and "scum."[91] It portrayed Mother Jones's past in a most interesting light. She had once been "a well-known character, not alone in the 'Red Light' district of Denver, but in Omaha, Kansas City, Chicago, and far off San Francisco." Mother Jones had been an "inmate of Jennie Rogers house on Market Street, Denver," "a procuress of girls for a house of prostitution in Omaha," and a "vulgar, heartless, vicious creature, with a fiery temper and a cold-blooded brutality rare even in the slums."[92]

This news about Mother Jones's past originated in the Denver office of the Pinkerton Detective Agency, an office directed by none other than James McParlan[d], our old friend of Molly Maguire fame. The procompany newsletter in which the story was printed was hardly a reliable source.[93] Publicly, Mother Jones ignored the accusation; privately, she gave conflicting accounts.[94]

Her real past remains buried under a mass of her own contradictory statements. But the truth of the matter may have been its least important aspect. Mother Jones's effectiveness as an organizer was built upon a carefully nurtured maternal image; the opposite idea presented by the company side could be quite damaging. Not even the charismatic Mother Jones could transcend the two boxes of "good" versus "bad" woman imposed by the larger culture. For years afterward, the story had a way of turning up in newspaper accounts, especially during coal strikes. But the campaign to destroy Mother Jones's reputation failed, because the miners believed it was slander.[95]

In addition to traducing particular organizers and the union in general, operators conducted a campaign of violence against the strikers.[96] In one incident, mine guards entered the house of a pro-union saloon keeper, turned over his furniture, broke open boxes of private papers, threw his pregnant wife on the floor, and arrested him without a warrant. He was "marched down through Delagua, his wife running along and begging for his release."[97] In another incident, unknown persons blew up the houses of five union activists in New Castle, Colorado. These perpetrators knew what they were about, leaving behind five piles of rubble.[98]

Week after week, strikers and organizers were badly beaten.[99] One of the worst cases involved Chris Evans, the financial manager of the strike. Three masked men boarded the Pueblo-bound train at Trinidad and began clubbing him with the butt ends of their rifles. They left him on the floor unconscious and bleeding profusely. As the train began to move, the attackers ran to the back of the car and jumped off. The passengers observing the incident were too dumbfounded to move, "and did not regain their presence of mind until the masked men had completed their job, leaving Evans for dead." At that point, the passengers washed Evans's head wounds with water from the water tank and forced whiskey down his throat to revive him. At the first stop, he was taken to a hospital.[100]

Such methods had their effect. Gradually, the mines filled with new employees. Precisely when the union should have called off the strike is debatable. But as early as the New Year the future looked dark.[101]

The more strikebreakers the companies brought in, the more union funds were drained in recruiting them, feeding them, and transporting them out. Yet coal production steadily increased. For the union, financial mismanagement of the strike may have aggravated the problems. Pro-operator newspapers gloated over announcements to this effect, and at least some national officers concurred.[102]

By March, district and national officers alike agreed that the Colorado strike was lost. Both groups were so impressed with the futility of continuing it that they called a district convention for the purpose of calling it off. But on 23 March, before the convention could meet, Governor James Peabody ordered the Colorado National Guard into the southern Colorado coalfields. The delegates to the convention, President Mitchell recounted, "were so incensed at the . . . uncalled for action of the governor that instead of calmly considering the strike and declaring it off, as they undoubtedly would have done," they voted to continue the strike. They decided that returning to work now would be a cowardly surrender to Governor Peabody, that "tool of the mining corporations of Colorado."[103]

Ostensibly, Peabody had ordered the Colorado National Guard to preserve law and order, but in fact the strikers were conducting themselves with remarkable restraint. The governor sent the troops to stand guard so the companies could import strikebreakers en masse and get production back up to normal. The governor remained quite indifferent to enforcing the state laws violated by the coal companies.[104]

On 23 March, four hundred troops and one hundred horses of the

Colorado National Guard, commanded by Major Zeph Hill, arrived at the Trinidad Depot. Major Hill immediately declared martial law, allowing only those persons able to convince him that they were "good, law abiding citizens" to retain their weapons. The strikers did not qualify, but most mine guards did. He canceled freedom of assembly, announced a nine o'clock curfew, and established press censorship. No messages were to be transmitted by telegraph or telephone without his "O.K." (The technology of the day required calls to be placed by the local operator.) He forbade telephone conversations in languages other than English. He closed down the strikers' Italian-language paper and confiscated the current edition, which (not incidentally) claimed that 200,000 dollars had been raised for the maintenance of the troops: 80,000 dollars from the Colorado Fuel and Iron Company, 70,000 dollars from the Victor Fuel Company, 30,000 dollars from the Citizen's Alliance, and 20,000 dollars from the railroads.[105]

Supposedly but not actually a neutral force between the two sides, the troops joined company gunmen in harassing the strikers. On 26 March, three days after their arrival, they began deportations. Mother Jones was one of the first to go. After being expelled from the strike region, she made her way to Denver and wrote the governor, "I am right here in the capital. . . . I want to ask you, governor, what in Hell are you going to do about it?"[106]

A week later, eight strikers, mostly officers of their local unions, were taken by train to the New Mexico border, put out, and warned never to return. No charges were placed against them. Upon questioning, Major Hill asserted that he believed their absence was better than their presence.[107]

Martial law escalated the violence against the strikers. Beatings, arrests, and deportations became daily occurrences. On 30 April, William Wardjon encountered his second beating, this time sustaining a severe head injury.[108] Incapacitated in the hospital, he developed a morbid fear of being assaulted by company thugs while lying in bed. So great was his terror that Chris Evans urged the national office to remove him from the state. This was done, ending Wardjon's participation in the strike.[109]

On 7 May, a seventy-year-old striker, Joe Raiz, was caught by three masked men at the back of the camp at Sunlight, Colorado, and castrated. He died three days later.[110]

On 19 May, mounted troops marched eighty Italian strikers from Berwind to Trinidad, a distance of eighteen miles. The strikers had refused to register their descriptions for future reference, reasoning that they had committed no crime. These Italians were not newcomers to the state but were "old miners [who] have property in the county." They told the press through an interpreter that the troops had driven them like cattle in the burning sun. "They repeatedly struck us," one miner said, "and several times when men would lag behind they would run their horses against them and compel them to run or stagger out of the way."[111]

The women endured sexual harassment from the troops as they no doubt had from company mine guards. In one incident shortly after the Guard arrived, a private "invaded the house of a miner at Segundo during his absence." He was drunk and "guilty of offensive conduct toward the miner's wife." Her husband returned "opportunely." In the ensuing scuffle, the private was shot through the hand with his own weapon. Subsequently, he was court-martialed and fined seven dollars for his offense. Major Hill remarked that the private "may have frightened some women, but the story of attempted assault I do not believe. A lot of Italians and strikers make the charge, and I will believe my own men before I will them." The following night, he heard a similar case involving a different soldier.[112]

Agnes Smedley, an American writer who grew up near Trinidad, recalled (in *Daughter of Earth*) the fear the troops inspired in the townspeople for their daughters. After the troops arrived, families did not allow their working daughters to return home alone at the end of the day. One night, a girl's father made the mistake of neglecting to fetch her from the laundry where she worked. When she did not appear at the usual time, he set out to search for her:

> He found her in the possession of two soldiers, away down between piles of lumber in the lumber yard. One standing on watch had warned the other and the shouts of the father had not led to their capture. . . . News travels by air beyond the tracks, and before the father and his daughter had passed a dozen houses along the street, men and women emerged from their doors or stood in groups watching. It was a silent march those two made, with heads bent and eyes that did not see, the girl's blouse torn from her and her eyes red and swollen from weeping.[113]

In comparison to such attacks, the strikers' offenses were insignificant, but they did occur. In March, an Italian striker fired at strikebreakers at the Pryor Camp. Returning his fire, mine guards shot him dead. On 18 April, strikers stoned strikebreakers on their way to work. Two days later, eight Italian strikers seized a tipple and attempted to force strikebreakers off the job. Strikers were also accused of starting several fires.[114]

By spring, corporate victory was complete. Strikebreakers were producing coal in the quantities the companies required. Most had come from outside the state, but the strike had also suffered an erosion in its own ranks. As the president of the Pictou local reported: "Some of our men that we supposed were good union men got under the influence of cowardice . . . and after being supported for three months broke away from our ranks, and today are working as scabs. One of them . . . is John Gardner by name, and claiming to be a good union man from Scotland. . . ."[115]

In late April, the UMWA executive board concluded for the second time that the strike could not be won. Again the board instructed the district to call a convention, this one to be held in June, to discuss the best methods for closing it up. While also supporting strikes in Pennsylvania, West Virginia, Ohio, Tennessee, and Kentucky, the union had poured 500,000 dollars down the Colorado drain and could afford no more.[116]

As instructed, District 15 called a convention to discuss terminating the strike. Instead, on 23 June the delegates voted unanimously to continue it without the financial support of the national office. They did so on moral grounds, and some believed that if they could keep going until cool weather increased the demand for coal, they could win. District President Howells explained: "The operators absolutely refuse to treat with us in any way and leave us no alternative than to continue the strike."[117] But these moralistic arguments mentioned nothing about a strategy for winning.

A district financial committee set itself to raising money to replace the substantial strike relief withdrawn by the national. Now the split between the two became a chasm.[118] "There is certainly some grave misunderstanding between District and National officers," wrote district secretary John Simpson to the national office in July. "I do not believe that the rank and file of the district should be punished for the wrongs of two or three, neither do I believe that the same rank and file should be forced to accept something they do not want and is not just. The process of starving men into submission is what we are in the organization to fight against."[119]

Angry feelings toward the national office came from the rank and file as well. Speaking of the separate settlement in the north, a Lafayette miner wrote to the *Journal:* "There's a lot of hard feeling all over Colorado, especially in the north, toward men who went to work at the advice of the national." He accused the national of going further than merely withdrawing from the strike. "Now comes word from one of the men who went East to solicit funds for the miners who are on strike in Colorado that the national is advising the locals in the East not to give anything to the Colorado miners. Is this true or a d——— lie?"[120] Still, no concrete proposals were put forward to suggest how a strike that had already failed with the support of the national union could succeed without it.

In September, the District called the strike anew, in the hope that Colorado's now fully functioning coal industry could be halted once again. Fewer than one hundred miners responded. On 12 October, the District officially gave up, authorizing the handful of stalwarts to return to work.[121]

Inevitably, defeat settled on hundreds of strikers and their families in the form of a personal tragedy. For instance, an Italian miner from Scofield, Utah, had spent years putting every extra cent into a home he had built on company land. He joined the strike at the cost of losing his house. To the superintendent who offered him the opportunity to go back to work as a strikebreaker, he said: "For eight years you have stolen nearly all my wages, but what little you have not stolen I have saved and built a home with. Now you are going to steal my home. You can take the wages you have stolen for eight years, you can take my home from me and from my wife and children but you cannot take my manhood."[122]

Unionists claimed that women and children were going around with gunny sacks (feed bags) on their feet, that superintendents and mine bosses had adopted "very mean methods to prevent individuals from securing work at mines in which they were formerly employed," and that "self-appointed posses" were seizing individuals and driving them out of the coal towns.[123]

A Lafayette miner confessed his deep discouragement: ". . . [A]fter being in a coal mine since I was 9 years old and am now over 40 and am almost in the same place as to finance I am almost tempted to ask, 'Is it any use?' "[124]

Within the union, controversy raged over responsibility for the defeat. At the UMWA convention of January 1905, a delegate from Deitz, Wyoming, Robert Randall, denounced John Mitchell in a long and bitter speech. He accused the union president of many treasons, including collaboration with the Northern Coal and Coke Company to force a settlement in the north. Randall called Mitchell the "little tin labor god of the capitalistic class" and proclaimed his greater respect for Governor

Peabody, because he was "at least true to his class and served his masters faithfully, but I cannot find words to express my contempt for a man who, having raised himself to power by the sufferings of the working class, falls a victim to the flattery of his capitalistic masters and proves himself false to the working man. In the name of the men of the West, whom you have proved a traitor to, I denounce you."[125]

Mitchell followed Randall to the podium to castigate the dissidents of District 15. He condemned the district convention because, in refusing to end the strike, it had acted "contrary to our expectations, and I unhesitatingly assert, contrary to the best interests of their constituents." He refuted Randall's various accusations at great length. He lectured the delegates that "when we are convinced beyond peradventure that a strike cannot be won, it becomes our solemn duty, regardless of criticism or condemnation, to recognize the inevitable and declare that . . . the strike should be brought to a close."[126]

Above all, Mitchell accused Randall of being an agent of the radical and increasingly revolutionary Western Federation of Miners. "I know how much Mr. Randall cares for the UMWA!" Mitchell shouted to the delegates. "I know where his heart is! His heart is where the documents he has read come from and you do not need to tell us where you got them Mr. Randall!"[127]

Others besides Mitchell voiced suspicions that Randall was not speaking as an individual but as an agent of the WFM. Chris Evans accused the western union of being behind the vote in Pueblo to continue the strike instead of calling it off: "I . . . happened to be in Pueblo when they decided to continue the strike, contrary to the wishes and advice of the national officials. What did I find there? I found the Representatives of the WFM there. . . ."[128]

Here was the old split between trade unionists seeking to reform the system and revolutionaries seeking to replace it. Yet the split was not a simple one of socialists versus nonsocialists. The conservatives were backed by a large group of moderate socialists, including the powerful Illinois socialist bloc, who criticized the WFM for dual unionism and who, when the battle lines were drawn, gave their personal loyalty to John Mitchell.[129]

The defeat and the altercation that followed it had consequences for the prime movers that were both political and highly personal: As victory could make a labor career, so defeat could end one. Randall's attack on John Mitchell at the convention of 1905 was the first in a series of blows that ended Mitchell's career at the head of the nation's largest union. Increasingly under attack, particularly for his participation in the National Civic Federation, Mitchell declined to run for reelection in 1908. By then, according to his biographer, "[h]e had for some time, lived on a social and financial scale approximating that of the small businessmen of the country."[130]

Randall himself paid for his attack. On Mitchell's ultimatum that "either this man or I shall not be a member of this union," the convention voted to expel Randall from the coal miners' union unless or until he retracted his statements, which he refused to do. The resolution to condemn him was moved by the Illinois miners' leader John Walker, a socialist, and seconded by William Wardjon, also a socialist.[131]

In the midst of the storm, the leadership of District 15 went down. At the end of 1904, the national office announced the names of new district officers. That the change was made under duress was indicated by Chris Evans's warning to the miners of the district. "[T]he new district officers . . . are the only officials . . . that have authority to issue circulars or appeals of any kind on behalf of this district and coal miners everywhere should govern themselves accordingly." William Howells, who first came into public view during the 1884 strike, returned to his home in Coal Creek. This significant figure was not heard from again in the context of the Colorado coalfields.[132]

The executive board recalled Mother Jones to inform her that her services would be dispensed with if she did not conform to the union's policies as set by the board. Although Mitchell later reappointed her as an organizer (she was not dismissed as one dissident claimed), she left the union by her own choice to work for socialism and for the WFM. Not until 1910 did she return to the UMWA fold.[133]

In the fight over who was to blame, the antagonists nearly forgot that it was the corporations, not any faction of coal miners, that defeated the Colorado strike of 1903–1904. Given the will of the coal operators to fight the union at almost any cost, the interlocking unity of the region's coal and railroad corporations, and the Colorado government's support of the companies, it would have been difficult for the miners to win even with a totally unified organization.

William Wardjon recognized this, and, at the 1905 convention, it was he who had the last word on the strike. He opposed Randall's attack on Mitchell. "Although I am a Socialist," he said, "I am a loyal trade unionist. As one of the men who was connected with the Colorado strike I will say that nothing was done by the national officers that caused the strike to be lost; it was lost through the influence of the mine owners, the Citizen's Alliance and the machinery of the state. That is what lost it. It was also lost because of the tens of thousands of non-union men that were sent into the state."[134]

"Corporation-cursed Colorado," as the *United Mine Workers Journal* had called it in the bleak month of June 1904, was now unorganized territory.[135] As delegate Randall had rightly said, the Colorado coal companies had "closed upon the western miners the door of hope for years to come."[136]

11

WHO LIVE LIKE RATS: MANAGERS AND MINERS REORGANIZE, 1905-1913

. . . considering these foreigners who do not intend to make America their home, and who live like rats in order to save money, I do not feel that we ought to maintain high wages in order to increase their income and shorten their stay in this country.

L. M. Bowers, Colorado Fuel and Iron Company manager, to John D. Rockefeller, Jr., 1909[1]

I am not an organizer, but I tell a couple of fellows to combine, like the big corporations. I says, Here boys, we must join the union. It's time for the working man to tie to something.

Mike Sekoria, southern Colorado coal miner, 1914[2]

The defeat of the 1903–1904 strike left the wreckage of UMWA District 15 in its wake. Union membership plummeted from some 9,000 strikers in November 1903 to exactly 425 paid-up members in January 1905. The Western Federation of Miners was in no better shape, having sustained a devastating defeat of its Cripple Creek strike.[3]

For the coal corporations, victory was costly. In June 1904, the Colorado Fuel and Iron Company informed its stockholders of a deficit of more than a million dollars. The worst damage had been inflicted by the necessity of shutting down the Pueblo steel mill in December 1903, due to the lack of coal. The company had decided to distribute its severely reduced tonnage to its outside customers in order to keep their business, and to prevent a coal famine in Colorado, which, the company felt, "would have affected the successful termination of the strike."[4]

MANAGEMENT REORGANIZES: THE COLORADO FUEL AND IRON COMPANY

The deterioration of Colorado Fuel and Iron Company earnings spurred the Rockefeller directors to action. In order to gain closer scrutiny and control, in 1907 John D. Rockefeller, his son, Rockefeller, Jr., and their ally, George Gould, placed their own manager at the helm. "The turn things have taken," explained Frederick Gates, a Rockefeller executive, "rendered the condition of the CF&I sufficiently acute to make it very important that we have a man on the inside, who can keep his finger on the pulse all the time and keep us informed as to exact conditions and at the same time serve in every possible way and remedy whatever he finds at fault."[5] Gates, the overseer of the company from New York, was writing to his uncle, Lamont Montgomery Bowers (1847–1941), a successful manager of other Rockefeller enterprises. Rockefeller appointed Bowers vice president of the company at a salary (twelve thousand dollars a year) exceeding all others in the firm except that of the newly promoted president, Jesse F. Welborn.[6]

Though the strike had shaken the company to the core, to outsiders it still seemed impervious. Its power was symbolized by the Denver suite out of which Bowers began to rearrange its business practices. A newspaper revealed that his new offices were being fitted with magnificent oriental rugs, hand-carved hardwood walls, and hand-painted ceilings. The occupant "will be surrounded with luxurious apartments that would befit a king."[7]

Thus settled, Bowers went to work. He inspected every corner of the company, and peered under every stone. "Mismanagement," he informed Frederick Gates, "is the only word that exactly fits the condition of operations . . . another year would have found the company bankrupt. . . ."[8] He told Rockefeller that "Osgood and his clique" had disgracefully managed the firm, which was "filled with dead wood, extravagance and lack of interest."[9]

Bowers accused the Osgood management of gutting the Fuel and

L. M. Bowers.

*Photo courtesy L. M. Bowers Papers, Special Collections,
Glenn G. Bartle Library, State University of New York at Binghamton.*

Iron Company in the interest of the Denver & Rio Grande Railway.
(The directors of the two firms overlapped considerably.) The railroad,
it seemed, charged high freight rates to the Colorado Fuel and Iron
Company, which on its part sold rails to the Rio Grande for practically
nothing. Gates responded that if Bowers' accusation was true, "It
means . . . that the directors . . . are looting the company in the interest of the railroads which they represent, and we, as very large bond-
holders as well as stockholders in the CF&I Co. are being robbed right
and left."[10]

Bowers uncovered prodigious waste in every department. He found
gross lack of detailed supervision in the handling of supplies, and con-
ferred with each department head on how best to plug this drain on
funds.[11] At the steel mill, he reduced the cost per ton "to a point that
causes us to look back with little less than disgust at the incompetency
of men who permitted such conditions to exist right under their noses."
The economical use of supplies was enforced by monthly statements and
records, a system also put into effect at the coal mines.[12]

Through "a general house cleaning from cellar to garret in all departments," Bowers had by 1909 reduced operating expenses by some two

million dollars, more than ten percent, while the gross earnings remained the same.[13]

Reducing waste was one strategy; reducing wages another. In 1909, Bowers cut wages at the Pueblo steel plant by ten percent. Although competitive labor conditions that year prevented him from doing the same at the coal mines, he plainly stated his opinion on the subject. "I always regret cutting the wages of laborers who have families to support . . . ," he confided to John D. Rockefeller, Jr., "but considering these foreigners who do not intend to make America their home, and who live like rats in order to save money, I do not feel that we ought to maintain high wages in order to increase their income and shorten their stay in this country."[14]

Bowers was no humanitarian. He did not value the lives of those who "lived like rats" and launched no campaign to improve safety in the mines. This was unfortunate, considering that conditions in Colorado Fuel and Iron Company mines were horrendous almost beyond belief. The Starkville mine, to give one example, extended underground for three miles (in 1907) but had exactly one opening—the entrance. Because the air traveled in what was essentially a loop, some miners worked from four to six miles away from fresh air. That year, the company authorized the sinking of a second shaft, since "neither men nor mules can do as good work as would be possible in reasonably fresh air."[15]

Local managers attended to production, not safety. Take the case of a gas explosion in the Primero Mine. One day in 1907, an experienced Irish-American miner reported gas to his pit boss. The boss told the miner to shut his mouth lest the foreign workers learn of the condition and "stampede." The miner refused to continue working, and the boss discharged him. On his way back down to retrieve his tools, the miner was caught in a local explosion and severely injured. Three days later, the Primero Mine blew up again. This time, twenty-four men died.[16]

Many Colorado mines were high, dry, and quite dusty. After the 1907 Primero disaster, the state Commissioner of Mines established new safety regulations that mandated sprinkling with water all underground entries, rooms, and traveling ways, "to purge the air as much as possible of suspended dust."[17]

The Colorado Fuel and Iron Company disregarded the sprinkling regulation. "I cannot find any evidence that the company paid the slightest attention to the recommendations of the Mine Commissioner," reported state Labor Commissioner Edwin Brake in 1910.[18] The occasion of his denunciation was another disaster at the Primero Mine. On 31 January 1910, the mine blew up for the second time.

State mine and labor officials rushed to the scene to find the small community in turmoil.[19] "The camp is a scene of indescribable horror tonight," a committee of coal miners reported from Trinidad:

> While every able-bodied man is taking his turn with pick and shovel to clear the shaft, the women and children, kept back by ropes, have gathered about the shaft weeping and calling wildly upon their loved ones who have not been found. . . . [F]rantic women and children surged against the ropes. . . . [S]ome of the women attempted to join the workers below and had to be restrained by force. As each body was brought to the surface, the women gathered about it with shrieks and prayers, but the bodies were so charred and disfigured that they could not be identified.[20]

A disputed number of men (the company said 75, the miners said 150), mostly Slavs and Hungarians, died in the explosion, which resulted from a lack of sufficient sprinkling to wet the dust.[21]

In fact, reported Commissioner Brake, the mines were practically never sprinkled, "except in the main entry . . . when the dust became so deep as to interfere with the efficiency of the mules in hauling the coal out of the mine." He accused the company of "cold blooded barbarism."[22]

Dust was only one of several problems. "Primero Mine No. 4 was known as rotten," stated the local committee of coal miners. "It had too much gas and dust and dirt. Water is unknown. The super did not order timber, the rock dropped constantly to cripple the men. Men who never worked in a mine received dynamite. From about 60 mines in Las Animas and Huerfano Counties, employing more than 8000 men, not one is safe, most are nearly as rotten as Primero No. 4."[23]

On the other hand, L. M. Bowers blamed the miners. On the day after the disaster, with most of the bodies yet to be removed, he set the record straight for Gates: "The disaster was probably caused by some miner smuggling in pipes and matches, the use of which is prohibited. The mine was thoroughly ventilated, but, like most soft coal mines, has pockets of gas that are struck which cause explosions, and then the dust ignites and havoc follows. The latest reports indicate that the mine is not damaged, and work will be resumed as soon as the miners get over the excitement."[24]

Bowers did not mention sprinkling, although the coal mine inspector, David J. Griffiths, found the lack of it to be the cause of the explosion. The evidence was overwhelming and Griffiths, a coal-mining expert, knew it. Yet not six weeks before he had reported the mine to be in "satisfactory condition," even going so far as to praise the coal operators for the "promptness and readiness" with which they complied with "recommendations for improvements made by this department."[25] His praise for the company, however, is suspect. Griffiths was one of those inspectors who interwove his public career with a private one as a Colorado Fuel and Iron mine superintendent.[26]

In some ways it was the women who caught the brunt of company indifference to human life. State law required coal companies to compensate widows, but only if the accident occurred through the fault of the company. Political corruption guaranteed that coroner's juries in southern Colorado never found fault with the Colorado Fuel and Iron Company. The jury that ruled on the 1910 Primero disaster could find no cause for the explosion. "[I]n their anxiety to protect the interests of the company against the widows, mothers, and surviving relatives of the dead miners," exclaimed the Commissioner of Labor Statistics, "[the jury] rendered a verdict without making any kind of investigation that would determine the cause."[27]

A mother could lose her children in the economic calamity resulting from her husband's death. In Las Animas County alone, in the five years following 1907, two hundred married coal miners were killed, leaving widows responsible for the care of nearly seven hundred children.[28] Many of these women had no means of support whatever. Annually, seventy or eighty of their children were sent from the county into various state institutions. "Mine disasters have left scores of children unprovided for," a Denver newspaper reported. "Many of them have been adopted and in this way entire families have become separated. . . ."[29]

Here was the dark side of keeping down costs. Improving efficiency was the least of it. Disregarding the most elementary safety practices did save the company money. In addition, the Colorado Fuel and Iron Company used standard methods of shortchanging the miners, particularly short-weighing the coal. "I was unable to test the scales at Primero," the Commissioner of Labor reported in 1910, "as the building was locked, but at Sopris I found that my weight, which is one hundred and fifty-five pounds (155) added to a car of coal only increased the weight 70 pounds. At Starkville my weight added to a car of coal only 35 pounds. At Ingleville [Engleville] my weight added to a car of coal increased it 92 pounds."[30]

LABOR REORGANIZES: THE CAMPAIGN OF 1907

In southern Colorado, the need for a union was plainly present. Nor had defeat of the 1903–1904 strike obliterated the union spirit. As early as 1905, organizers had once again infiltrated the southern coal camps. From 1905 to 1907, they worked on behalf of the increasingly radical Western Federation of Miners.

Since the defeat, politics had added bile to bitterness in the relations between the Western Federation of Miners and the United Mine Workers of America. In June 1905, the WFM joined with other socialist unions and organizations to establish the Industrial Workers of the World (IWW), a revolutionary union that declared the great facts of the age to be ". . . the displacement of human skill by machines and the increase of capitalist power through concentration. . . . Class divisions grow ever more fixed and class antagonisms more sharp."[31] In December, eighty percent of the membership of the WFM (including both coal and metal miners) made their anticapitalist sentiments explicit by voting for affiliation with the new organization.[32]

The fight over jurisdiction continued; it may have been the thorniest issue dividing the two unions. From 1905 to 1907, WFM organizers actively recruited coal miners in the Rocky Mountain coalfields.[33] For two years, the WFM had the field to itself, but in 1907 the Denver office of UMWA District 15 began receiving two hundred dollars per month from the national office "to try to do something for the miners in southern Colorado."[34] "During this period conditions were badly mixed," a UMWA committee later reflected, "because of the fact that the W. F. of M. and the I. W. W. were making a strenuous effort to control the field. The coal company's thugs and corrupt politicians took advantage of this and every member and official of the United Mine Workers was persecuted to the limit; many brutal assaults were committed on our people."[35] During the futile campaign of 1907, one organizer was shot dead.[36]

The threat of corporate violence was an ever-present reality. "You can not get into some of the camps," reported a UMWA organizer, "because there are deputies who meet you and follow you around until you leave the camp and if you stay too long they will help you out."[37] Organizer

John Lawson, who was emerging as an important regional leader, began walking from Trinidad out to the coal camps by night. Lawson knew company tactics firsthand; during the 1903 strike, he had been shot, and his house had been destroyed by dynamite. These experiences had only hardened his commitment to the union cause. "I had to go over a certain road in Colorado every night . . . and I always carried a six-shooter," he recalled. "I made up my mind I might as well be killed as to have my head beaten off."[38]

Lawson would insinuate himself into a coal camp by avoiding bosses and superintendents he knew would recognize him. On a good day, he could spend a whole day or more inside, contacting miners and signing them up. "Later the new little local would meet at night, out in the hills, giving new members the 'obligation' of the union by the light of a bonfire." In this clandestine manner, one local grew to seventy-five members. But this system, in which the members of a large secret local were known to one another, produced the inevitable result of the best men of the local being discharged and blacklisted out of the southern Colorado coalfields. Spies reporting to the companies were invariably the first to join the union.[39]

But in the end the economy did a better job of destroying the 1907 campaign than the coal operators could. Late in the year, a depression pulled down the American economy and both union drives collapsed with it. On 17 December, the WFM secretary recorded for posterity a picture of hard times. In careful handwriting, he entered into the minute book a scene of "crumbling trust companies, bank failures and the withdrawal of legal tender from circulation," which money was replaced by "all kinds of so-called paper." The human cost was in "the weary tramp of the vast armies of the unemployed, the alarming increase in crime, the enormous increase in the number of suicides. . . . The wails of the hungry, destitute, and miserable assail our ears on every hand, all combining to make up an earthly hell such as only the pen of a Dante or a Milton could fitly describe."[40] On its part, the UMWA found hundreds of miners desperate for work, and soon closed the doors of its Trinidad office.[41]

Company directors also viewed the depression with alarm. In December 1907, Gates comforted Bowers, "I suspect it will take all the skill and resources both [you and President Welborn] possess to steer the craft successfully away from the rocks toward which the present adverse winds are blowing it." Indeed, the Colorado Fuel and Iron Company avoided catastrophe only through its close association with Denver banks. During the 1907–1908 panic, the company met its payroll by vastly overdrawing its accounts.[42]

ROCK SPRINGS: A NEW DIRECTION

By 1907, the opposition of southern Colorado coal operators to the UMWA had hardened into an ideological fury that excluded strategic considerations. The Colorado Fuel and Iron Company and the Victor Fuel Company viewed the UMWA as the devil in a holy war. But in southern Wyoming, where the Union Pacific had successfully kept workers divided by mixing Asians with Asian-hating whites (and through other tactics), a profound change was taking place.

The effectiveness of using racial mixing to keep workers divided depended absolutely on the racism of the white miners. The miners themselves guaranteed its success. They had only to set aside their hatred to make the policy shatter like glass. Coming to this conclusion in 1903, the Western Federation of Miners became the pioneering labor organization in the United States to oppose Chinese and Japanese exclusion. In doing so, the western union reversed its own exclusionist rules and began to welcome Asians as members.[43] White coal miners at Rock Springs, long open to WFM influence, undoubtedly began to rethink their self-defeating position at this time.

It took four more years for the UMWA to climb on the antiexclusion bandwagon. As late as 1906, the coal miners' convention reiterated its slander of people of the Orient. But on this occasion, some delegates jammed the wheels of business as usual. A resolution submitted to the convention would have excluded Asian mine workers not only from America but from the UMWA. The Resolutions Committee concurred, a step that normally would have caused it to pass on the floor of the convention. The vociferous objections of several prominent delegates caused it to be tabled instead.[44]

A year later, the UMWA reversed itself on exclusion. The executive board titled its new policy of admitting Chinese and Japanese mine workers "The Japanese in Southern Wyoming." In Rock Springs, where Jay Gould and his successors had defeated the union movement by aiming at its Achilles' heel, the miners finally realized that the road of racial harmony was the only road to union wages and conditions.[45]

The Union Pacific had historically approached its labor problem pragmatically. The railroad needed coal to run its locomotives; this took priority over ideology. Perhaps in 1907 controlling director E. H. Harriman realized just how costly it was to fight the union at any cost. For whatever reasons, on 1 September the Union Pacific Coal Company signed its first contract with the UMWA.[46] The agreement reduced the race differential in pay, a differential that the company eliminated by 1911.[47] In one stroke, the Union Pacific bypassed the intense and costly labor strife that would erupt in the region in the years to come.

To the surprise of organizers, the Japanese entered the union readily.[48] At the founding convention of the new Wyoming district (the union had separated Wyoming from District 15 for the sake of this contract), two Japanese miners were the first Asian delegates ever to be seated at any UMWA convention. "An unusual feature connected with their presence as delegates," one newspaper reported, "is that Americans predominate in their own locals and their election was practically unanimous. The other delegates are exceedingly friendly towards them and the least intimation of a slight . . . is received belligerently."[49]

THE NORTHERN COLORADO LIGNITE FIELD, 1908–1910

In July 1908, the UMWA negotiated its second contract in the Rocky Mountain region, this one in the northern Colorado lignite field. (The famous separate settlement of 1903 had been signed by employees only, not the union. Technically, it was not a UMWA contract.)[50]

English, Scottish, and Welsh miners, or their sons, still predominated in the northern Colorado coal towns. In 1908, the U.S. Immigration Commission found that of 528 mine workers enumerated, all but 69 were either foreign born or the sons of foreign-born fathers. The vast majority were Britons. Added to the ethnic mix was a sizable and strongly pro-union Italian minority, many of whom had been blacklisted out of southern Colorado during the 1903 strike. Other minorities were virtually absent because, the Commission speculated, "[d]iscrimination to the extent of refusing employment has been constantly exercised toward Chinese, Japanese, and Negroes, and except in times of stress, toward Greeks. Doubtless the operators have been guided in this position by the known antipathy of their workmen toward the races mentioned."[51]

E. L. Doyle (1886–1954), who became an important figure in the Colorado miners' movement, typified the working miners of the northern lignite field; his biography illuminates the life pattern of the group. Doyle was born in Illinois to an Irish miner and his wife and began his working life at age twelve when his father became too ill to work. In 1907, already an experienced unionist and a skilled miner, he moved to Lafayette with his mother and four younger siblings. The Lafayette local elected him check-weighman. In 1908, he became chairman of the pit (grievance) committee established by the new contract.[52]

In 1909, Doyle, at age twenty-three, began keeping a diary of his experiences as chairman of the pit committee at the Capital mine. This remarkable document opens a window on the world of the lignite miner of northern Colorado.[53] Doyle attended to complaints and problems relating mainly to safety. For instance, the mine had no ladder in the escape shaft. For two weeks, he worried about the ladder and the pit boss who refused to provide it.[54]

Another day, the cage hit a miner standing at the bottom of the shaft, and the man died later that day. Doyle agitated for safer procedures, such as warning bells to be rung while the cage was going down, and a chain barring miners from accidentally standing under the cage while waiting for it. He also pressured the management to cease the dangerous practice of sending explosive powder down with the men in the cage.[55]

As the weather grew colder, Doyle began to needle the manager about sinking a well near the air shaft to prevent water from running down the shaft, freezing, and cutting off the air supply. The manager refused to do it. One day in January, the worst happened—ice blocked the air shaft. The cagers, following the pit boss's orders, refused to let the men up and (Doyle recorded) "[y]es, there is lots wanting up." Doyle leaned over the shaft and hollered that they must be let up. The boss, his mind on production, ordered him away, "[s]aying that I had no business telling those men to do anything and that I had gone too far already and he would put me off the place. That I should work there no more. To this I said, 'Put me off if you can, I am going to have those men out.' "[56]

Doyle's diary portrays far from ideal conditions in the northern lignite field, but it also shows the day-to-day benefit of having a union, in this case in the form of a pit committee with a competent miner at its head. Most of the details to which Doyle attended were petty annoyances, yet any one of them could have quickly developed into a matter of life or death.

Despite the 1908 contract, the northern lignite miners lacked the strength to prevent conditions from deteriorating. Lignite was used mainly as a home heating fuel: In the summer, there was little demand for it. At that time particularly, the locals lost members "on account of poor work." The northern field was not economically important, but it was the center of union strength. Just how fragile the center was came out in June 1909, when the UMWA executive board, noting that District 15 was unable to sustain itself financially, voted two hundred dollars a month into the district, with the request that it "discontinue all unnecessary expenses."[57]

THE NORTHERN COLORADO COAL STRIKE, 1910

It is therefore astonishing that less than a year later, in April 1910, the 2,200 miners of the northern lignite field struck. The Northern Coal and Coke Company had refused to renew the contract under pressure, it was widely speculated, from the Colorado Fuel and Iron Company.[58] Despite their weak position, the lignite miners voted to strike in order to comply with a national UMWA wage policy set at a stormy, shouting convention that broke into fistfights more than once. The Cincinnati Wage Convention of March 1910 stipulated that when contracts expired on 1 April all districts should negotiate a five-percent increase in wages as well as improved working conditions. If they failed, they were to strike, regardless of local conditions.[59]

Fragmentary evidence suggests that the fight in Cincinnati was fought again in Colorado, preceding the vote to strike on 1 April 1910. Former Senator George McGovern, an important student of Colorado coal mining, suggests that "a considerable number of the 2200 miners who laid down their picks on April 4, 1910 actually believed the strike was a mistake."[60]

Mother Jones concurred. In a 1911 speech, she bluntly stated: "Now you have a fight of the miners in Colorado. You have got to call a strike in the southern field and lick the Colorado Fuel and Iron Co. out of its boots. . . . You cannot win in the North until you do. You are wasting money. I know that field thoroughly. I was up against the guns too many months not to understand the situation."[61]

In May 1910, the northern operators erected high fences of barbed wire and hog wire around their properties. Inside, they built houses; before long, Mexican, Bulgarian, Greek, Rumanian, and Serbian strikebreakers were living in them. A debate began over whether or not the strikebreakers were imprisoned inside.[62] The composition of the work force in the north began to resemble that of the south.

The dispute over whether or not to strike did not prevent the northern coal towns from supporting the action once the decision was made. Doyle's diary preserves a picture of the everyday life of the strike community. One Sunday in October, a parade two thousand strong headed by a band marched through the streets of Lafayette to the ballpark, where speaking and singing were the features of the day. Doyle's comment: "Note: Banners with short statements were carried in the parade, such as, Welcome All, We Are Here To Stay, Union Forever Open Shop Never, Six Months on Strike and Never Missed a Meal; the U.M.W.A. the Miners Greatest Friend." The next day he observed, "Everyone wore a smile to think we had such a fine showing on Sunday."[63]

The women took a strong part. Consider their resolution submitted to a District 15 convention held three years after the northern lignite strike had begun: "The Ladies Auxiliary now organized," it read, "have . . . fought gallantly to help win the battle by humiliating scabs and doing picket duty where it was dangerous for men to go. . . ." The women also declared they had done other "daring deeds," not mentioned specifically but "known to many of the delegates." They asked the convention to officially approve their actions and to formally recognize their part in the organization.[64]

The fighting spirit of the strike community waxed large, but it was insufficient to overcome the continual arrival of new strikebreakers.[65] To make matters worse, in November 1910, Denver Judge Greeley W. Whitford enjoined the strikers from picketing, congregating in groups, posting notices addressed to strikebreakers, or using "profane or obscene language" in the presence of strikebreakers. He forbade them from using "any language, sign or gesture, banners, cards, or badges calculated to intimidate any person or persons in the employ of the plaintiffs [the coal operators]."[66]

In the winter of 1911, the UMWA assigned Mother Jones to Colorado. Although she believed the fight was already lost, she put her shoulder to the wheel of the struggle, writing to a friend, "I am on the warpath here for the miners."[67]

The strike continued in the face of satisfactory coal output, and it sucked up union funds in amounts that eventually exceeded a million dollars.[68] The Whitford injunction escalated expenses due to the legal costs of keeping strike leaders out of jail, where they began to be regularly deposited for violating the injunction. In June 1911, the executive board considered a resolution from the Lafayette local requesting more strike relief because "our wives and little ones are in need of shoes, clothing, and other necessaries of life." At this time, the board privately conceded that the northern Colorado strike was lost. Yet the union's governing body would not publicly admit defeat, and for two more years continued to send five thousand dollars per week to northern Colorado in strike relief.[69]

THE SOUTHERN COLORADO COALFIELD, 1910

Gradually, the hopes of the northern lignite miners shifted southward. They were too few and too fragile to win a strike in isolation from miners working in the domain of the Colorado Fuel and Iron Company and the Victor Fuel Company.[70]

By 1910, a new anti-union force had joined the formidable opposition unionists faced in southern Colorado. The Rocky Mountain Fuel Company, originally a small Wyoming firm operating near the Union Pacific mines at Almy, steadily expanded southward, first into the northern Colorado lignite field where it absorbed the Northern Coal and Coke Company, and then into the southern Colorado bituminous field. It became one of the three largest coal operators in Colorado.[71]

In April 1910, Rocky Mountain Fuel hired detectives to observe the extent of union activism in the southern Colorado coal camps. The *Rocky Mountain News* reported that southern miners felt considerable sympathy for those about to strike in the north.[72] The daily reports submitted by the detectives confirm intense union activity.[73] Inspector D-85 noted "a feeling of unrest" and "just enough excitement in all the camps to keep everyone on edge."[74] Another detective reported that, with the exception of a few Mexicans, he had not encountered a single miner who did not talk in sympathy with the UMWA. In his view, they were at least two-thirds organized.[75] Indeed, union organizers were telling miners they "must keep down and not start trouble until the strike order is given from headquarters."[76]

"The organizers feel they can depend on the Italians better than on any other nationality," Inspector D-3 commented, "knowing that they are 'stickers' when it comes to organization work; this is borne out strongly by their actions in the southern field during the last strike when the Italians stuck to the very last."[77] He noted that the District 15 office in Denver employed its own "spotters" (detectives) in every southern coal camp. He learned from two detectives working for the union side (who took him for a union miner) that they "tried to keep spotters on the move by exposing them."[78]

Company detectives reported on two UMWA organizers, one paid, the other not, "doing their missionary work together at night."[79] Another organizer, "a fellow by the name of Frey," worked as a painter and paper hanger. While renovating boarding houses in the coal camps, he quietly distributed union cards and collected initiation fees.[80] Still another organizer posed as an insurance agent. He inadvertently confided to a company detective that he had been driven out of nearly every Colorado Fuel and Iron camp.[81]

According to the detectives watching them, organizers used the town of Trinidad as a base of operations.[82] Organizer Frank Morrissy met field workers by appointment at the Commercial Hotel, where "no less than 15 men from different camps called on him, arriving—some in hacks, some by train, and others by street car . . . each man remaining from 20 to 40 minutes."[83]

By 1910, extensive surveillance by the companies had persuaded the union to modify its organizing practices. Four years later, district officers believed their ingenious methods were still unknown to the operators.[84] Neither organizers nor the organized were extensively known to each other. The Denver office hired different sets of organizers who worked in different ways. Some forwarded membership lists to the Denver office while others collected dues and gave out cards directly. When an organizer was exposed in one camp, he was switched to another. Organizers were kept floating, as one detective observed, "getting in a lick wherever they could." Miniature locals (with fewer than ten members) collected their own dues and knew the identity of only one organizer.[85]

Most cleverly, the UMWA fooled superintendents into discharging procompany workers and replacing them with union activists. Two organizers would work together, one inside the camp, the other outside. The inside organizer would feign a grudge against the union; he wanted the superintendent to hire him as a spy for the company. After acquiring this position, he would report anti-union workers to the superintendent as union activists. The superintendent would then discharge them. Their places would be taken by new applicants, who also feigned bitter antagonism against the union, but who in reality were union stalwarts.[86]

No matters of importance were ever brought up at public union meetings. One detective complained to his superior that the meetings at Trinidad were more or less a farce. This practice helped to foil the operators' campaign against the union, but of course it also made it more difficult to organize. "Organizers say they are up against it in the southern fields," one detective claimed, "because they cannot hold meetings which would be the means of holding the membership together. It would be useless they say to hold meetings because the operators would know everything that took place a half an hour afterwards."[87]

Despite the union's various methods and strategies, many unionists lost their jobs. "If a miner was thought to be a union man," a miner testified years later, "if he lit in any of the Colorado camps, it was goodby for that union man. He would have to move. They would not leave him light." Asked how the company would know, the miner replied, "You take these big camps, and there are always a lot of sleuths."[88]

Another miner recalled that at the Colorado Fuel and Iron Company's Berwind camp, "One fellow reported that [three or four men] had been having a little union meeting at the wash shanty and they all got to leave the next morning."[89] At Sopris, yet another man got along "fine as silk" for about a month until he was suspected of associating with "that hot bunch from Kansas." He was fired.[90]

"I am not an organizer, but I tell a couple of fellows to combine, like the big corporations," testified Mike Sekoria, a Slavic miner. "I just talk with a couple of fellows; and I says, Here boys, we must join the union. It's time for the working man to tie to something." Sekoria made these remarks in a saloon within earshot of a spy. He was discharged. "Gentlemen," he concluded to the assembled Congressmen, "you think I don't know the corporation. They are tight, just the same as a chain."[91]

Besides the coal companies, organizers faced the legacy of the 1903 strike: anger over the separate settlement in the north. "So embittered are the mineworkers in the southern field against the treatment they received six years ago," explained an organizer in 1910, "that the very mention of the name John Mitchell precludes any possibility of securing any success." Miners felt that if the northern lignite miners had stayed on strike in 1903 to support the workers in the south, "no strike would be in vogue now. . . ."[92] In 1913, E. L. Doyle elaborated: "Every time an organizer or officer of this District goes to the southern fields, he is confronted with the criticism of the northern miners' action at that time. It is true the northern miners acted under the advice of their national officers but the feeling against the north for having returned to work has not yet died out in the southern field, and is now one of the greatest factors which prevents complete organization of that district."[93]

Company detectives posing as union activists took every opportunity to inflame such discontent. "I never fail to criticize certain acts of union officials," boasted Inspector D-87. "I call . . . attention to the fact that union officials take boat rides . . . and dine with some mine operators, and this I notice makes the rank and file jealous. . . . This sentiment is quite strong among the Italian element. . . . One of these union men . . . said that if it had not been for the crookedness of the union officials seven years ago in 1903, that the union could easily have won out in Colorado."[94]

Company opposition and internal dissension combined to make the work of organizing a dangerous and knotty proposition. But in 1910 one of the chronic difficulties evaporated. In late 1907, the WFM began moving away from its radical past; as a result, increasingly friendly feelings developed between the WFM and the UMWA. There was even talk of a merger. In October 1910, the WFM ceded jurisdiction of western coal miners to the UMWA and offered to aid the organizing effort. A major obstacle to the success of organizing the southern Colorado coalfield had been removed.[95] With the transfer of three hundred coal miner members of the WFM into the UMWA, "the men began to flock into the organization," an organizer recalled, "with the hope that a contract would be made for them that Spring. . . ."[96]

In the midst of the 1910 organizing campaign in the south, Colorado was shocked by its second coal mine disaster of the year. On 8 October, the Colorado Fuel and Iron Company's Starkville mine exploded, killing fifty-six men. According to miners, the mine was unsprinkled and virtually unventilated. The workers literally had waded in coal dust.[97]

Early the next year, President Welborn reported to the company's executive committee that the two disasters of 1910 (at Primero and Starkville) had cost more than 60,000 dollars. Otherwise, the six months under review would have shown a surplus of one million dollars. The costs included 5,000 dollars to replace a ruined fan and another 5,000 to buy twenty-seven new mules. At this time, company officers were finally moved to invest more than 35,000 dollars in sprinkler systems for seventeen Colorado Fuel and Iron Company mines.[98]

A month after Starkville, the Victor American Fuel Company's Number 3 mine at Delagua blew up, killing seventy-nine miners. These catastrophes raised the death toll in Colorado coal mines to 319 mine workers slain out of less than 15,000 working, or one out of every 46 miners. Colorado's death rate was twice as high as that of the United States as a whole. Even with the new Colorado Fuel and Iron Company sprinkler systems, other matters were not improved; the state's death rate remained double that of the United States.[99]

The explosions of 1910 drove hundreds of men from the camps. A labor shortage developed, particularly in the south. The miners who stayed, according to the coal mine inspector, were mostly natives of Italy, Austria, Japan, and Korea.[100]

Many of the eight thousand mine workers in the southern coalfield waited anxiously for the UMWA to call them out on strike. Because the northern Colorado miners were already out, the time seemed opportune. In October 1910, district officials went before the national executive board to convince it to authorize a strike because "at this time there is the best opportunity of winning it." The board decided against taking action at that time, pointing to the substantial funds that such a strike would require.[101]

Organizing continued, but with greater difficulty. Membership dropped away, and organizers had trouble collecting dues. Miners felt that union membership without hope of a strike was not worth the risk. Yet they were "more or less impatient" with the desire to organize.[102] Organizers stationed at Trinidad, Walsenburg, and at Florence (near Cañon City) periodically rushed to the site of an unauthorized strike to persuade the men "to remain as they are for the present at least, and that eventually the organization will begin a vigorous campaign of organization."[103]

THE STATE OF THE UNION, 1910

What eluded many coal miners was the state of affairs existing in the United Mine Workers of America, at both the national level and in their own District 15. Unbeknownst to the membership or to the public, the union was staggering under a debt in excess of 300,000 dollars. Some districts were stronger by far than the national. The union was surviving on loans received from the powerful District 12 (Illinois), as well as others.[104] But the worst news was the extent to which coal operators had gained influence within the union. The evidence makes it hard to avoid the conclusion that by 1910 UMWA President Thomas (T. L.) Lewis, a longtime activist who had served as John Mitchell's vice president, was surreptitiously working in the interests of the operators.

Defeated for reelection in the midst of a near riot at the January 1911 convention, Lewis offered this in his final Presidential address: "Whether or not I succeed myself as president of this organization, you can rest assured of one thing, the knowledge, the information, the training and experience I have acquired will not be sold to the operators or employers of this country at a fixed salary . . . because I am not too old to earn a living mining coal and I haven't too much pride in my makeup to swing a pick."[105]

Shortly after this masterful deception, Lewis joined the ranks of the union's most formidable foe: the West Virginia Coal Operators' Association. The union's new president, John P. White, upon taking office in April, discovered that Lewis had taken the office records with him to the other side.[106] This and other examples of Lewis's treasonous behavior embarrassed the union intensely; public revelations would have damaged the cause. They were not made, but board members discussed the various distressing matters resulting from his perfidy behind closed doors. For instance, in the last year of his administration Lewis had sent encouraging letters to both sides in a faction fight that had nearly destroyed the important District 5 (Pittsburgh). This fight had resulted in the defeat of a major strike.[107]

Other examples of Lewis's betrayal extended back at least to the January 1910 convention. On that occasion, President Lewis proposed a highly peculiar wage resolution. All districts would demand an advance in wages, but miners would continue working after their contracts expired. Further, no district would sign a new contract until all other districts had finished negotiating, including those which had never before succeeded in bringing employers to the negotiating table. This arrangement consisting of no contracts and no strikes was an operator's dream. It would have destroyed the union, and quite naturally the scale committee rejected it.[108] This suggests that Lewis was already working for the operators at that time and also, therefore, at the Cincinnati Wage Convention that required all districts to strike unless certain conditions were met, regardless of local conditions. In Colorado, the effect of this policy had been to push the northern lignite miners out on their ill-advised, expensive, and inopportune strike.

Corruption and incompetence had also seeped into the affairs of District 15. E. L. Doyle was elected district secretary-treasurer in 1912. When he took office on 1 April, he found the financial records in such chaos that he could find no way to audit them—this in a District that had been pouring five thousand dollars per week into strike relief for the past two years. Worse, Doyle became convinced that Frank Smith, district president from 1910 to 1912, was a detective working for the operators.[109]

CAMPAIGN RENEWED: 1912–1913

The 1912 elections swept the UMWA house clean. John P. White, formerly president of the Iowa district, now took the reins of the presidency. John McLennan, a Scottish-born miner long active in Colorado, was elected president of District 15.[110]

The northern Colorado miners were still out on strike, by now hopelessly so except for the hope they fixed on the south. "In the southern field," Doyle wrote to President White, "lies our hope to win the struggle . . . it would seem that a battle lasting as long as this one could never be won, but to my mind there is still a chance of winning if we put forth a more vigorous campaign." He noted that membership was increasing in various parts of the state.[111] On New Year's Day 1912, the UMWA publicly opened an office in Trinidad for the first time since 1907.[112]

In the spring and summer of 1912, a local minister observed many threats of a strike in southern Colorado. Unauthorized strikes were frequent. On each of these occasions, an organizer would hurry to the scene to convince the men to return to work until such a time as "the organization was ready to get behind the movement." District President John McLennan was convinced that but for the efforts of union organizers, southern Colorado miners would have struck in the spring of 1912.[113]

Company managers were also worried. To counter "unusual activity on the part of spies sent out by the union," the Colorado Fuel and Iron Company took the precaution of raising the tonnage rate. "You know our mines are non-union," Bowers explained to Gates, "and I know of no better way than to anticipate demands and do a little better by the men than they would receive if they belonged to the unions."[114]

Among northern lignite strikers, pressure mounted for a district convention to discuss grievances. A convention, miners knew, would explode into a statewide strike as surely as lighting a squib would produce a blast. In June 1912, a committee representing all northern locals on strike had "considerable talk, pro and con," on the question of calling a convention. The committee decided the timing was inopportune, "for various reasons which cannot be divulged at present, and also the good advice of our National and District officers."[115] Union executive board minutes reveal that shaky finances (never a matter for public discussion), and costly strikes in West Virginia and Vancouver Island as well as in northern Colorado, prevented a districtwide strike.[116] Colorado workers would have to wait. Mother Jones, speaking to a crowd of West Virginia miners, said accurately enough, "Even now those fellows [in Colorado] are begging me 'When are you coming back Mother? We are ready to strike the blow.' "[117]

In the summer of 1912, the national office sent a committee to Colorado to thoroughly canvass the situation. Noting the fact that seventy percent of Colorado's nearly fourteen thousand mine workers spoke a non-English language, the committee urged the union to hire foreign organizers out of the southern coal camps who were leaders of their fraternal organizations.[118] English-speaking organizers from other states should be dispensed with. These organizers couldn't get near the coal camps once they were known, "as their lives are in constant danger, but their hotel bills and railroad fares would keep many foreign organizers in the mines." Moreover, the English-speaking miners in the south couldn't be trusted: "They are hired by the company to play union man and as a rule they have a union card in their pocket and their business is to betray and have discharged good union men." They were company sympathizers, "if not spotters, detectives, and man beaters."[119]

The committee recommended secret methods of organizing. Before a strike was called, there should be a great demand for coal. At the appropriate time, coal operators should be invited to a joint conference, where they would be confronted by their employees. Organizers from the camps would act as delegates to the conference, because the necessity for secrecy before the strike precluded an open election. The union should lease land outside the coal camps on which to set up tents. Lastly, the national office should have "absolute jurisdiction over the strike."[120]

Welborn's analysis was that if the UMWA committee's report reflected true conditions, "there would be little reason for calling a strike . . . as there is no question but that the conditions are more satisfactory and the men better contented in southern Colorado than they have been for years."[121] Two weeks later, he reported to Bowers, who had gone east for a rest, that the profits of the previous year came to a gratifying two million dollars.[122]

Any appearance of malcontent on the part of miners resulted in discharge. The UMWA estimated that in 1912 the coal companies fired twelve hundred Colorado mine workers on suspicion of their having union sympathies.[123] A minister approaching the Victor-American's Hastings camp, observed two deputy sheriffs on horseback herding three miners out of camp. It was between four and five in the afternoon and "they still had their pit lamps in their hands, they had not been allowed to get anything to eat, in fact, they were unwashed. I made inquiry as to what the trouble was because I thought that there must be a terrible charge to have them herded out of town in this way, that they must have been guilty of murder or some other terrible crime, and I found that the trouble was that they were suspected of being union men."[124]

Outsiders could not get inside the coal camps unless they could give an unquestionable account of themselves.[125] The companies banned peddlers, according to Bowers, in order to protect the mine workers from "being swindled by unscrupulous Jews, Italians and other cut-throat dealers who would control the business if we should withdraw and leave the field open."[126]

Vincent Militello, a blacklisted coal miner who had turned to selling fruit after the 1903 strike, was "chased out like a rabbit" from Berwind, Morley, Sopris, Primero, and Segundo. "They said they knew I was a union man, a spy," he testified. On one occasion, a camp marshal dragged him from his niece's house in Coaldale—he was delivering bananas—and threw him on the ground, took the whip from his buggy, and broke it over his body.[127]

The repressive atmosphere weighed on the entire community. Gunmen were everywhere. "There was never a time," district President John McLennan recalled, "that I wasn't followed constantly by one to three guards, not only in Trinidad, but all over the state. . . . If I went to eat in a restaurant they stood at the door until I came out. . . . [M]any times five or six of them would stand for half a day in front of our office or across the street from our office for the purpose of intimidating men from going up there. Every man who was seen going up there was . . . trailed until they found out what camp he went to, and immediately he was discharged."[128] State officials were not exempt. "One does not need to be a sleuth," a deputy Labor Commissioner reported, "to become acquainted with the elaborate and complete system of surveillance that is maintained by the fuel corporations of this district, a system that has Trinidad as a center and radiates to all camps in the southern field. Under this system of espionage it is an easy matter for one of the Trinidad agents to notify a mine when to expect a visit from the [State Coal Mine] Inspector. . . . I myself have been trailed around the streets of Trinidad."[129]

To the women of the coal camps, company gunmen posed the constant threat of rape. During the workday, the coal camp above ground was populated chiefly by women, children, and gunmen. To what extent the gunmen raped the women is of course impossible to know, but there is no question of its occurrence. An official of the Bureau of Labor Statistics wrote: "I have been told by women in Primero that there was no privacy in their home life, that whenever a representative of the company or deputy sheriff desired, they entered the house unannounced." A coal miner wrote to President Woodrow Wilson about an incident which occurred in 1910 in which a "campmarshall . . . was trying with a revolver . . . in the daytime, when man was in mine[,] a wife to unjust porposes, and the mining foreman and Superintendent was clearing him the way to flee befor[e] he could be arrested."[130] One particular mine guard, Bob Lee, was a known rapist. A former undersheriff described him as "a brutal man, very brutal. . . . The miners wives . . . who used to come up the hill to do washings for the American women, told hard stories on him of how he terrified certain miners wives into submitting to him by authority of his star and threatening the loss of their men's jobs . . . if a miner's woman saw him coming she'd get up and hike over to a neighbor's house to keep out of the way. . . . Myself, I've heard him brag as some young Italian wife went by, 'She's a peach. I'm a goin' to get her.' "[131]

Under such conditions, in the fall of 1912 organizers again began walking the roads out of Trinidad to the coal camps. One was the Slavic organizer, Mike Livoda, a twenty-five-year-old miner who in 1904 had left the eastern European farm of his childhood to follow his father to Ohio. From there, he drifted west, looking for an easier life than that provided by loading pig iron for thirteen hours a day. He began digging coal in Montana; in 1910, he turned up in Trinidad. According to the custom, he deposited his union card with the Trinidad (coal miner's) local of the WFM. By the time John Lawson hired him as a UMWA organizer in 1912, he was adept at talking union in the "Slavic," Russian, Polish, and English languages. Livoda worked at night. Under the cover of dark, he would walk six to ten miles from Trinidad into a coal camp, visit with one or two friendly miners, and stay overnight.[132]

The Greek organizer Louis Tikas grew up in Crete and emigrated to Denver in 1906 at age twenty. By 1910, he and a partner were operating a Greek coffeehouse in Denver. The records show that he had taken out papers declaring his intention to become a U. S. citizen. In 1912, Tikas received his introduction to coal mining when he took a job as a strikebreaker in the northern lignite field. After a few weeks, he walked out to join the strike with some sixty other Greeks. He soon became one of the interpreters through whom organizers spoke to the Greek mine workers. Before long, he was doing the talking himself.[133]

Colorado Fuel and Iron Company managers were well aware of these activities. Bowers pontificated to Rockefeller:

> [T]he lawless agitators are on the alert and we can never tell when trouble will come, their efforts to unionize our coal miners being unceasing. Our men are well paid, well housed, and every precaution known taken to prevent disaster. So far as we can learn, they are satisfied and contented, but the constant dogging of their heels by agitators, together with the muckraking magazines and trust-busting political shysters, has a mighty influence over the ignorant foreigners who make up the great mass of our ten thousand miners.[134]

Winter approached. The companies suspected union activity everywhere. Mine guards broke up meetings of fraternal societies, disrupted socials, and routed church meetings. Indeed, any of these gatherings might have included clandestine union business. As time went on, the community seemed to turn to religion. One Sunday, a minister sympathetic to the union observed miners going in twos and threes to the Ludlow Depot, a little station out on the prairie just beyond company property. A miner informed him that in fact they were going to a miners' meeting. But inside the camp he found no evidence of agitation.[135]

The companies did find evidence, and they fought back. One night in June 1912, Mike Livoda woke from his sleep in the bachelor's quarters of the Ravenwood camp with a bright light shining in his face. Mine guards yanked him out of bed and ordered him to dress. They pulled him out of the house and began beating him. When he cried for help, they responded, "Shut up you son of a bitch or we'll kill you." After severely beating him, they picked him up by the arms, and pushed him out of the camp, firing shots over his head. Injured and bleeding, he could walk only with extreme difficulty. It took him four and a half hours to walk the three miles to Walsenburg, where he then lay on a mattress in a coal shed for two days before a doctor got to him.[136]

Neither did the companies forget the carrot option. On 1 March 1913, the Colorado Fuel and Iron Company inaugurated the eight-hour day in its coal mines. Later, President Welborn reassured his executive committee that this had not reduced productivity, because along with shortening hours the company had increased the supply of coal cars to the miners. Spending less time waiting for cars, they could easily load out as many cars per day as before, and still go home early. No one mentioned the irony that ten years before the company had spent more than a million dollars to fight what had been nothing more than a demand for this very thing.[137]

In August 1913, Louis Tikas visited fourteen southern Colorado coal camps to ascertain the situation of the Greeks. He learned that out of 350 Greeks working in the southern field, 13 had been killed and from 20 to 25 injured in the past seven months. One injured Greek miner lay in the company hospital in Trinidad filthy and untended. According to this man, the company doctor had called him names and had refused to treat him. Whether true or not, this was one of many stories reflecting the miners' fear and loathing of the company doctors.[138]

While organizers like Tikas and Livoda visited the camps, secret organizers worked on the inside, identifying themselves to the Denver UMWA office with numbers.[139] Number 52 reported from a camp in Huerfano County: "Most diggers clame they are getting very short weights. The Big four [coal operators] compel all single men to board at their boarding house which is very bum most of the time. The dinner pail they don't fill is a poor lunch. Friday they only put in the dinner pail two fat meat sandwiches and a piece of poor rhubarb pie and a little green apple. All the Boys kicked. But its take what they give you or leve the Camp. . . . The wash house is fair but not sanitary."[140]

From Superior, Colorado, Number 94 reported, "the rooms are 40 feet and not a timber in sight. When the boss comes around he never tells a man to set up timber and the men won't do it because they don't like to do all that work for nothing. . . . [T]here is always plenty of smoke in the mine, for the men shoot any time of day they get ready, no air to speak of and in some places . . . every time a man shoots about two feet of rock come down with no extra pay for loading it out. The men lay their own track and its awful hard to get rails."[141]

By July 1913, the UMWA prepared to move. The strike in West Virginia was over. The fact that the Colorado Fuel and Iron Company had raised wages and initiated other improvements led some union officials to think that the company's position toward the union was softening. Others saw this anti-union tactic for what it was. In any case, in late July the national office issued circulars to the local unions in Colorado announcing the UMWA's continued support for the strike of lignite miners in the north. Union officers were authorized to call out on strike any part or all of the district at their discretion.[142]

Mine guards from the notorious Baldwin Feltz Detective Agency of Bluefield, West Virginia, began to appear in the Colorado coalfields. The miners and their families waited. Organizer Number 94 reported twenty-one Austrians, "all ready when the word is said."[143] Tikas waxed eloquent: "Conditions in the Southern Coal Fields of this state are so tyrannus, the injustices, brutalities and cohersions heaped upon the miners are such, that I found the spirit manifested among my three hundred and fifty countrymen working there to be that of war. They are ready at any time . . . to engage in an industrial war and to fight, just as their fathers and brothers in the fatherland have fought the Turks until their freedom has been obtained. . . ."[144] Adolph Germer, the national organizer in charge of Huerfano County, gave the word that the miners were "ready at the word to drop their tools to the man."[145]

On 16 August, Gerald Lippiatti, a sandy-complected man with a slight limp, walked slowly down a Trinidad street. An Italian organizer from the lignite field, he had been working in the south for two months. As he walked along the main street toward his hotel, two Baldwin Feltz detectives watched him approach and then drew their guns. Lippiatti drew his. He shot one detective in the leg before he himself died of a gunshot wound. Unionists considered him "the first to pay the death penalty for fighting corporate greed in Colorado."[146]

Mother Jones returned to Trinidad on 2 September. She immediately began receiving threat letters. "They have my skull drawn on a picture and two cross sticks under neath my jaw to tell me that if I do not quit they are going to get me. Well, they have been a long time at it."[147] Although ill, "almost broke down," Mother Jones began making the rounds of the coalfields, always accompanied by one of the foreign-speaking organizers, so the miners would, as Mike Livoda put it, "understand what she's driving at, what we wanted to do for the miners. . . ."[148]

At Rockvale, organizers openly billed a meeting for the evening of 12 September. They had no idea what to expect. On the appointed night, as the sun sank, they were greeted by the extraordinary sight of miners' lamps coming over the hill. Four hundred miners were marching behind the Coal Creek Brass Band. They were joined at an intersection by the Brookside local, led by its own brass band. The Brookside miners openly carried their union banner for the first time since 1903. For twelve years they had met secretly, keeping their local intact.[149]

Inside the camps, organizers continued an intense whispering campaign. At Delagua, two couples, Charles and Cedi Costa and Tony and Margo Gorci, took a leading role. One of their techniques was to arrange Saturday night dances. The people would string up paper lanterns and dance to music provided by a miners' dance band that included Tony Gorci, an expert fiddler. Under the sound of the music, organizers would mill in the crowd recruiting for the union in full sight of detectives. In August, these union stalwarts gained an unexpected asset. A miner's wife, Mary Thomas, arrived from Wales with her two small daughters. The organizers befriended her and soon discovered that in Wales she had won trophies for her beautiful singing voice. She began singing at the dances; her nostalgic songs from the old countries and her lovely voice attracted more and more people. Only later did her friends tell her that they were using her in the union cause. By then, she herself was a vocal advocate for the UMWA.[150]

In September, the union appointed a policy committee to direct the strike. The committee immediately began addressing letters to the coal operators requesting a conference. On 5 September, it appealed publicly to the miners of District 15, urging them to join the UMWA, "which has advanced the interests of its members in a hundred different ways and has brought sunshine and happiness into thousands of homes."[151]

The policy committee then called a Special District Convention, to be held in Trinidad on 15 and 16 September. The operators were invited, but no one expected them to come. One delegate would be present for each local in the state. Louis Tikas went immediately to Trinidad and spent four days writing 125 letters to the Greek members, informing them that they must elect delegates. Newspapers began to announce daily that the "angel of the coal camps" would speak in Trinidad on the day before the convention opened.[152]

On that day, excitement rippled through the coalfield. In Walsenburg, three hundred cheering miners listened to UMWA Vice President Frank Hayes speak, after which they formed a parade behind a brass band and marched through the streets. Out on the prairie near the Ludlow Depot, Mike Livoda, John Lawson, E. L. Doyle, and organizer Adolph Germer spoke to a large crowd. In Trinidad, hundreds of miners milled on the streets and talked in groups on the street corners. That evening, many were turned away from the West Opera House where Mother Jones was to speak, "owing to the lack of seating capacity in that large hall."[153]

The audience of two thousand included Mary Thomas, Tony and Margo Gorci, and Charles and Cedi Costa. Thomas observed a woman giving orders in a deep husky voice. It was Mother Jones. When it came time for her to speak, she "strode to the center of the stage, took the long pins out of her hat and threw them over for the men in the front row to catch, smoothed down her lovely pure white hair, straightened the shawl over her floor length black dress, and put up her hand in a command for silence."[154]

She held the audience spellbound for two hours, the *United Mine Workers Journal* reported, "mercilessly scoring the operators and the Baldwin Feltz guards." She also pleaded for arbitration, and urged the governor to use his influence to force the operators into a conference with the union. Then she built to the crescendo of her address. With fists clenched, she cried, "If it is strike or submit . . . why for God's sake, strike—strike until you win!" The cheers that greeted this utterance, reported the *Rocky Mountain News*, "ended in a veritable pandemonium. Springing to their seats men shouted until they dropped back exhausted."[155]

On Monday, Mother Jones and other organizers held open meetings directly outside the coal camps, now enclosed in high barbed wire fences. In Trinidad, 255 delegates to the convention paraded through the streets behind a band and a banner that read "We Are Fighting for Our Homes."[156]

The convention opened. One by one the delegates got up to detail their grievances.[157] "The conditions at Tolberg are very bad," said delegate Fernandez. "We cannot travel through the man way, and have to risk our lives going through the haulage way. We are not paid for room turning, for rock or dead work."

"The men work for practically nothing," said delegate Morson from Tobasco. "They have to look around the mine for rails and ties, and thereby lose a half day's work."

"There is a Doctor Cole in that camp," said delegate Miller, "who has made a number of cripples."

"The boss has the privilege of changing a man every day from place to place," said Lawrence Losey, "and he gives the good places to the fellows who put up the most money to buy him drinks."

"I tell you conditions are outrageous," said Robert Harley. "I know of four men who asked for a checkweighman and were fired."

"I am charged $5.00 a month for a shack; $2.00 a month for coal whether I burn coal or not" said Chas Costo. "House is in bad condition, and when it rains we have to get under the bed to keep from getting wet."

"They get very poor food," said delegate Lamont from Cokedale, "and some of the children are dressed in clothes made of gunny-sacks [feed bags] and their fathers are working every day."

"We are expected to trade at the company store," said delegate Cornwall. "We were compelled to buy shoes for $4.00 that my wife got elsewhere for $2.50. We were charged 45 cents for a bottle of 30 cent oil."

The following day, the convention reduced the complaints to seven demands. Of these, four were already incorporated into state law. They asked for a raise, for freedom from repression in the company town, and for the eight-hour day. Above all, the miners demanded union recognition.[158] To outsiders, this demand often seemed arbitrary and unreasonable, but the fact was that without it victory was hollow. Companies merely waited for an opportune moment to retract any concessions granted. Both miners and operators understood, if the public did not, that only with union recognition did the collective work force have any real power.

As expected (because the entire effort had been aimed toward this conclusion) the delegates voted unanimously to strike, setting the date for the following Monday, 23 September 1913. During the week, organizers intensified their efforts. Louis Tikas went out in rain, wind, sleet, and snow to call out his countrymen. "I went to El Paso County and saw that thirty-six Greeks who worked in the Pikeview Mine walked out on the strike. I then took the train to Walsenburg on the 23rd, and also went to La Veta, where I called out twenty two Greeks. . . . On the 24th I went to Ludlow and called out about two hundred and fifty of my Greek countrymen who were working in that locality. . . . I managed by using an automobile day and night to reach all of my countrymen and advise them of the necessity of sticking together." In the process, Tikas caught a severe cold.[159]

During the convention to call the strike, a week before the strike actually began, coal operators began evicting the residents of the coal camps.

On Monday, 23 September, a freezing rain fell over the canyons, prairies, and rolling foothills of the southern Colorado coalfield. On that day, the motley and bedraggled inhabitants of the coal camps struggled through the mud to begin one of the greatest industrial struggles the United States has ever known.

12

THE VOICE OF THE GUN:
THE COLORADO FUEL
AND IRON STRIKE,
1913-1914

The only language common to all, and which all understand in Colorado, is the voice of the gun.

Ethelbert Stewart, Chief Statistician, U. S. Bureau of Labor Statistics, 1913[1]

. . . You're not to think we could do any differently another time. We are working people—my husband and I—and we're stronger for the union than before the strike . . . I can't have my babies back. But perhaps when everybody knows about them, something will be done to make the world a better place for all babies.

Mary Petrucci, miner's wife, after the death of her children on 20 April 1914[2]

It rained and it snowed. Strikers abandoned furniture in wagons stuck in the mire, and teamsters returned their teams to the stables.[3] Prior to the strike the mining people were in a shabby condition; the week it began they were shabby, cold, and wet. A Denver reporter called the procession he saw leaving one camp an "exodus of woe."[4]

At Starkville on 24 September 1913, Mother Jones held up the mirror of her wit to show the strikers to themselves. "There was a lot of poor wretches on that wagon," she said of one family, "their life earnings were piled on that wagon. . . . [A] child of about ten years old, and the

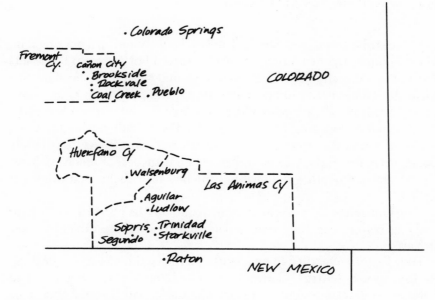

So. and central Colorado coal towns, 1914

mother—she had a babe in her arms and they did not have clothing enough to shelter them from the cold. . . ."[5]

"Dozens of single men were compelled to tramp across the prairies at night through the rain and snow," reported E. L. Doyle, "but in spite of this fully ninety-eight percent of the miners responded to the call."[6]

Federal observers estimated that ninety to ninety-five percent of the Colorado work force went out. Even L. M. Bowers had to admit the absence of forty to sixty percent of the miners from their jobs. "It is safe to say," Bowers told Rockefeller, Jr. "that . . . [most] of them have quit from fear of the black-hand and similar organizations who . . . threaten to kill the men, do violence to their wives and daughters and practice all of the hellish villainy that these creatures possess."[7]

Strikers either departed without delay for other coalfields or gathered with their families at tent villages set up by the UMWA on land leased just outside of coal company property. The Ludlow Tent Colony was the largest of eight. Located eighteen miles north of Trinidad at the Ludlow Depot, the colony was strategically located at the intersection of two roads: One climbed the steep hills of the Delagua Canyon to the Delagua and Hastings camps; the other climbed the Berwind Canyon to the Berwind and Tobasco camps. Tent colony residents could plainly see all who came or went from these camps.[8]

As the strikers straggled into the Ludlow Tent Colony in a pouring rain, Mary Thomas, the first woman to arrive, served hot coffee to the adults and milk to the children. They were cold and soaked to the skin. She recalled: "Some of the miners came in wagons piled with their cheap rickety furniture, their small children perched on top. Others had carts, the husband pulling and the wives pushing, their children trudging along behind through the sticky mud. Hundreds walked. . . . As each family would arrive the Slavs, Germans, Russians, Portuguese, French, Italians, Greeks and many other nationalities pitched in to help each other get settled."[9]

The Hungarian and German women set up their stoves and began cooking; before long they were serving hot food to the exhausted new arrivals.

Tony Gorci was carried in on a mattress, the victim of a severe beating. At the Special District Convention to call the strike, detectives had watched while clandestine organizers like Gorci had revealed their true union colors. Mine guards hit Gorci repeatedly until his face looked like "a blubber of black and blue jelly mixed with blood."[10]

The people came out of the coal camps hungry: These were America's working poor. Union regulations stipulated that relief could not begin until the second week of the strike, but the union suspended the rule on 24 September, when several hundred strikers presented themselves at the union office. "We are actually starving," they said. "We must have assistance."[11]

Difficult beginnings. Yet life in the tent communities governed by the United Mine Workers of America contrasted starkly with life under company rule. The Ludlow Tent Colony was an elaborately organized village of twelve hundred inhabitants speaking twenty-four languages. Strikers and their families lived in white tents arranged in numbered streets. At the center stood a speaker's platform on which was mounted an American flag and a bell to be rung in emergencies. Next to the outdoor platform, a large tent flying its own stars and stripes served as a meeting tent and for church services in cold weather. Americanism was important to the strike community. The flags were ceremoniously raised and lowered each day, and all meetings began with the singing of the national anthem.[12]

The union supplied coal, water distributed in barrels, and strike relief (three dollars per week for each miner, one dollar for his wife, and fifty cents for each child). Of the twenty thousand persons on strike relief, four thousand were women and nine thousand, or nearly half, were children. The miners supplemented the food supply by fishing and by hunting rabbits and quail—at least until the onset of an unusually severe winter.[13]

John Lawson (1871–1945), executive board member from District 15, served as camp boss of the Ludlow Tent Colony. He, like E. L. Doyle, represented the archetypal miner of the previous century. Born in anthracite Pennsylvania to a Scottish mining family, he began his working life at age eight as a breaker boy. He came West in 1893, working in Rock Springs before taking a job in the Colorado Fuel and Iron Company's Walsen mine in Huerfano County. By the time of the Colorado Fuel and Iron strike, he was a seasoned organizer who had lived and worked in the Colorado coalfields for more than twenty years. At forty-two, Lawson was nearly twice the age of the vast majority of strikers and of most organizers. Edward Doyle, Louis Tikas, and Mike Livoda were in their mid-twenties. Most strikers and their wives were like Mary Thomas, in their early twenties or even younger.[14]

Louis Tikas took second command at Ludlow; he was resident camp boss since Lawson worked out of Trinidad. Both men actively helped to organize the life of the community. After the tents were set up, they made lists of necessary tasks for the approval or disapproval of all. These included cleaning, guard duty for day and night, emergency squads, and the picket committee. Tony Gorci was elected head musician, and Mary Thomas, "greeter-singer." A visiting journalist marveled at how thoroughly organized the tent colony was. It even had its own police officers. The chief of police wore a "pie-plate badge with 'chief of police UMWA' emblazoned on it."[15]

The Ludlow Tent Colony evolved into a close-knit, multinational community. Mary Petrucci, a miner's wife, later testified that the strikers enjoyed themselves, preferring the life at Ludlow to that in the coal camps.[16] Mary Thomas had never seen so many people get along together so well as these did.[17] For entertainment, the strikers played baseball, and one visitor watched the Croatians playing a game of bolo.[18] Another saw "lots of little youngsters," noting that the women shared the work of caring for the children.[19] The Italian organizer Charlie Costa "kept everyone in stitches with his jokes. The children followed him around like Pied Piper."[20]

Mike Livoda lived in Trinidad but often stayed overnight at the big tent colony, paying a family for his meals. In the evenings, he would walk around listening to music played on accordions, violins, mouth harps, guitars, and tambarizzas. "I used to get out and listen the songs," he reflected, "different songs, Spanish songs, Mexican songs, Slavic song, Italian song, at night. . . . Sometimes they put on a dance, they used to polka. You just begin to feel like that even though they're out on strike, they're happy, because they're singing and enjoying themselves."[21]

"Everything is getting along fine in our camp," a striker wrote from one of the other tent colonies. "We had a social on Tuesday last. Our hall was filled with members and their wives and children. The ladies of course took charge of the entertainment. They served coffee and cake. Music was furnished by a graphaphone and everyone had a jolly time until midnight." He concluded that "our local is among the happiest in southern Colorado."[22]

The strike was a family affair. Picketing was the women's specialty and they did it with aplomb. Mary Thomas led the effort at Ludlow, making sure that whenever a manager passed by on his way to the Berwind or Delagua Cañons the women and children were at the fence singing the union song while the men held up their signs and picketed.[23]

The women demonstrated their antipathy for the companies in other ways as well. On the second day of the strike, for instance, a predominantly female crowd waited outside the barbed wire fence of the Walsen mine (in Huerfano County) for a tracklayer who had refused to join the union. When he emerged, "[t]he infuriated women rolled him in the mud and kicked him." His cries produced several mine guards, who chased away all but one woman. She jumped on the tracklayer and beat him over the head with a heavy bucket, breaking his nose.[24]

The women's own lives in the coal camps gave them ample experience from which to form strongly antagonistic feelings against the companies. Yet in their activism the women followed the leadership of the UMWA and of their husbands. "Margo [Gorci], Cedi [Costa], and I continued our gossip sessions," Mary Thomas recalled. "Their husbands tried to bring a little good news to cheer us up. They knew more about things than we, and daily we women awaited their opinions regarding the strike."[25]

The wives did not seek positions of leadership, yet for some Mother Jones was a compelling role model. The elderly organizer was "on the firing line as always," a labor paper reported, "cheering the women and inspiring the men."[26] Mary Thomas recalled her "thrilling talks," and one day invited her to her tent for tea. On the way, the white-haired organizer chatted with miners and their wives, and distributed candy to the children.[27]

The operators saw Mother Jones's activities in a different light. On 25 September, according to an operators' advertisement in the *Rocky Mountain News,* she delivered an "incendiary speech" at Walsenburg. The next day, she attacked Governor Ammons in "a violent speech."[28] Repeatedly, the coal operators portrayed her as the instigator of the violence that began to erupt in the southern Colorado coalfield. The evidence suggests the opposite: Until 20 April 1914, she and other organizers consistently cautioned strikers not to use violence. Indeed, as a union leaflet pointed out, the worst example of Mother Jones's rhetoric that even the operators could produce was tame: "[W]e will lick the hell out of the operators," she declared early in the strike. "We are not going to take any guns; we are going to take picks along, and we will take the mines and own them."[29]

The fanatics were the coal operators, particularly L. M. Bowers, who informed Rockefeller, Jr., that the company would close any mines they could not protect, and work the others with non-union labor "until our bones were bleached as white as chalk in these Rocky Mountains."[30]

Bowers' antipathies extended to government officials he perceived as pro-labor. A stream of these began arriving from Washington to attempt to settle the strike.[31] One was Ethelbert Stewart, Chief Statistician of the Bureau of Labor Statistics, sent by President Woodrow Wilson. Bowers met with Stewart and that evening wrote to his son: "Part of the afternoon was spent with a man sent from Washington. . . . [W]e have no trouble with our men and the thing is to force us to employ union men only: Not while LMB is in the saddle: I wound him up into kinks and beat him at every point: all his arguments were silly and amounted to no more than a busted bubble: Gosh he was a big chap. . . ."[32]

Stewart was equally disgusted with Bowers. "For some time the interview, or monologue," he reported, "dwelt upon the ancestral stock and industrial experience of Mr. Bowers. . . ." Coming to the matter of union labor in the mines, Bowers waxed bitter. The company "had never recognized the existence of the union and they never would." Bowers informed Stewart that the company consciously mixed nationalities and that "when too many of one nationality get into a given district . . . they would so adjust their men that no very large per cent in any mine could communicate with the others." He closed the interview with a tirade against labor leaders.[33] Mediation was hopeless.

The voice of the gun, as Ethelbert Stewart later commented, was the only language common to all in Colorado.[34] During the strike, the Fuel and Iron Company spent 25,000 to 30,000 dollars for guns and ammunition, while the UMWA spent 7,500 dollars for the same purpose.[35] "[O]ur folks are getting pretty peeved," one Rocky Mountain Fuel manager wrote to another in late October. "If anyone comes around here with a gun looking for trouble, we contemplate giving them all that is coming to them and hope to prepare a few subjects for the hospital."[36]

The union claimed that Gerald Lippiatti, shot on 16 August, was the first death of the strike. The second, then, was the hated mine guard Bob Lee, widely regarded as a rapist. Lee was shot, reported the *Rocky Mountain News*, "in the first flame of outlawry which sprang from the smouldering fires of class hatred in the southern coal field. . . ." The bullet that killed him on 24 September emanated from a group of Greek strikers who then vanished in the direction of New Mexico.[37]

The first gun battle occurred on 7 October at the Ludlow Tent Colony. That morning, John Lawson and Mother Jones had delivered speeches. In the early afternoon, shooting broke out and continued until dark. Each side accused the other of starting it.[38]

Ten days later, on 17 October, a drenching rain fell on the strike field. Mine guards gathered near the Forbes Tent Colony and fired six hundred rounds of ammunition into it, destroying the tents. A Slav striker was killed, and a mine guard wounded. A boy was shot in the legs and lay in the cross fire for hours before he could be rescued. He was crippled for life.[39]

Around this time, a "feeling of terror" began to pervade the tent colonies. According to reporter Don MacGregor: "[A] machine gun had been turned on it [the Forbes colony] in a most ruthless fashion. What if this were done at other colonies? What if machine guns were turned on Ludlow with its 150 women and 300 children? The strikers swore by their various gods that neither armored automobile nor force of deputies would ever be permitted near enough Ludlow to allow this to happen."[40]

A newspaper reporter, MacGregor became increasingly involved on the strikers' side. Before the strike was over, he abandoned his career to pick up a gun in their cause. The written account he submitted to federal investigators in 1914—shortly thereafter he disappeared into the Mexican Revolution—remains one of the most vivid documents of a strike that produced thousands of pages of documentation, little or none of it "objective."

In the autumn of 1913, the companies installed electric searchlights that swept the hills and prairies for six miles around. At Ludlow, the strikers slept in a rhythm of sweeping light.

Operators began to pressure Governor Elias Ammons into ordering the Colorado National Guard into the strike zone. Governor Ammons, elected in part with labor's vote, was a Democrat, not the party of the companies. His reeducation now began. "There probably has never been such pressure," Bowers boasted to Rockefeller, Jr., "brought to bear upon any governor of this state by the strongest men in it, as has been brought to bear upon Governor Ammons." The Colorado Fuel and Iron Company secured the cooperation of every Denver banker, of real estate and other business people, and of fourteen newspaper editors, who visited "our hesitating governor" in a steady stream to urge him "to drive the vicious agitators from the state."[41]

The UMWA began its own campaign to oppose the calling out of the guard. On 18 October, the Trinidad Trades Assembly sponsored a mass meeting for the purpose.[42] Don MacGregor reported that the strikers believed that

> the militia was coming to help the mineguards break the strike; they would do as the deputies had done and raze their tent colonies; that they would do as the militia had done in 1904, tear them from their families, load them on trains and dump them on the prairies of Kansas, Texas and New Mexico, threatening them with death if they ever returned to Colorado. This feeling was caused by the . . . Americans among the strikers, who remembered the events of 1903–04. . . .[43]

Late in October, Governor Ammons traveled to Trinidad to inspect conditions for himself. In response, the UMWA organized a giant parade to demonstrate the union side of the question. On the morning of 21 October, some four thousand strikers arrived in Trinidad by streetcar, by wagon, and on foot. They first gathered in front of the union office, where food was served, then began the march.

"Never in local history," reported one newspaper, "has there been a more spectacular demonstration." The parade was led by a band, followed by the women and children: "Hundreds of women carried small babies at their breasts and toddling youngsters hardly able to keep up walked sturdily in line . . . women with baby buggies followed . . . shouting and singing." Behind the women marched more than one thousand coal miners, many of them also carrying children. The marchers carried great banners that read "We Will Not Be Whipped Into Citizenship by the Sheriff and Gatling Guns," "The Democratic Party Is on Trial," and "Some of Mother Jones's Children."[44]

The marchers sang and chanted their way through town. When they reached the governor's hotel, Mother Jones, "in her foghorn voice," shouted up to the governor that there were some women who wanted to see him. The governor declined to descend to the street. The demonstrators shouted union songs, and Mother Jones "delivered a tirade of abuse upon the Baldwin-Feltz detectives who also had their headquarters in the same hotel."[45]

Three days later, strikers milling the streets of Walsenburg had devised the game of surrounding any isolated mine guard and disarming him. At midafternoon, mine guards were assisting a family of strikebreakers on Seventh Street to load their things into a cart. Harassment had persuaded them to prefer the coal camp to Walsenburg. As the mule moved the cart out onto the street, women and children threw nut coal and tin cans at it. A large crowd including many children began to gather. At 4:15, the mine guards opened fire. Four strikers were killed.[46]

The following day, strikers killed a mine guard. Gunfire had become nearly continuous out on the prairie and in the hills.[47]

Don MacGregor arrived at Ludlow just after the residents had learned of the Seventh Street incident at Walsenburg. The news "set the tent colony to seething. Taken on top of Forbes, the strikers believed the mine guards planned to wipe them out utterly. Many of the women at Ludlow became half hysterical." Rumors now combined with actualities to feed a growing fear. Someone in a saloon overheard a company doctor boasting that the mine guards were going to clean out the tent colonies, including men, women, and children. "This threw the entire colony into the wildest excitement. Men ran from tent to tent shouting the news. A dozen nationalities gathered in jabbering groups. Hysterical women demanded that their men take steps to protect them. A woman whose time was at hand began shrieking in most horrible fashion."

Despite a bitterly cold night, the men took the women and children out to the arroyo north of the colony, out of the line of fire if the deputies did come. "There was despairing talk of cleaning out the two canyons, of killing every mine guard in them," MacGregor wrote. But in the end, nothing at all happened, and the people came straggling back to the tent colony.[48]

By this time, the strike had worked its way onto the national agenda.

President Woodrow Wilson wrote to company President Jesse Welborn to say that he regretted the failure of federal mediator Ethelbert Stewart to resolve the distressing situation in the Colorado Fuel and Iron Company mines. He asked for management's views before recommending a congressional investigation of the strike. Bowers answered the letter. The employees were prosperous and happy; only the utmost friendliness pervaded the relations between them and the company. Further, the "coming of these agitators to stir up strife was almost, if not quite a crime." Bowers also delivered a tirade against Mother Jones. "You will pardon me, I am sure," he wrote, "in stating that the Scotch-Irish Presbyterian blood coming down from my ancestors, becomes somewhat heated to know that such disreputable creatures as the self-named 'Mother' Jones can secure the attention and cooperation of statesmen." Bowers enclosed a 1904 clipping on the subject of Mother Jones's morals.[49]

Rockefeller, Jr., also wrote to President Wilson to inform him that the failure of the men to come to work was "due simply to their fear of assault and assassination."[50] President Wilson made three more attempts to persuade the Fuel and Iron Company to compromise. He failed.[51]

On October 28, Governor Elias Ammons ordered the Colorado National Guard into the southern Colorado strike field. Since the governor had assured the union that the troops would be impartial, the strikers welcomed them in good faith. "With banners flying, band in the lead, men, women, and children of Ludlow marched a mile down the country road to meet the Colorado State Militia," reported a labor paper.[52] But the warm welcome did not signify trust. Ordered to relinquish their weapons, the strikers at Ludlow offered the commander, General John Chase, thirty-seven rusting rifles and one child's popgun.[53]

Had they freely handed over their weapons, they would have been handing them to their enemies, the mine guards. Within two weeks, these hired gunmen emerged in the uniform of the Colorado National Guard. Don MacGregor reported that at Sopris the entire force of mine guards was enlisted in one day. The mine guard who had led the attack on Forbes, whom the strikers regarded as a murdering thug, now appeared in uniform. "The bitterness of the miners," MacGregor reported, "was intense."[54]

Supposedly, the Guard was to stand as a neutral force between opposing sides: The governor had ordered the troops not to escort strikebreakers into the mines. Yet not a week had passed before a miner indignantly reported to the *United Mine Workers Journal* that troops were bringing in strikebreakers. His letter was printed under the rubric "Crooked Work Laid Bare."[55] A month later, on 26 November, the governor rescinded his order. That which had been done covertly could now be done in plain view.[56]

Troops began to ride around openly in Fuel and Iron Company automobiles. Strikers regularly saw them going in and out of the firm's Trinidad office.[57]

They began daily weapons searches. At Forbes, strikers accused soldiers of stealing things, vandalizing, and dumping possessions on the floor. At Ludlow, the women considered it their duty to harass the soldiers. Lieutenant Karl Linderfelt, a uniformed mine guard whom strikers loathed, later testified: "[T]he women in the colony would take these little children and line them out when we were searching. . . . 'Now tell them what they are [repeat] after me.' And I have heard an American mule skinner in the Philipine Islands drive eight head of mules, but I have never heard anything equal to it."[58]

The children could be horrible indeed. On a visit to the strike region, State Senator Helen Ring Robinson reported that when a militia man appeared, the child who a moment before had looked like one of Raphael's cherubs, would mutate into a fiend, taunting the soldier with "scab herder" and similar epithets.[59]

Bitterness smouldered and flared easily into violence. On 8 November, a group of strikers ambushed three mine guards and a strikebreaker and shot them to death.[60] Two weeks later, George Belcher, the Baldwin-Feltz detective who had killed organizer Lippiatti in August, was himself shot to death.[61]

As the Guard began publicly escorting strikebreakers into the camps, the picketers grew more aggressive. One morning in late November, a rumor circulated in the Ludlow Tent Colony that strikebreakers would be brought through the Ludlow Depot on the 8:30 A.M. train. The women came out to block the train. An officer of the Guard described the scene: ". . . [T]he women carried clubs . . . many had spikes driven through them. Others were . . . limbs of trees with sharpened branches. Some were just plain boards and billiard cues. . . . [O]pposed to ten sentries [guardsmen] was a solid mass of strikers, with their club swinging women in the front rank, giving vent to all manner of profanity. . . ."[62] But the rumor proved to be baseless, and strikers returned to the tent colony without incident.

The troops were not neutral, but, what was worse, their behavior could be violent and unpredictable. Discipline was lax, and it gradually deteriorated. As time went on, the strikers found it increasingly difficult to distinguish them from a collection of rowdy, dangerous mine guards. A governor-appointed committee investigating their conduct collected testimony on abuses that ran to over seven hundred pages.[63] A former guardsman who regretted his own enlistment as a youthful mistake recalled that "plain holdups on the street and armed robberies of the saloons became common. Once a crime had been committed, the uniformed thugs disappeared into a pool of 2000 militiamen, confident no one would betray them."[64]

In early January 1914, Mother Jones was deported from the strike region.[65] Soldiers drove her to the depot in an automobile, while "cavalry men charged through the streets clearing them against possible disorder."[66] Mother Jones was put on the northbound train under guard.

Her deportation caused a flurry of comment across the United States. Telegrams questioning the legality of the procedure flooded Governor Ammons's office. In response, the governor remarked that the commander of the strike region had done the right thing, as three-fourths of the strike violence could be blamed on Mother Jones's "incendiary utterances."[67] The procompany Pueblo *Star Journal* editorialized that "Mother Jones was a disturber employed for no other reason than to incite class hatred."[68]

In Denver, the Equal Suffrage Association denounced her deportation on the grounds that the action was unconstitutional and abridged the right of free speech. This quickly developed into a "hornet's nest" that revealed the disparate values and class loyalties among the suffragists of 1914. At the next meeting, the "Mother Jones resolution" triggered a dispute, with some of the women seeing Mother Jones as a dangerous agitator while others stood by their principles regarding free speech. Governor Ammons stirred the hornet's nest by "delivering a castigation over the telephone" upon two of the women who had sponsored the resolution.[69]

A week later, Mother Jones returned to Trinidad, where she was again arrested. This time, troops placed her in an automobile, which "whirled rapidly through the streets with a cavalry escort galloping at full speed" in front of and behind the car. Strikers lined the streets and cheered wildly as she waved. The twenty-eighth person to be detained without charge that week, Mother Jones was taken to the San Raphael Hospital, a Catholic institution, and kept there for nine weeks.[70] Although she was held incommunicado, her imprisonment drew national attention. As lawyers for the union began the process of petitioning the courts for a Writ of Habeas Corpus, indignant telegrams began raining on the White House.[71]

Women took the forefront in the protests against Mother Jones's incarceration. On 15 January, they marched two hundred strong from the union hall to military headquarters, chanting, waving flags, and singing union songs. Once there, they confronted General John Chase and demanded her immediate release. Chase informed them that Mother Jones was a disturber of the peace and an inciter of riots, and under no circumstances would she be released. His remarks were greeted with jeers.[72]

A week later, the UMWA organized a women's demonstration to protest her detention. General Chase outlined the union's strategy to the governor: "The strikers had evinced a disposition to cause disturbance and disorder through their women folks. They adopted as a device the plan of hiding behind their women's skirts, believing, as was indeed the case, that it would be more embarrassing for the military to deal with women than with men."[73]

The demonstration began peaceably but did not end well. On the appointed day, one thousand wives, mothers, and sisters of striking miners, dressed in their Sunday best, began marching with their children. The men brought up the rear; many of them also carried or walked with children. The procession moved down Commercial Street and then turned onto Main Street, where it confronted one hundred mounted troops. General Chase ordered the women to halt. Instead, they continued to advance, slowly. Then, according to sixteen-year-old Sarah Slator, "General Chase's horse became frightened at something . . . and it ran into a horse and buggy . . . and he fell off the horse. . . . He had been treating us so mean that everybody screamed and laughed at him and that made him angry." What followed caused this women's demonstration to be called "The Mother Jones riot."

General Chase lost his temper and ordered the troops to charge. They obeyed, swinging their rifles and cutting several women with their sabers. The demonstrators scattered into porches and yards. Transformed into viragos, they began hitting the troops with sticks and throwing bottles.

Sarah Slator recalled women's hats lying in the mud. General Chase remounted close to her and from his horse kicked her in the breast. She crossed the street and heard a woman asking a soldier "what right they had to chase women away like cattle?"

Miners's wives demonstrating. Photo courtesy Denver Public Library, Western History Department.

"When women sink beneath our respect," the soldier replied, "they need to be treated like cattle."[74]

Mary Thomas was arrested while beating a soldier with an umbrella. An officer of the Guard later remarked that she seemed to be the ringleader of a mob of women. General Chase reported that she was a "vociferous, belligerent and abusive leader of the mob. She forcibly resisted orders to move on, responding only with highly abusive, and to say the least, unwomanly language. She attacked the troops with fists, feet, and umbrella."[75] She passed the next fortnight in the county jail.

For those not in jail, the day closed with a meeting to protest the presence of the militia, and to organize the defeat of Governor Ammons in the next election. Calling themselves the Women's Voting Association of Southern Colorado, these coal mining women had invited a suffragist to speak. The guest speaker, the prominent member of the Denver Equal Suffrage Association who had authored the "Mother Jones resolution," failed to appear. She had apparently caved in under pressure. The assembled women greeted the news of her absence with cries of "Quitter" and "Deserter of the Cause."[76] All was not sisterhood in the women's movements of 1914.

"One cannot help extending sympathy to these women," wrote an "Eyewitness" from the Trinidad *Chronicle-News*, "when this banner came in view: 'Mother Jones has not done anything we would not do ourselves.' Surely, surely, these poor misguided women did not think that." This reporter (who spoke only English) went on to ridicule the women who rose to speak in their native languages. "The next woman on the program was a Polish woman from Walsenburg. . . . She stepped up to the front of the stage, rolled up her sleeves and from her build and actions one was not sure whether she was just preparing for the wash tub or to enter the prize ring. . . ." Reporting that the woman had decided to telegram President Wilson, the journalist concluded, "Did they suppose our President would notice a petition from a few women, many of whom are foreigners . . . and laboring under a load of misapprehension . . . ?" The writer was moved to "a feeling of compassion."[77]

By this time, January 1914, the strike had settled into a standoff. This was bad news for the UMWA. The longer the strike lasted, the more difficult it became to hope for victory. The strike drained enormous sums and required a high level of continuous organizing. Aggravating the difficulties of an already Herculean task, the union's organizational structure was flawed. Like an old canker sore, jealousy flared between men loyal to the district and those loyal to the national office. Wrangling, bickering, and faultfinding accompanied the inevitable tightening of the purse strings.

The national treasurer held district Secretary-Treasurer E. L. Doyle responsible for the management of the strike funds. In December, the national office ordered him to practice retrenchment.[78] This was easier said than done. Doyle distributed relief checks to the officers in charge of each subdistrict, but the UMWA had devised no uniform system for handling such funds, an astonishing lacuna considering the sums involved. Although Doyle worked incessantly to devise a system, he lacked the authority to impose it upon men who allegedly worked under him. Subdistrict officers assigned to their posts by the national office sometimes refused to follow his instructions. The national office refused to back Doyle's authority.

William Diamond, the national organizer in charge of the important Trinidad office, ran his subdistrict without regard for the opinions of Doyle. "You mention that there seems to be several heads from which this strike is being conducted," he charged in an angry exchange with Doyle: "I again ask you for your approval or disapproval as to me remaining in charge of this office? I would like to have an immediate reply, because I have had enough misunderstanding on the part of the District Office as to my standing in here. . . . I am either in charge of this office or I am not in charge. In either case I want to know it."[79]

Funds were not the only problem. Doyle assured Diamond that he had nothing against anyone delegated to work in Colorado by the national, "with the exception of men who get drunk, and this is one thing that cannot be said of you." But he criticized Diamond for not dealing with strikers' complaints "whether real or imagined" in a proper manner: "Often they go away feeling that they didn't receive sufficient attention. I'm old enough in the organization to know that hundreds of these complaints are most unreasonable, and in a great many cases injust, not only to other members but to the officers in charge, but I look upon it that it is our place as officers to act in a fatherly way with these poor fellows, encouraging and advising each individual. . . ."[80]

Other complaints from other organizers reveal the organizational malaise. Louis Tikas, working in Diamond's Trinidad district, sent off a bristling letter to the national office that began by detailing his own contributions to the work. The people, he insisted, had regarded him with love and respect. It was jealousy at his success, he thought, that had caused certain national organizers "to plot and connive against me. These men, seated around a large table in comfortable chairs in the Trinidad office, smoking their good fifteen cent cigars, riding in automobiles at the expense of the organization and their work limited to a few speeches . . . plotted to discredit me with the people. . . ." Tikas accused them of trying to undermine his influence by refusing to recognize his orders for supplies. "They know my power over the people and know that the quickest and surest means to discredit me . . . was to show the people that I no longer had influence with those in authority and that when . . . in crying need of the necessities of life, they must look elsewhere for supplies from headquarters." He claimed that during a blizzard while he was trying to keep his thirteen hundred strikers at Ludlow from freezing, he was "given as supplies to take to the people this, 'You can go to hell.' "[81]

A local official from Fremont County, Davis Robb, directed his irritation at Doyle: "On several occasions there has been a little 'butting in' that does not savor of good judgement, like the cheque question, *as you know* on which I should have been consulted. . . ." Responding to Doyle's request for retrenchment, he spat out that, although he had been preaching economy, "I would like to draw your attention to the fact that the abuses of the funds in other places is well known here and I am getting hell all the time about what the men are getting elsewhere. If this strike is lost it will be lost through carelessness."[82]

Underlying these squabbles was the hard truth that by March the strike was disintegrating under its financial burdens. National Secretary-Treasurer William Green scolded that the union could not furnish the constantly increasing amount of money required in Colorado: The money was not there.[83] Doyle accused Green of being penny-wise, pound-foolish by not supplying him with competent auditing assistants. "[W]ith the lack of experience on the part of many Local Union Secretaries, I am satisfied that the amounts wasted through ignorance or otherwise will more than double the amount it would cost the organization to have these accounts examined regularly, as was planned in the beginning of the strike."[84]

In the winter of 1914, the UMWA ran a monthly deficit of between thirty and fifty thousand dollars. Supposedly, the Colorado strike (and simultaneous strikes elsewhere) were being financed by a monthly assessment of fifty cents per member. Actually, the assessment didn't cover the costs. As in the 1903 strike, substantial loans from the solvent districts were fueling the union motor.[85]

Opponents of the strike reveled in the opportunity to sling mud. The Trinidad *Chronicle-News* printed the union's published financial report under the headline "Report Shows How Union Organizers Spend Thousands While Strikers Suffer." The article showed District President John McLennan receiving "$2683.55 with an expense account of $2469.55." Of course, McLennan's expense account went to pay the bills of the strike, but the paper claimed that he was paid grandly for his services, while the strikers eked along on three dollars a week.[86]

No doubt detectives fanned the flames of the same fire. In the winter of 1914, the district office warned the strikers: "Believe not the poisonous statements or stories being whispered throughout your camp about our officials . . . we find that many such stories are being circulated at the behest of the enemies of our organization, all for the purpose of destroying the confidence of the rank and file in their elected officials . . . we know you will discard them, without ever repeating any of these slimy whisperings. . . .[87]

One of the few hopeful signs to cheer unionists on was the arrival in early February of a congressional delegation to investigate the strike. This subcommittee of the House Committee on Mines began taking testimony in Denver amidst national publicity.[88] The antagonists went on good behavior: Not a single violent incident occurred during the committee's tenure. In mid-February, the five congressmen toured the Ludlow Tent Colony, where the strikers gave them the grand welcome. "An orchestra of strikers played strike songs, while the Congressmen applauded and scores of children plied the candy counter at the tent colony store with nickels distributed to them by the visitors. When the automobiles chugged away the strikers waved their hats and set up a rousing cheer."[89]

Unionists hoped the effect of the investigation on public opinion would force the operators to compromise. They were too optimistic. Against the power of the coal operators, public opinion had no force. The congressional investigation produced one of the rich historical records of the strike, its only accomplishment.[90]

If the presence of the five congressmen had inhibited violence, their departure seemed to set it loose. On Tuesday, 11 March, twenty-three mounted troops of the Colorado National Guard tore down the tents at Forbes. A woman who had just given birth to twins was thrown out into the sleet and snow along with everyone else.[91] Doyle feared this was the beginning of a reign of terror designed to drive the miners back to work.[92]

Troops began surrounding the tent colonies every night after dark. The miners and their families lay awake in fear.[93]

The Guard was going from bad to worse as more and more of the better class of recruits returned to their homes. The remaining troops had begun receiving I.O.U.s from the state instead of paychecks. The state was bankrupt, the governor inept. For lack of a better idea, he ordered the withdrawal of some but not all of the troops. One Lieutenant Conners revealed their pathetic condition in a letter to his superior officer. "Things are in an awful mess here," he wrote from Starkville. "You would almost have cried to see the men when they started home last night, ragged, dirty, and with only a few nickels left after paying their bills, or as much of them as they could."[94]

In mid-April, two key persons left the state. One was Mother Jones. Released from her incarceration in a Trinidad hospital on 16 March, she had returned to the southern field a week later only to be rearrested. This time, she was put in the Walsenburg County jail. On 3 April, the Colorado Suprème Court issued a Writ of Habeas Corpus, ordering her captors to produce her before the court within ten days. To avoid an important test case of the statute under which she was being held, the state released her. She boarded an eastbound train. The second person who absented himself from Colorado at this critical time was Governor Ammons. Feeling the need to consult with President Wilson, he, too, boarded a train for the East.[95]

In the coalfield, mine guards increasingly joined the remaining companies of the Colorado National Guard. A new company, Troop A, was formed of "mineguards, pit bosses, mine superintendents, mine clerks, and others in the employ of the company."[96] Troop B, commanded by the hated Lieutenant Karl Linderfelt, consisted entirely of mine guards. "All that is left now," fumed one striker, "are the gunmen, the scum of the earth, barrel house bums, professional killers from every part of the country who think nothing of human life."[97]

L. M. Bowers saw the world through a different pair of glasses. He saw the latest troop formations as a "favorable feature" of the situation.[98]

The troops went fearfully out of control. One officer of the Guard wrote to another: "The detachment at Sopris turned out just as I thought. They had a drunken brawl the first night and raised cain in general, shot holes in the walls and ran each other all over the place."[99] On a visit to the strike zone, State Senator Helen Ring Robinson observed militiamen openly entering the office of the Colorado Fuel and Iron Company to receive paychecks. Both soldiers and strikers informed her that they expected the Ludlow Tent Colony to be wiped out.[100]

At Ludlow, the senator found an atmosphere of waiting, of dread. The women showed her pits or cellars their husbands had dug beneath the tents, for them to run into with the children in case of attack. An employee of the railroad who lived on the prairie nearby noticed that the strikers "seemed to be kind of restless and scared all the time, for fear something was going to happen."[101]

But Senator Robinson also discovered something else at Ludlow, an atmosphere unusual in an age so saturated with public and unchallenged racism. "I found," she testified before the Industrial Relations Commission, "among the women particularly, and many of the children, that this long winter had brought the nationalities together in a rather remarkable way. I found a friendliness among women of all nationalities—22 at least. I saw the true melting pot at Ludlow."[102]

The community had survived the winter. To Senator Robinson, the people seemed "pretty comfortable" and "rather happy." "The sun was bright and warm," Mary Thomas recalled, "which felt so good after seven months of winter. . . . We emerged like groundhogs our clotheslines were full of washings and bedclothes."[103]

On Sunday, 19 April, the Greeks celebrated the Greek Orthodox Easter. Everyone took part in the service. A baseball game followed, attended by all, including the troops. The militia had come to ballgames before, but never with their guns. This time, according to one woman, they entertained themselves by training their rifles on the players.[104]

After the game, the community shared a large Easter dinner. Mary Thomas, Cedi and Charlie Costa, and Margo and Tony Gorci spent the evening together. They passed the time singing hymns, "thinking of the Easters in the lands we came from," Thomas remembered, "we women in tears. Cedi, Margo and I before retiring . . . had our coffee and talk. Little did I think, this is the last coffee."[105]

The next morning around nine o'clock, the attack on the Ludlow Tent Colony began. At that hour, not one-half of the tent colony residents were up and dressed. Ometomica Covadle looked out her tent window to see three of the Greek bachelors still celebrating—singing, dancing, and playing on a mandolin, a violin, and a flute. Mary Petrucci had just started to wash. Mary Thomas's two girls sat eating their oatmeal. Margaret Dominiske had decided to let her children sleep late. After putting wash water on the stove, she was walking to another tent to get some postcards to "send away from our Ludlow Easter." Mrs. Ed Tonner, in an advanced stage of pregnancy, was sweeping her front room tent. Leyor Fylor, aged ten, was out playing ball.[106] Louis Tikas was engaged in a dialogue with the commander of troops who seemed to be moving into positions relative to the tent colony. Major Hamrock wanted Tikas to produce a man supposed to be in the tent colony. Tikas did not have him. Suddenly, three explosions went off. They may have been signals.

The shooting began. "Suddenly the prairie was covered with human beings running in all directions like ants," Mary Thomas recalled. "We all ran as we were, some with babies on their backs . . . not even thinking through the clouds of panic.[107] Many women with small children ran into the cellars dug beneath the tents. Others ran out of the tent colony. The men went for the rifles they had managed to keep through successive weapons searches. They ran into rifle pits dug purposely away from the tent colony to draw fire away from the tents.

Pearl Jolly, a twenty-one-year-old nurse, spent the day in a rain of bullets. She and Tikas ran from tent to tent trying to provide for people— making sandwiches and tending the wounded. "Shots were flying all around me and a bullet hit the heel of my shoe and shot it off," she later testified. "I thought my foot had been shot off but I didn't have time to stop and see whether it had or not, I had to keep on the run . . . we would go from tent to tent and make the women and children in these tents as comfortable as could be. We went about dodging bullets all day."[108]

Hour after hour, the shooting continued. By midafternoon, the children in the cellars were becoming restless and hungry.[109] Eleven-year-old Frank Snyder, the eldest of six, came up into the Snyder tent to get something to eat. He was shot in the head and killed. His father came up and lay down beside him. He then began running hysterically from tent to tent telling people to make their children lie down rather than have them killed. After this, it seemed to Mrs. Tonner that "the machine guns turned loose all the more. My tent was so full of holes that it was like lace, pretty near."

Meanwhile, the strikers shooting back from the rifle pits dug to one side of the colony were faring badly. At approximately 4:30 in the afternoon, they ran out of ammunition.

At dusk, the attackers entered the colony with war whoops. They looted the tents of musical instruments, quilts, and clothes. At 7:10, a Colorado Fuel and Iron Company manager in Trinidad received and recorded a telephone message: "Tent colony on fire."[110]

Louis Tikas and Pearl Jolly ran from tent to tent, getting out the women and children. As they were leaving with a group of fifty, they heard a rumor that one of Mary Petrucci's children had been shot. According to Jolly, Tikas couldn't see how she could get out with her small children and one of them dead. He went back. On the way, troops led by Lieutenant Karl Linderfelt captured and killed him.[111] Troops killed two other prisoners just as the tents were starting to burn. One of them, John Bartoloti, had run back into the tents for his wife and children. He had been crying for them all day. The other was Charlie Costa.

In the light of burning tents, gunmen escorted a few families up to the Ludlow Depot. There, according to Juanita Hernandez, one of the soldiers took an accordion and "played on it before us." Sometime after midnight, soldiers came upon the Snyder family with their five living and one dead child. One soldier ordered them out of the tent, saying to the father, "You redneck son-of-a-bitch I have a notion to kill you right now."

The light of Tuesday morning revealed whiskey bottles strewn across the battlefield. Sometime after dawn, Mary Petrucci regained consciousness. Lying beside her in the cellar were thirteen corpses—the remains of her own three children, Cedi Costa and the two Costa children, Mrs. Patria Valdez and her four children, and two other children. Mary Petrucci staggered out of the cellar toward the Ludlow Depot, not entirely in her right mind. "I suppose I was like a drunken person," she later said. "The road was all mine." Her only thought was to go to Trinidad to see her mother. Nine days passed before the information that her children were dead entered her conscious mind.

On Tuesday, other miners' wives, cold, hungry, and half-dressed, straggled into Trinidad. They had spent the night on the prairie. One woman had given birth; mother and child came into town half-clad and freezing cold.[112]

Pedro Valdez, husband of Patria, was in El Paso, Texas, when he heard rumors of trouble at Ludlow. He immediately jumped a train for Trinidad. He arrived too late for anything except the news that his entire family was dead.[113]

"For God's sake," Doyle wired UMWA President White, "urge the chief executive of this nation to use his power to protect the helpless men, women and children from being slaughtered in southern Colorado."[114]

Bowers wired John D. Rockefeller, Jr.: "[A]n unprovoked attack upon small force of militia yesterday by two hundred strikers forced fight resulting in probable loss of ten or fifteen strikers, only one militiaman killed. Suggest your giving this information to friendly papers."[115]

As the news hit the nation's front pages, an armed rebellion gripped the southern Colorado coalfield. The mayor of Trinidad left town as district union officers openly distributed guns.[116] Doyle, Lawson, and McLennan, as well as officials of the Colorado State Federation of Labor and the Western Federation of Miners, issued a "Call to Rebellion" requesting donations of guns and ammunition. The call asked volunteers to organize themselves into companies "to protect the workers of Colorado against the murder and cremation of men, women and children by armed assassins in the employ of coal corporations. . . ."[117]

On Tuesday, strikers set up camps in the hills and elected military officers.[118] On Wednesday, Doyle wired the national office: "Battle raging now. Five large tipples said to be on fire. Five guards killed, two strikers wounded. . . ." The next day, he wired again: "Hell is loose in this state."[119]

"Campfires gleamed along the ridges at Aguilar, scene of the burning of the Empire Mine Property," reported the *New York Times* on Thursday, 24 April. "Fifteen hundred men were gathered for what was supposed to be the first concerted attack. . . . At the Southwestern Mine, which broke into flames early this afternoon, a battle is raging between guards and a strong force of strikers."[120] In the meantime, according to a government investigator, women were still wandering the hills, "frantic with anxiety over the fate of their husbands and children." Many spent as long as forty-eight hours huddled in "ditches, gulches, and similar shelters."[121]

By Wednesday, the strikers controlled an area eighteen miles long and five miles wide, embracing the town of Trinidad as well as numerous mines. Trapped inside this territory were superintendents, mine guards, strikebreakers, and the two companies that had attacked the Ludlow Tent Colony.[122] A minister ventured into the battle zone in an attempt to rescue strikebreakers and commented of the armed strikers: "[T]here were men there who had their children killed at Ludlow just two days previous. . . . [I]t is almost impossible for anyone to appreciate the intense excitement that prevailed among these men; in fact it would almost be right to say that some of them were insane in their grief."[123]

13

WE ARE UP AGAINST
A STONE WALL:
THE ROAD TO DEFEAT

*I know the conditions you have to bear . . . I know from my own personal
experience . . . I know what it is to suffer. . . . It is not new to me.*

John Lawson, December 1914[1]

*We will find out if the government is in Washington or New York, and
I feel that we have found that it is in New York yet.*

John Lawson, December 1914[2]

The coalfield war competed for space on the nation's front pages with
news of another fiasco: the invasion of Mexico by American troops on 21
April 1914. President Wilson had converted a minor incident involving
a few American sailors into a national insult to the American flag. The
invasion of Veracruz left nineteen Americans and more than three hundred Mexicans dead. In the public mind, this and the war in Colorado
became irrevocably linked.[3]

Telegrams rained on the White House. A typical message from New
York contrasted the trivial Mexican insult with the "far greater blot upon
the honor and flag of this country" caused by the "outrages against miners
and their families."[4] A Denver garment workers' union wired: "You are
sending thousands of our citizens into Mexico to resent an insult to our
flag yet there were three flags waving over Ludlow when fifteen innocent
babies and five defenseless women two approaching motherhood were

brutally murdered and cremated by the demons wearing the uniform of the U.S.A. in the employ of the Rockefeller interests."[5]

Demonstrations and rallies protesting the murder of women and children in Colorado erupted in cities across the country. Denver, of course, felt the effects the most strongly. On the Saturday after the event, one thousand women gathered at the State House to insist that the governor wire the president to request federal troops. They sang hymns for hours while waiting for President Wilson's answer.[6]

The next day, five thousand union supporters stood on the lawn of the capital building in a pouring rain "and called upon the justice-loving citizens of Colorado to arm themselves so that if law and order are still defied, we may be able to protect our homes." The protesters denounced Governor Ammons as a "servile tool of special privilege."[7]

In New York, on 29 April, five people left an immense protest meeting at Carnegie Hall and made their way to the Rockefeller offices at 26 Broadway. Donning black armbands, they began a silent picket on the sidewalk outside. Among them was the novelist Upton Sinclair, who was arrested with others and taken to the Tombs. (In the next few months, he would write *King Coal*, based on the Colorado Fuel and Iron strike.) After Sinclair's arrest, his wife marshaled fresh forces and the picket was resumed. The silent mourning picket continued for two weeks. It was joined by the preacher for St. Marks Church, who one day read the service for the dead from the Book of Common Prayer.[8]

In Chicago, reformer Jane Addams presided over a mass meeting to protest the Ludlow Massacre. A silent picket appeared in front of the Standard Oil building. In San Francisco, hundreds of silent demonstrators paraded through the business district "to the measured beat of the dead march," ending with a vigil at the Standard's local offices.[9]

Coal miners across the country were profoundly enraged. An Illinois UMWA local wired President Wilson that if he did not "stop the bloodshed of our members while they are standing for their rights," they would call out the entire Illinois membership of eighty thousand members, march to Colorado in a body, and protect them at all costs. In Rock Springs, Wyoming, a mass meeting of organized labor requested Wilson to take action "in the name of humanity" or, if not, they themselves would defend the strikers "with our lives if necessary." Miners at Arnot, Pennsylvania (in Tioga County), ordered the President to "stop this wanton carnage which is not only a blot on the state of Colorado but on the government of the United States."[10]

For once, coal operators and coal miners, Denver Bankers and union officials, and the governor of Colorado and the president of the UMWA could agree on a single idea: that Woodrow Wilson should order the United States Army into Colorado. The vice president of the First National Bank of Trinidad added his telegram to the hundreds accumulating at the White House: "Thousands of people have investments here. Please stop war while you can."[11]

If anything, the messages pouring into the UMWA national office in Indianapolis conveyed more hysteria than those addressed to President Wilson. One letter written in Italian at Johnston City, Illinois, translated: "[T]hree hundred Italians of this place are ready at any time to leave for Colorado to defend their brothers against the murderous attacks. . . . [W]e need only a few firearms, we will attend to the rest. . . . [O]ur blood is boiling, and we are ready to sacrifice our lives for their just cause."[12]

On 25 April, three officers of the Colorado National Guard, appointed by Governor Ammons, convened to investigate the "Battle at Ludlow." All witnesses insisted that the strikers had started the incident. "[T]he conflict," the report stated, "was contemplated, prepared against, deliberately planned and intended by some of the strikers." Neither strikers nor unionists were interviewed. The Guard's investigation of itself concluded that the tent colony inhabitants were "almost wholly foreign and without conception of our government. A large percentage are unassimilable aliens to whom liberty means license." But the remote cause of the battle lay with the coal operators, who had established in Colorado "a numerous class of ignorant, lawless and savage South-European peasants."[13]

In southern Colorado, the fighting raged on. Under intense pressure from Welborn and other operators, Lieutenant Governor Fitzgerald (Ammons had not yet returned) ordered the Colorado National Guard back into the strike field. Now there was no pretense of neutrality: The troops reenforced the mine guards, the mine superintendents, and the troops of Companies A and B who had attacked the tent colony.[14] Company C refused to go. Its eighty-two members hissed the troops departing Denver "to engage in the shooting of women and children."[15]

The companies in the field had taken prisoners and forced them to work long hours without sleep. One prisoner recounted how a guard "called us rednecks and waps, and told us if we made a move he'd kill us. I asked him what a wap was. He said I was a wap. He said foreigners were waps. I told him I was raised in this country the same as he was."[16]

On Friday, UMWA attorney Horace Hawkins negotiated a truce with the lieutenant governor. State troops would not move past Ludlow, would not try to retake Trinidad (which strikers controlled), and would not attack. Doyle wired all the leaders, ordering them to stop fighting but to keep their weapons. Jesse Welborn denounced the agreement as "nothing short of a conspiracy."[17]

That day, the strikers buried their Ludlow dead, except for Tikas, in a massive outdoor funeral held in Trinidad. "The manifestations of grief on the women's part was pathetic," reported the *Rocky Mountain News.* "But for the most part, the men watched the services grim-jawed and silent." The trucks bearing the caskets also carried "a large quantity of guns and ammunition." Among the mourners was the aging father of Charles Costa. As the caskets of his son and his son's family were being removed from the morgue to be taken to the church, Costa "raised his hand and vowed vengeance upon his son's slayers. His voice rose higher and higher, tinged with hysteria, until some person . . . caught him by the arm and quieted him."[18]

The truce lasted with some success through the weekend, after which intense fighting resumed. On Monday, 27 April, Doyle wrote in his diary: "Shooting in north is reported by phone 10:30 P.M. Refused to offer help and when asked for ammunition said 'You cannot get it from me.' "[19]

On Monday, the strikers buried Louis Tikas and two men killed the previous Wednesday. "Fumes of incense filled the chapel," a newspaper reported, "as the Greek burial ceremony progressed, and the mourning strikers responded in chant to the priest's solemn intonations." Outside, 578 armed coal miners representing a dozen nationalities waited to accompany Tikas to his grave. Music was omitted "in light of the intense feeling here." After Tikas was buried, fighting resumed, spreading to the northern lignite field, where "in the hills around Louisville 20,000 to 30,000 shots were fired."[20]

Meanwhile, hundreds of UMWA locals across the country were petitioning the national office to call a nationwide strike to protest the killings at Ludlow. In all the union's history, there had never been a better time. The sympathy of the nation was behind the miners. Money to finance a nationwide strike would have poured in. Under the circumstances, coal miners everywhere, union or not, would have laid down their tools. But in 1914 the national UMWA officers lacked the visionary boldness that taking such a dramatic step required.

Executive board minutes reveal that "hundreds" of locals had requested a general strike. A circular issued in response understated: "We are in receipt of a number of resolutions calling for a general strike on account of the situation in Colorado. . . ." After surveying the situation "from every angle," the board had concluded: "[I]t is not the better part of wisdom for the miners in the organized states to engage in a general strike at this time, believing that we can better aid our gallant brothers in Colorado by remaining at work, thus insuring in a financial way our fullest support to the men, women, and children who have been so long engaged in the great industrial struggle in Colorado."[21] With these words, the union passed up its last opportunity to emerge from the Colorado Fuel and Iron strike victorious.

So far, some fifty persons had died. (Still, this was fewer than the seventy-five dead in Colorado from underground accidents that year.) Matters finally reached the point that Woodrow Wilson overcame his reluctance to interfere with Colorado's state's rights. On 28 April, he ordered the United States Army into the Colorado coalfield. The troops arrived on May Day during a heavy rain. The coal war was over.

Immediately, the army ordered the Colorado National Guard out of the field and required both sides to relinquish their weapons. District 15 officials cooperated, urging strikers to put away their guns. Army troops followed instructions to treat strikers with courtesy: They were not searched, and many kept their weapons. But the fighting stopped. On 6 May, Doyle wired UMWA President John White that the strikers and the federal troops were on the best of terms.[23]

Still, the strike was not over. The outcome, as usual, depended on whether or not the companies could hire enough strikebreakers to bring production back up to normal. The army had orders not to escort out-of-state strikebreakers into the coal camps. Federal troops were to allow coal mines already operating to remain open, but to prevent those not operating from opening. This policy was short-lived. Colorado Fuel and Iron Company President Welborn later testified that strikebreakers came freely to Colorado from other states, and were protected by the army as they took jobs in the coal camps. He asserted that the policy of the army in regard to strikebreakers was "not materially different" than the policy of the Colorado National Guard had been. Doyle agreed, producing compelling evidence.[24]

On 18 May, an entourage of strikers' wives and supporters set out from Denver to publicize the union side of the story. Its leader was Denver Judge Ben Lindsey, the founder of the country's first juvenile court, who was familiar with the coal industry mainly through its numerous orphans.

The party, including Pearl Jolly, Mary Thomas, Margaret Dominiske, and Mary Petrucci, traveled to Chicago, Washington, and New York, speaking at rallies and giving interviews.[25]

Mary Petrucci, in the clutches of grief over the death of her children, broke down in midtour and had to return to Colorado. Before going home, she agreed to talk to a Washington reporter. The interview stands as a poignant document of the strike. It is a rare window into the worldview of a miner's wife—a fusion of class, female, and ethnic consciousness. "Perhaps it seems strange to you that I want to go home," she said. "But I do."

> My man is there and my children are buried there. . . . I have been so happy there. Why, there wasn't a happier woman anywhere than I was. . . . You see, I'm Italian, although I was born in this country, and our people are gay of heart. I used to sing around my work and playing with my babies. Well, I don't sing any more. And my husband doesn't laugh as he used to. I'm twenty-four years old and I suppose I'll live a long time, but I don't see how I can ever be happy again. But I try to be cheerful on account of my husband. It is so hard for him when he comes home from work to find only me in the house, and none of the children. . . . But you're not to think we could do any differently another time. We are working people—my husband and I—and we're stronger for the union than before the strike. . . . I can't have my babies back. But perhaps when everybody knows about them, something will be done to make the world a better place for all babies.[26]

This sort of comment could not reach L. M. Bowers. So fanatically did he oppose the union that he seemed untouched by the events of the year. Bowers denounced Judge Lindsey and the "ignorant women" who accompanied him. In his view, Pearl Jolly was a notorious prostitute who had been living with an equally notorious assassin, "Louie the Greek." Ben Lindsey, Bowers informed Rockefeller, Jr., was a "political demagogue and a religious hypocrite. . . ."[27]

Incredibly, considering all that had happened, Bowers insisted as late as the summer of 1914 that no disagreements had marred relations between the company and its employees since his management had assumed control in 1907. "The word 'satisfaction' could have been put over the entrance of every one of our mines."[28]

By summer, Colorado Governor Ammons had abandoned any pretense of independence from the coal operators. In June, he began allowing Ivy

Lee, a publicity agent employed by the Colorado Fuel and Iron Company, to compose his letters addressed to President Wilson. These compositions were also the work of Rockefeller, Jr., who wrote to Lee on 10 June: "Several points in my memorandum . . . could well . . . be used in the letter from Governor Ammons to President Wilson which you are proposing to prepare. . . ." The actual letter sent over Ammons' signature went through several drafts passed between Rockefeller, Jr., and Lee.[29]

During the summer of 1914, the federal government became as involved in the conflict as the state of Colorado had been. One federal investigator reported that the operators "become wild men when they begin to discuss the strike. They are a long way from sanity. . . . They fly into a rage, curse the federal government and froth at the mouth when investigations or investigators are mentioned."[30]

In June, two federal mediators who had arrived on the scene in the late spring, proposed a settlement. It would waive the right to strike for three years, did not raise wages, and did not provide for union recognition. For three years, there was to be no "picketing, parading, colonizing or mass campaigning by representatives of any labor organization of miners that are party to this truce which will interfere with the working operation of any mine." Operators were not to employ mine guards, "but this does not preclude the employment of the necessary watchmen." The mining and labor laws were to be enforced. All strikers not "found guilty of violation of the law" would be employed by their former employer. Each mine was to have a grievance committee elected by the miners, but its members had to be employed more than six months and more than half had to be married men. A three-man commission appointed by the president of the United States would arbitrate disputes that the grievance committees were unable to settle. The UMWA national office supported this proposal. The district officers violently objected.[31]

For the national office, the proposal represented a face-saving way to withdraw, but district officers were not prepared for defeat.[32] Doyle insisted that if the proposal was accepted the operators would have every advantage and the miners would "be compelled to endure conditions very similar to, if not the same or worse, as existed in the mines of Colorado previous to the strike." He rejected clauses that seemed advantageous to the strikers. The state mining laws were to be enforced, but no enforcement machinery was provided. The grievance committees would be ineffectual because the operators would use the six-month employment condition to insure that only strikebreakers served on them. The condition requiring that the majority on the grievance committees had to be married meant the committees would be filled with men who could least afford to displease their employers. The reemployment of any striker not guilty of violating any law was a hollow promise considering the operator's control of the legal apparatus in the southern coal counties. Operators could arrange to make any striker "guilty of violating a law."

Again district and national officers were feuding. According to Doyle's detailed notes, at one meeting the federal mediator opened the agenda with an argument that the Colorado strike was lost. "He emphatically stated that the production was so great there was no chance of winning the strike." District men disputed this conclusion. Later, UMWA Vice President Frank Hayes, John Lawson, and E. L. Doyle went over the same ground. Doyle recorded: "Other statements during the argument were 'I know there will be some who will holler no matter what kind of agreement we get.' 'We can't carry this thing on forever.' By God, John, we are up against a stone wall.' "

In August, Mother Jones returned to Denver with Frank Hayes to argue for the settlement. They told Doyle that the strike had to be settled in three weeks for the simple reason that the money was gone. Doyle reiterated his position. At this, Mother Jones told him not to expect too much.

On 6 September, district officers were shocked to find the hated proposal announced in newspapers as "President Wilson's proposal." That morning, Vice President Hayes walked into Doyle's office. "I asked him what he thought of the dirty double crossers . . . ," Doyle wrote. "Hayes replied that we couldn't help it now and would have to accept it since the President offered it." A week later, the UMWA executive board accepted the proposal, "subject of course to the approval of the miners in Colorado." A Special District 15 convention was called.[33] Before it convened, Doyle went to Omaha to talk to national officers. Although they claimed that the president's action had come to them out of the clear blue sky, one officer let it slip that he had discussed certain clauses with a mediator in Washington. Indeed, the mediator's reports indicate that the "Wilson Proposal" had been discussed at length with national officers. Doyle was correct to suspect that they were lying to the district.[34]

On 15 September 1914, the Special District 15 Convention met to consider President Wilson's proposal. Mother Jones set the tone of the convention: "Thank God that we have another great man, another Lincoln, in Washington today in our President." The federal mediator took up the same theme: "Are you going to throw cold water on the United States?" Against a move to bring the question up in a referendum of striking miners, Vice President Hayes sneered, "If you don't feel big enough or brave enough to decide this question then a motion is in order to refer it back to the rank and file." In the end, Doyle and Lawson also spoke for acceptance. Doyle said, "[W]e are up a tree and must get down the best way we can."[35]

The Wilson proposal to end the strike was a humiliating defeat for the strikers, with a few face-saving arrangements that would affect appearances but not the conditions of the mine workers. The Colorado operators wanted more. They turned the Wilson proposal down.[36]

Wholly without funds to continue, in late November the national executive board recommended the closing up of the strike.[37] The district convened once again. "I do not want to make a great speech in the matter," John Lawson told the delegates. "I don't want to tell all the things that are deep down in my heart concerning these affairs. I have worked in the Colorado mines for quite a few years. I know what you men struggled and rebelled against; I know the conditions you have to bear, poor wages and so forth; I know from my own personal experience. . . . I know what it is to suffer because of my activity in 1903 physical and mental pain. . . . It is not new to me." Thus Lawson, representing not the national but the district, urged the delegates to end the strike. He told them that as a member of the executive board he had not only voted in favor of the recommendation but had helped to draw it up, "because I knew the time had come when the organization without any more pretenses, without any more pomp, the time had come when the organization was no longer able to finance the strike." A voice interrupted: "That is what we want John Lawson, a clear statement."

A strong note of disillusionment with the American government accompanied the closing up of the Colorado Fuel and Iron strike. "The national government is the newest thing we've had to contend with," one executive board member confessed. Frank Hayes, who had pushed the Wilson proposal from the beginning, said, "Before this civil war had went very far the President of the United States was asked to intervene. He responded by sending federal troops. The war and strife then ended. We thought that the federal troops would not permit wholesale importation of scabs but again we were disappointed, and under the administration of federal troops these mines have been filling up from day to day until the present time they have a force at work to produce all the coal they can sell." John Lawson concluded that at Trinidad, "[W]e will find out if the government is in Washington or New York, and I feel that we have found that it is in New York yet."[38]

14

ORGANIZING CONSENT: THE RISE OF CORPORATE LIBERALISM

This is no local brawl in the foothills of the Rockies. The commanding generals are not here, the armies are not here—only the outposts. . . . It is not local and moreover it is not "western." You can not dismiss the bleeding here with that old bogus about the wild and woolly west.

Max Eastman, in The Masses, June 1914[1]

What can businessmen do to clean up the rot that these muckrakers and demagogues have dumped upon our door step?

L. M. Bowers, Leslie's Illustrated Weekly, 1914[2]

The corporations defeated the strikers by sheer superiority of resources. But once again, victory was not cheap. Two days before the attack on Ludlow, Bowers reported to Rockefeller, Jr., "We have already lost in nine months . . . $835,351.15, which will reach a million by the end of our fiscal year, and besides this an entire year's profits are lost."[3] Class conflict killed profits. Simple force could break a strike but could not provide the womb of social stability in which profits could grow.

Colorado Fuel and Iron directors were well positioned to understand the costs of class conflict, but the problem exceeded the boundaries of any one firm, and a new world situation lent a certain urgency to solving it. By the turn of the century, American capitalism had replaced that of Britain as the leading economic force in the world. Previously, the American West had provided room for expansion, but by 1900 corporations were turning to world markets, to the cheap labor supply and rich natural resources of Cuba, Puerto Rico, Mexico, China, and the Philippines. Continued economic expansion could not proceed under conditions of class war at home.

During the great coalfield war, Colorado operators won the battle on the ground but lost the propaganda war. Repeatedly, Colorado miners had put forward their view of the coal firms as undemocratic and un-American, seeing themselves as embodying American values. To them, Americanism meant democracy. "Colorado: America or Russia, Which?" a union broadside had asked, referring to the notorious czarist regime.[4]

Ludlow exploded the strikers' view of the company across the United States. From New York to Chicago, from Boston to San Francisco, thousands now perceived the Colorado Fuel and Iron Company and its owners as brutal and un-American. Many saw John D. Rockefeller as cartoonist O. E. Cesare portrayed him in *Harper's Weekly,* a vulturelike figure standing over the shambles of Ludlow. The caption read, "Success."[5]

On the other hand, businessmen and others saw the specter of revolution on the western horizon, chimera though it may have been. Even in armed rebellion, the strikers were not revolutionaries: They were asking for simple decency, not the overthrow of the government. Nevertheless, a Trinidad banker who asserted that a revolution extending to all corners of the nation had only narrowly been averted was expressing a common opinion.[6]

Ludlow added points to anticorporate arguments, socialist or otherwise, that were freely and frequently discussed in intellectual and working-class circles throughout the country. And American socialism was a force to be reckoned with. In 1912, the Socialist Party itself could claim some eighteen thousand members. It had elected twelve hundred public officials to office, including the mayors of seventy-three towns and cities. By 1913, the combined circulation of 323 socialist newspapers reached over two million. One of them, the *Appeal to Reason,* circulated to more than 750,000 subscribers. It was one of the most widely read weekly newspapers in the world.[7]

Socialism was a significant political current in American life, and the socialists pounced on Ludlow as a perfect example of corporate greed. Castigations reached artistic heights in *The Masses.* Editor Max Eastman took the train west to write "Class War in Colorado" and "The Nice

People of Trinidad." He had tea with the leading ladies of Trinidad (all connected with the coal firms) and reproduced the ensuing conversation with devastating wit. One woman called the strikers "nothing but cattle and the only way is to kill them off." Another insisted that the strikers had sealed the women and children into their Ludlow death trap themselves in order to reduce expenses to the union. "The subject of the native iniquity of every person not born on American soil," Eastman wrote, "was . . . tossed from chair to chair for the space of about an hour."[8]

Yet socialists like Eastman represented only the furthest left of widespread anticorporate opinions. Adding its voice to the general criticism was a group of writers and investigative reporters who came to be known as the muckrakers. The muckrakers, including writers like Upton Sinclair and visual artists like John Sloan, lit up the darker corners of American society to expose poverty, political corruption, and the varied evils of unrestrained power. Their politics ranged from socialist to a moderate reformist bent, but their work did nothing to improve the corporate image.[9]

Even before Ludlow, the muckrakers had sharpened their claws on the image of the Colorado Fuel and Iron Company. A writer in *Pearson's Magazine* described the 1910 Starkville disaster under the title "How Coal Owners Sacrifice Coal Workers." "Officially these men were killed by an explosion of coal dust," he wrote, "but they were really killed by greed."[10] In 1912, a writer in *The Survey* deplored the atrocious conditions in the Colorado Fuel and Iron Company's Pueblo steel mill. Theodore Roosevelt echoed these sentiments in a *Century* article that referred to conditions at the mill as "appalling."[11] None of this was lost on L. M. Bowers, who asked in *Leslie's Illustrated Weekly*, "What can businessmen do to clean up the rot that these muckrakers and demagogues have dumped upon our door steps?"[12]

The labor and socialist press had been publishing anticorporate sentiments for decades. But the muckrakers were different because they linked working-class concerns to a middle-class audience. Many of these writers were themselves middle class in origin and outlook. They could sympathize with the miners and publicize their plight, but they could fall quite short of communicating the miners' experience to the outside world. Colorado State Senator Helen Ring Robinson, an accomplished journalist and a strike sympathizer, could write (in the *Independent*): "There was also the demand for a check-weighman of their own—a bitter subject among the miners who believe to the depths of their dumb minds that the 'Rockyfeller fellers' are sneak thieves who rob them of the results of their digging by crediting them always with underweights. A monstrous charge, no doubt, but quite natural under the circumstances which have been described."[13]

Ludlow elevated the policies of the Colorado Fuel and Iron Company to the center of national debate. Day after day, newspapers like the Chicago *Tribune* and the New York *World* carried the news on the front page. In the spring of 1914, the *New York Times* ran so much news on Colorado events that by itself the index of articles for three months runs to six pages of small print. In May 1914, there was hardly a newspaper or magazine that did not feature the Colorado Fuel and Iron Company and its faults or virtues. Even the *New York Times,* generally a procompany paper, editorialized that at Ludlow "somebody blundered."[14] The *Wall Street Journal* maintained several days of silence before observing that "a reign of terror" existed in southern Colorado.[15] Throughout the country, procompany editorials countered anticompany editorials. Two views of the company or, more broadly speaking, two views of industrial capitalism and its chief institution, the large corporation, were competing for the allegiance of the American people.

THE RISE OF PUBLIC RELATIONS

On the side of the corporations, at least one Rockefeller executive understood the seriousness of the situation. In late May 1914, Jerome Greene wrote to John D. Rockefeller, Jr.:

> . . . [W]e must consider afresh the whole situation looked at as a great social and economic problem. . . . It sounds grandiloquent, but it is perhaps not overstating it to say that we must think first of all of saving the country from a great danger. That we must beat the unions in the wicked game they are playing goes without saying. There are two ways of doing it. We may stand by our present position—stand pat and wait for somebody else to move. . . . The other way is to work through public opinion until the pressure becomes so strong that the United Mine Workers' Union will have to slink off and acknowledge itself beaten.[16]

This private expression of opinion foreshadowed a momentous change. The giant corporations had reorganized industry, and they had organized finance. Now, as the events of 1914 demonstrated, they must organize consent.

The man to mastermind the job was the pioneer of corporate public relations: Ivy Ledbetter Lee (1877–1934). By late spring 1914, Lee had already broken the ground of public relations in the employ of the Union Pacific and Pennsylvania Railroads.[17] His largest opportunity was yet to come.

Before Ludlow, the Colorado Fuel and Iron Company had eschewed public relations, with the exception of the occasional purchase of a Colorado newspaper publisher.[18] L. M. Bowers, for one, did not care at all for the public's opinion, telling a major stockholder in 1913: "[W]e flatly refuse to furnish financial, commercial or any other papers with data about our business affairs."[19] Nevertheless, in May 1914 Ivy Lee hastened to New York in response to an invitation from John D. Rockefeller, Jr. The younger Rockefeller had concluded that he and his father were "much misunderstood by the press and the people of this country."[20] Shortly thereafter, Lee set to work refurbishing the image of the Fuel and Iron Company and of the men who controlled it.

In August 1914, Lee toured the Colorado coalfields because the first task, as he saw it, was to understand the characteristics of the audiences to be convinced. Here we have the beginnings of modern market research, as well as a definite ideological shift. For the first time, the public's opinion mattered.

Lee found several audiences; one comprised company employees, who were strikebreakers to the man. The campaign seeking to influence them would thank them for their loyalty during the strike, assure them that the strike was broken, and "tell of the desire of the company that every man shall be treated fairly and inviting him to send his complaints. . . ." Lee shrewdly observed that leaflets and posters should be placed in the homes of the miners "where the women will see them. Women are voters in this state and their influence is important in every way."

A second important audience comprised the general public in Colorado. Lee discovered that although Coloradans generally felt hostile toward the companies, they were not at all pro-union, feeling, rather, extremely patriotic about their state. Accordingly, the propaganda strategy to win them over would show how the Rockefellers had built up the state rather than milking it dry.[21]

The centerpiece of Lee's campaign was a series of pamphlets entitled *The Struggle in Colorado for Industrial Freedom.* These bulletins were similar to the advertisements coal operators had been running in regional newspapers throughout the strike: They simply maligned UMWA while defending the company. A typical issue contained an operator's letter justifying their position to President Woodrow Wilson. Another claimed that union organizers earned fantastic salaries while strike relief was meager. This particular issue embarrassed Lee intensely when it was exposed as an outright lie perpetrated by the operators. Lee's biographer insists upon Lee's quest for truth, but in fact he never questioned any of the operator's opinions that he reproduced. His bulletins contained standard pap from the operators' side; his truth was, as one federal investigator described it, "the truth as the man you were serving saw it."[22]

Lee's propaganda, unlike the paid advertisements, carried no indication of its source. It was distributed nationally and repeated in friendly newspapers and from pulpits and podiums across the country. His campaign received the helping hand of all manner of prominent persons. Charles Eliot, president emeritus of Harvard University (and a trustee of the Rockefeller Foundation), wrote to Rockefeller Jr. to suggest that the bulletins be sent "to all professors of Economics, Government, and Sociology in the American Colleges and Universities. . . . Presidents of Colleges and Universities might also be included to advantage."[23]

The publicity campaign brought the morals of Mother Jones back into the fray. In the summer of 1914, Colorado Congressman George Kindel reiterated the old Polly Pry allegations into the *Congressional Record*. A Rockefeller executive confided to a supporter that this disclosure of the "disreputable and criminal career" of Mother Jones was the best achievement so far in the way of publicity. "We can now quote chapter and verse in referring to the true character of that energetic person."[24]

Under Lee's guidance, by March 1915 the Colorado Fuel and Iron Company had spent nineteen thousand dollars on publicity, excluding his own salary. This was almost as much as the company had spent on guns the previous year. By Number 15, the bulletin had reached a printing of 37,000 copies. A wide range of readers received it, from the six hundred subscribers of *The Survey,* a magazine for social workers, to every clergyman in Colorado.[25]

The Rockefellers rewarded Lee with a full-time position the following year. Thereafter, he grounded his career in Rockefeller companies, occasionally going outside to consult for other major interests, contributing "Betty Crocker" to General Mills and "Breakfast of Champions" to the Wheaties box.[26] His career ended, however, in an ignominious but revealing fashion. In the late 1920s, the Rockefeller-owned Standard Oil of New Jersey formed a cartel with the German petrochemical monopoly I. G.

Farben, and sent Lee to Germany to polish the image of the Nazi firm. In 1934, before he died that year of cancer, Lee was exposed (by the Special House Committee on Un-American Activities) as a press agent for Adolf Hitler, work he had undertaken for the Standard. He died in disgrace, while the Rockefeller firm continued its relationship with I. G. Farben up to and through the time it was operating with slave labor.[27]

Ivy Lee pioneered corporate public relations; his own career illustrated their ambiguous character. But in 1914 publicity, however sophisticated, was insufficient to stem the tide of the new journalism. One problem was that the mass-circulation muckraking magazines virtually monopolized the country's best writers. Readers did not want the ruminations of reactionaries, they wanted the muckrakers. So the large firms turned to nonverbal methods of combat. In *Leslie's Illustrated Weekly*, Bowers suggested the use of financial clout, particularly withdrawing advertising.[28] Using this and other methods, "[t]he destruction of the magazines," as Louis Filler writes, "was deliberately planned and accomplished in short order."[29]

Corporations bought the magazines in order to control the editorial policy. The West Virginia Pulp and Paper Company acquired the pioneer of the lot: *McClure's* promptly lost its fire. The Butterick Pattern Company purchased *Everybody's* for three million dollars. Its reforming zeal fizzled immediately. A banker called in the loans of a financially successful muckraking magazine called *Success*, and the publication folded. *Hampton's* went down on the same boat. Bankers told the editors of *Colliers* to change the editorial policies. The distributor of *The Arena* informed the publisher that hereafter newsdealers could not return unsold copies. An advertising boycott of the magazines also took a heavy toll.[30]

THE RISE OF THE COMPANY UNION

Public relations reflected the new idea that a firm's social and political setting could contribute to or detract from its power, that, like a government, it must seek legitimacy in the eyes of the people. Lee, with his constant attention to audience, was one of the first to understand this. It is not surprising, then, to find him suggesting real, if extremely moderate, concessions to employees. "It is of the greatest importance," he wrote to Rockefeller, Jr., in August 1914, "that as early as possible some comprehensive plan be devised to provide machinery to redress grievances. Such provision would not only take the wind out of the union's sails, but would appeal, I am confident, to the soundest public opinion."[31] The goal—busting the union—was the same, but the method was undergoing a subtle change.

According to Lee, the miners lacked a safety valve to get petty grievances out of their systems. The mine superintendents and pit bosses had "all the faults of their kind"; the company had no way of knowing whether or not they were carrying out company policies; in practice, the men had no appeal from the decisions of the pit boss; and there was virtually no record of complaints, which demonstrated to Lee that miners were afraid to complain.[32] Give workers some sense that they had a voice, he seemed to be saying.

If corporate liberalism had a constitution, Ivy Lee would have to be considered one of its framers. A coframer, then, was W. L. MacKenzie King (1874–1950), a man John D. Rockefeller, Jr., brought into the thought process at about the same time. King would go down in history not for his role in the American industrial system but as Canada's Prime Minister for much of three decades (beginning in the 1920s).[33] But in the spring of 1914 the forty-year-old labor expert's first dazzling political career lay in the ruins of the recent fall of the Liberal Party. Despondent and near bankruptcy, King was ripe for opportunity when it came in the form of an invitation to meet with John D. Rockefeller, Jr. The two men took an immediate liking to each other; shortly thereafter, King became a Rockefeller employee.[34]

King's contribution to corporate liberalism was an influential form of the company union known as the Industrial Representation Plan. In early August 1914, he suggested that relations between employers and employed could be eased through a system of "easy and constant conference" between the two, "with reference to matters of concern to both." The proposal lay somewhere between "the extreme of individual agreements" on the one side and union recognition on the other. Here was Ivy Lee's safety valve. But now King came to the crux of the matter: the cost to the company. "Granting an acceptance of the principle outlined," he wrote to Rockefeller, Jr. "the machinery to be devised should aim primarily at securing a maximum of publicity with a minimum of interference in all that pertains to conditions of employment." At this stage of liberalism's development, the greatest change would occur in the arena of appearances. Coal miners' conditions would remain the same.[35]

But King's plan ran a foul of Lamont Montgomery Bowers, a man of principle if ever there was one. Bowers found the plan unwise because, in the first place, other operators would balk and, in the second place, "the socialistic papers would charge us with dodging and hiding behind this eleventh hour scheme to save our faces."[36] Jesse Welborn concurred that the plan looked like an admission of weakness.[37] Later, Welborn would change his mind.

THE ROLE OF THE FEDERAL GOVERNMENT

Whether or not the company union was a form of public relations, the corporate regard for the opinions of the public signaled the stirrings of a new outlook, one that saw a stable social environment in which to conduct business as a high priority. In this, the federal government became an important ally of the developing liberal wing of the capitalist class. As Gabriel Kolko has shown, Theodore Roosevelt, Woodrow Wilson, and other liberals once widely perceived as trustbusters were instead engaged in hammering out a social and political environment in which the large firm could thrive. In 1912, Woodrow Wilson expressed this quite explicitly: "[N]obody can fail to see that modern business is going to be done by corporations. The old time of individual competition is probably gone by. . . . We will do business henceforth . . . on a great and successful scale, by means of corporations."[38] "*We* will do business," he had said. In his mind, the national interest and that of the large-scale firm were identical. It was for the sake of the stable social environment business required that Wilson opposed individuals like L. M. Bowers and firms like the Colorado Fuel and Iron Company.

Sincere, issue-oriented reformers filled the ranks of the reform movements of the time, but the reforms themselves were indelibly marked by the fact that it was the liberal capitalists and their allies in government whose voices were heard when it came time for enactment. The founding of the United States Bureau of Mines in 1910 was a case in point. What miners needed was good and strictly enforced federal safety legislation. What they got was the Bureau of Mines. As William Graebner has persuasively argued, coal operators were instrumental in lobbying for legislation to establish the bureau. Once established, it did not threaten their interests (or increase their costs). The bureau gathered and published scientific information on coal mine explosions, and on other subjects of interest to the industry, but it lacked the power to enforce safer practices.[39]

But the Bureau of Mines represented the *idea* of a federal regulatory umbrella, even if in practice it did not regulate very much. It illustrated a change in the intellectual climate, a relatively new belief that government might legitimately constrain industry in certain areas. Another such idea emerged in the Industrial Relations Commission, established in the summer of 1912. The commission embodied the notion that the federal government had a legitimate role to play in the relations between capital and labor.[40]

The idea for the commission originated with a group of reformers including Jane Addams and Rabbi Stephen S. Wise. In 1911, they had petitioned President William Howard Taft to establish a government body with "wide powers of investigation into trade unions, trade associations and the economic and social costs of strikes." Among the reasons the petitioners put forward were the "profound restlessness" among large groups of workers, and the danger of a "larger lawlessness" that could extend "beyond the purview of the criminal courts." "A house divided against itself cannot stand," they warned. "We have to solve the problems of democracy in its industrial relationships and solve them along democratic lines."[41] They, too, feared the possibility of revolution.

In his 1912 State of the Union address, President Taft urged the formation of an Industrial Relations Commission; Congress passed the appropriate legislation that summer. It fell to President Woodrow Wilson to appoint the members, who were to represent, in National Civic Federation fashion, business, labor, and "the public."[42]

Headed by Frank Walsh, a Kansas City attorney and reformer, the commission began to investigate the Colorado Fuel and Iron strike in January 1915. In its third week, the highly publicized investigation moved from Denver to New York to put Rockefeller, Jr., on the stand. Rockefeller's testimony became the high point of the proceedings, and it ranks as the public relations coup of the decade. It was a tissue of lies, as Walsh later demonstrated through subpoenaed company correspondence. Rockefeller claimed that he had had little contact with the officers of the company during the strike, when in fact he had had constant contact. He claimed (incredibly) that he had always favored collective bargaining with labor unions. He claimed ignorance of details large and small pertaining to the Colorado Fuel and Iron Company. He did so pleasantly, firmly, regretfully.[43]

At one point during the sessions, he shook hands with Mother Jones and confided, "I wish you would come to my office and tell me what you know of the Colorado situation." Mother Jones, who was nothing if not

John D. Rockefeller Jr.,
1910.

Courtesy of the Library of Congress, Biographical File.

vain, succumbed. "Well that's nice of you," newspapers across the country reported that she replied. "I've always said you could never know what those hirelings out there were doing." The following day, she visited the younger Rockefeller in his office and subsequently announced to the world that he was a "much misunderstood young man." She retracted her statements under pressure from a delegation of labor and socialist leaders, but not before they had done their work of polishing the tarnished Rockefeller image.[44]

Despite his own and Mother Jones's assertions to the contrary, John D. Rockefeller, Jr., was the real power in the company. This was a fact keenly felt by Bowers in January 1915. Bowers' fanaticism had finally backed him into the corner of opposition to those who held the controlling interest. In late December 1914, the month the UMWA called off the strike, President Welborn announced an embryo of the Representation Plan to employees.[45] Concurrently, Rockefeller asked Bowers to resign all positions with the company except for membership on the board of directors. He invited him to take an extended vacation, after which he was to work more closely with the New York office in an advisory capacity.

Initially, Bowers felt pleased, thinking the change a sort of promotion. But when two weeks later Rockefeller, Jr. requested his resignation from the board of directors in order to make room on it for Ivy Lee, Bowers began to suspect a plot. "[L]et me say," he wrote to Rockefeller, "that I am not one of the old fossils who want to hold fast to such positions after my usefulness has passed. That there is some underlying reason for this move in asking my retirement from every position . . . after so many years of activity, is hardly to be doubted." Bowers told Rockefeller that he was unwilling to resign under criticism unless he was told what the criticism was and by whom inspired. Rockefeller answered evasively. Bowers correctly perceived that he was being eased out, being put, in his words, into cold storage because of his opposition to the scheme "of playing the soft pedal to suit the labor organizations." His bitterness was great.[46]

That John D. Rockefeller, Jr., was in the driver's seat was clear enough from inside the company. Only the public was in the dark. This was soon to change. For the three months following Rockefeller's testimony in January, Frank Walsh and his assistants quietly collected company correspondence that revealed his close familiarity with events in Colorado, his wholehearted approval of company policies as carried out by his managers, and even his work in helping to compose letters from Governor Ammons to the president of the United States. On 20 May 1915 Rockefeller, Jr., was recalled to the witness stand. Walsh cross-examined him aggressively and with great skill, exposing the fact that he had lied his way through his entire testimony earlier that year.[47]

The Industrial Relations Commission's final recommendations strongly resembled the New Deal, still twenty years away. But because they were not enacted, the commission's work resulted in no concrete reforms. Nevertheless, the investigation functioned to reassure Americans, including muckrakers, unionists, intellectuals of several political persuasions, and many socialists, that their government was the caretaker of all the people, not the strong arm of one class. The Industrial Relations Commission did more, as James Weinstein writes, "to conciliate workers and radicals [to the status quo] than any previous presidentially appointed body. . . ."[48] Chairman Frank Walsh, whose experience as a Kansas trial lawyer enabled him to skillfully grill the capitalists brought to the witness stand, was in large part responsible for the soothing effect. As Commissioner Mrs. J. Borden Harriman reflected, "For the first time labor felt that a branch of the Government was giving them a real say and a square deal. Walsh made them feel that way. . . ."[49] Although Walsh was a committed reformer who had intended to do more than smooth society's ruffled feathers, he was essentially powerless.

THE COMPANY UNION IN PRACTICE

The setback of John D. Rockefeller, Jr., on the witness stand did not discourage him from pursuing his new direction in industrial relations. In September 1915, he traveled to Colorado to discuss his Representation Plan with company employees, and to announce it to the world. Newspapers across the country chronicled his two-week visit. "He donned overalls and a jumper," one paper reported, "and trudged through two or three miles of narrow tunnels. . . . In one of the 'rooms' Rockefeller borrowed a pick and chopped away lustily until chunks of coal came rattling down to the floor, greatly to his delight."[50]

One day, he stood before a crowd and held up a small table to illustrate his new concept that capital and labor were partners. The four legs, he explained, each necessary to the others, represented stockholders, directors, officers, and employees. Then he put coins on the table to illustrate the idea that if one leg reached up to take more than its share, the coins would fall off. (He tilted the table, and they did.) Only if the four legs stood equally would the table remain level. Then he told his audience that for the past fourteen years the common stockholders had made no money whatever on the Colorado Fuel and Iron Company. This fit nicely with the partnership between capital and labor concept. Rockefeller did not mention that it was preferred stockholders like himself who had reaped handsome profits.[51]

But the point was not whether Rockefeller told the truth but that he was charming. He danced with miners' wives and bounced their children on his knees. In general, he impressed the nation as a nice fellow.[52]

On 2 October, he spoke to a large meeting in Pueblo to discuss the Representation Plan with employees.[53] The soul of the Plan was the grievance committee made up of representatives of workers and management. Discussions between the two sides were supposed to ease discontent on the workers' side. If a miner's grievance could not be solved in this manner, he had three levels of appeal: The first was the company president; the second was a committee above the president (also comprising workers and managers); and the last was the newly formed Colorado Industrial Commission. But there was a catch. The Colorado Industrial

Commission could not be resorted to except by permission of the grievance committee preceding it on the ladder of appeals. In turn, this body
could not muster a majority vote without company consent. In short,
power remained securely in company hands.

The first reaction to Rockefeller's talk (according to verbatim minutes
of the meeting) came from one Patrick from the Fremont mine. He rose
from his chair. "I think that the plan is great, and there can't be anything
added to it," he said. "I think you have covered everything." Next, a
man from the Walsen mine spoke up. "I have listened to your explanation. . . . I do not see any reason why capital and labor can't come
together if they follow the plan. . . . It looks very bright to me." Another
miner chimed in, "It looks like a square deal." A Sopris miner assured his
listeners that he was not a union sympathizer before proposing that the
outside men be granted an eight-hour day (which underground workers
had). In response, Rockefeller asked E. H. Weitzel, manager of the Fuel
Department, to explain why this was not possible. Weitzel did so at some
length. Following Weitzel's talk, Rockefeller repeatedly urged the delegates to withdraw and consider the plan among themselves, but the men
voted to remain and to continue the discussion with everyone present.[54]
These miners were in no mood to discuss their grievances among themselves. They were looking for ways to please the company.

Local managers, on the other hand, greeted the plan with pure anguish.
Especially distressing to them was the provision that miners could join a
labor union or not, as they chose. At first, apparently, superintendents
continued barring organizers from the coal camps. But in early 1916,
Rockefeller, Jr. received a visit from an organizer who asked whether or
not he and other organizers could visit the camps, since company publicity
now claimed they were open. "Were I to have said that I objected,"
Rockefeller explained to manager Weitzel, "organized labor would at
once have a platform upon which to publicly berate the Colorado Fuel
and Iron Company and me for hypocrisy. . . ."[55] The new rhetoric was
creating its own pressure to institute new practices.

Weitzel could not accept the idea of organizers openly coming into the
camps. "I have time and again told our superintendents . . . that they need
never fear a weakening on the part of the company," he explained to
Welborn, "that we had spent too much to keep out of the clutches of
these grafters [union organizers] to ever consider a back down." Welborn
instructed Weitzel to announce the new policy to superintendents. Weit-

zel responded that he "simply couldn't have the heart to do it, after what they have all gone through":

> . . . [I]f it has to be done, please write exactly what you want me to say and I will take it around and read it to them without comment; of all the discouraging and disheartening things that have occurred since September 1913, nothing has made me feel as this has, and I sincerely hope you can dissuade Mr. R from forcing this humiliation on the people who have stood by us so faithfully and whose opposition to the U.M.W.A. has grown to be a religion with them, men who have never considered their personal convenience and safety nor that of their families a sufficient reason for relaxing or avoiding their duty as they saw it through your instructions and mine, during the dark and stormy days and nights of 13 and 14. . . .[56]

Rockefeller held firm. "[E]ither the camps of the CF&I are open camps, which means that any individual may visit them . . . or they are closed camps, from which the company exercises its right to exclude certain individuals. . . . [I]t has been heralded broadly over the land that the CF&I's camps were open camps, and the provisions of the Plan make this perfectly clear."[57]

It was easy enough for Rockefeller to take this position now, but it was more difficult for local managers who had spent years at a high personal cost excluding organizers from the camps. These were lieutenants being asked to throw down their weapons in the midst of a war. Weitzel's anxiety over the new policy prompted several more letters. "Bert [another manager] and I are more disturbed each day," he wrote to Welborn, "and I can not concentrate my attention on any one thing for 5 consecutive minutes without thinking about this matter. . . ."[58]

His chagrin moved him to articulate with great candor the late company philosophy in regard to the working people. The mere fact of allowing organizers into the camps would be seen by miners as company approval of the union, since heretofore no one and nothing had entered the camps without company approval. "You may wonder why I think our men would be quick to join the union," he elaborated to Welborn. "They constantly try to adjust themselves to the policy of the company and the admission of organizers would be all that they would need as evidence that we wanted them to join. With the smooth talk of the organizer and the strong talking points they would have, few men would need a second urging to join."[59]

"I think you have not known how much reason we have given our men to expect us to direct them in every phase of their lives," he wrote to Rockefeller, Jr. "A majority of our people have had little business experience and are almost as easily persuaded as children, especially the foreigner when approached by one of their own tongue who has some education and is a fluent talker." To illustrate the childlike qualities of the miners, Weitzel explained that when the Slavic organizer Mike Livoda came to Walsenburg the Slavs from nearby camps stayed religiously away from Walsenburg. They were afraid to trust themselves to his influence. Weitzel believed they did not want to join the union but "could not cope with Mike in argument." He felt that without company opposition the UMWA could enroll the entire work force in a few months. Under union domination, "they would not be our men," and as a result the Representation plan would fail. He conceded that his views would undoubtedly be considered paternalistic, but he defended them on the grounds that it was the company's duty to protect the employees against exploitation by labor unions.[60]

"Paternalism is antagonistic to democracy," Rockefeller chided Weitzel in a lengthy response.[61] Yet Rockefeller's Plan hardly illustrated the workings of a democracy. In 1917, a friendly but critical article in *The Survey* stated: "In addition to the undemocratic origin of the plan, it is . . . undemocratic in its most essential provisions all the way through. The plan is the company's. Not a single act is done under it, not even the election of representatives in the camps, except under the direction of the president of the company." The writer had interviewed miners and found among them a low interest in the grievance procedure and a distinct aversion to criticizing the Plan or even talking about it inside the coal camps.[62]

Since the purpose of the Plan was to defeat the union, it is not amazing that the UMWA opposed it. E. L. Doyle believed that the miners regarded it as a joke. "[T]hey laugh at the promises made under it and attend meetings in the same way they would go to any gathering out of curiosity."[63] In the early 1920s, the UMWA put the Plan on its list of organizations considered "dual organizations" created for "the express purpose of destroying our Union." The union stipulated that membership on a company grievance committee was grounds for expulsion.[64]

Nevertheless, the Plan did introduce improvements. For the first time, organizers were permitted into the camps (at least until 1919, when the UMWA again became a real threat). Miners had some recourse from arbitrary rulings by pit bosses. Houses in the camps were repaired, the camps cleaned up, and trees, grass, and gardens planted: This may have been the most important change. In the first five years, employees received several raises, following closely behind wage agreements negotiated by the UMWA in the East.[65]

During World War I and after, Rockefeller's Industrial Representation Plan became an influential form of company union across the United States. The government itself through the War Labor Board ordered at least 125 companies to install some form of it in their plants. In the 1920s, major firms like International Harvester, Standard Oil of New Jersey, and Goodyear Tire and Rubber adopted it.[66]

For the Colorado Fuel and Iron Company, improvements introduced by the Plan were insufficient to squelch union sympathies among the miners. In October 1919, the UMWA issued its first postwar strike call for higher wages. In answer, ninety-one percent of the mine workers employed by the Colorado Fuel and Iron Company failed to appear at their jobs. More than half stayed out for the full six weeks of the strike. The company's response showed that the new liberalism was a tool that could be dropped on a moment's notice. Once again, organizers were barred from the camps; once again, the most active union members were fired. Further, in this new World War I era the company had discovered a new weapon. Early in the strike, the company posted on coal camp bulletin boards a lengthy statement opposing the strike. The conclusion: "Those who are disloyal to our country or the company, or who engage in efforts to disturb harmonious relations within the company, will neither be retained in our service, nor allowed on our property."[67]

Americanism, in less than five years, had changed sides. Castigated frequently in the past as un-American, the Colorado Fuel and Iron Company had successfully fused the idea of company loyalty to that of loyalty to America. It was potent anti-union ammunition. In 1920, a list of offenses for which an employee could be dismissed without further notice included "[t]alking or spreading propaganda disloyal to the United States or to the Company."[68]

Worse than the spread of UMWA influence, from the company's point of view, was the growing attraction of Colorado miners to the revolutionary Industrial Workers of the World (IWW).[69] In 1920, a group of IWW miners issued an open letter to John D. Rockefeller, Jr.: "For five years your scheme of Industrial Democracy has functioned without interference from any of the Labor Unions. . . . As a result of that we miners

Sopris mine family outside their shack, 1915. Photo courtesy of the Rockefeller Archive Center.

John D. Rockefeller Jr., center, and E.H. Weitzel, right, on a tour of a Colorado Fuel & Iron Company mine, 1915. Photo courtesy of the Rockefeller Archive Center.

of Colorado find that our wages have been cut. . . . When our wages were cut we were told that we should have more steady work, but the fact today is, that we work less days a year than when we had Industrial strife." They accused the company of short-weighing, and concluded: "If that is a result of your principles of Baptist Christianity, please let us have some heathen to rule us for a while. Or is your religion only for Sundays and not injected into the business of weekdays? We feel that you will be willing, as Tolstoy said, to do anything for us but to get off our backs."[70]

In 1921, the Colorado Fuel and Iron Company pressured employees to sign petitions requesting a twenty-percent wage cut. Some workers signed, but others struck instead, forcing the company to rescind the reduction. But in 1925 the company again dealt a twenty-percent cut. In August, managers circulated a petition in which employees were to request a further eleven-percent cut. At one mine, sixteen miners refused, and they were discharged. The company closed down two mines where every miner refused to sign.[71]

The Representation Plan represented no real compromise with labor, and therefore it provided no real salve to class conflict. In 1927, a major strike again gripped the Colorado coalfields. This time, it was led not by the UMWA but by the revolutionary IWW.

For the companies, company unions worked as long as labor was in decline, or defeated. But with the depression of the 1930s and the mass strikes and renewed labor militancy that came with it, they became as irrelevant to the companies as they probably had been to most workers. With a newly strong and quite radical labor movement again on the rise, conservative unionism began to look good to many employers.

THE RISE OF JOHN L. LEWIS

Liberal capitalists had their counterpart in the labor movement. John Mitchell, president of the UMWA from 1899 to 1908, was one of labor's pioneers in the move away from an anticorporate critique of society. He had gained prominence among coal miners through his organizing successes in the 1890s and through personal charisma, but these could not carry him through mounting opposition after the turn of the century. Even at the height of his influence, he and his supporters had coexisted with socialists, militants, and radicals of all stripes. By 1908, his days as president were plainly numbered. Accordingly, he left the UMWA to take a job at twice the salary with the National Civic Federation, that pioneering institution of liberalism. But coal miners did not approve. Mother Jones expressed the view of the majority when she asked the delegates to the 1910 convention, "What good is that Federation? . . . Where does the money come from that runs it? It comes from Morgan, Belmont, Harriman and old Oily John."[72]

Although the National Civic Federation was supposed to represent capital, labor, and the public, miners thought it served mainly capital. In 1909, they forced Mitchell to choose between his union membership (which he had retained), and his National Civic Federation job. Because the job was based on his standing in the labor movement, he had little choice but to comply by resigning from the Civic Federation.[73]

But John Mitchell had a spiritual heir, a man destined to tower over the American labor movement for fifty years to come. John L. Lewis (1880–1969) was shrewder and less ethical than Mitchell, a man emotionally and intellectually capable of a calculated and ruthless climb to power, a man who probably enjoyed presiding over the demise of his opponents. His outlook was highly similar to Mitchell's. Organized labor would not take its place as a partner with capital until the 1930s, but Lewis arranged for the UMWA to be ready when the time came. He marched in step with the times and stamped its soon-to-be dominant outlook on much of the labor movement. Within the UMWA, he eradicated each and every dissenting view.[74]

Lewis became acting president of the union in July 1919, his husky frame and forceful personality filling a vacuum created by a weak organizational center. His path to power was cleared by patronage and littered with election fraud. He was also a competent speaker, strategist, and organizer who was dedicated to his own rise, renouncing in the process all but the trappings of a personal life.

The son of a Welsh coal miner, he worked in his youth as a miner and a farm laborer. In 1908, Lewis and his wife, accompanied by Lewis's family, moved from Iowa to Illinois, the union's central pillar. By 1909, John L. Lewis was the president of the important Panama, Illinois, UMWA local; his father and brothers had attained positions of power in the union and in the town. As Melvyn Dubofsky and Warren Van Tine write, the local became a Lewis family political machine that utilized graft, embezzlement, patronage, and election fraud to insinuate family members into positions of local, and then state, power.[75]

Lewis's increasing visibility in Illinois led American Federation of Labor (AFL) President Samuel Gompers to appoint him in 1911 as an

AFL organizer. This position, which Lewis held until 1917, became his springboard to power within the UMWA, which was the largest and most important union within the AFL. His work involved making extensive contacts in the union movement and in politics. He also involved himself secretly but quite heavily in UMWA politics. In 1912, for instance, he composed pseudonymous broadsides against his enemies in the union. He maintained wide contacts with coal miners and their leaders, aligning himself with President John P. White, who was to raise the battle cry against socialists, militants, and other malcontents who formed a large bloc in the union.[76]

It was John P. White, not John L. Lewis, who began to eradicate dissent. But as early as 1912, Lewis was acting as White's silent partner. White's project extended over several years, and it was completed by Lewis. The purpose was to concentrate power absolutely in the union's national office. Under Lewis, this came to mean the consolidation of power absolutely in one man.

The means to the end included methods both legitimate and foul. President John P. White fought for and won the policy of holding the annual national convention only once every other year. Costly as it was, the convention was the union's only machinery for democratic decision making involving the entire paid-up membership.[77]

In 1915, John Walker, president of the powerful Illinois district, challenged White for the presidency of the union. White fought back. Lewis came to his aid by composing a set of fake telegrams from John Walker to coal operators, allegedly proving that Walker had appealed to the operators for funds. This "evidence" of class collaboration destroyed Walker's campaign.[78] In December 1916, massive fraud characterized district and national elections. A defeated candidate from Illinois reported that in a number of places "they had one or more men marking ballots all day long. . . . I am satisfied they marked the ballots of at least 8000 of our foreign speaking miners."[79]

In late 1915, Gompers assigned John L. Lewis to act as White's assistant at the January 1916 UMWA convention. His job was to help White quash dissenters. (Lewis was legally present in the convention as a delegate from Illinois.) According to Dubofsky and Van Tine, Lewis proved a formidable chair of the important resolutions committee, and of the convention during key debates.[80] One who felt his influence was E. L. Doyle. At the 1916 convention, Doyle took his last stand against the White administration in the matter of its handling of the Colorado Fuel and Iron strike. On 27 January, he took the floor for three hours in an attack that Lewis arranged to have expunged from the record. A fragment of the speech surviving in Doyle's papers reveals that he castigated the White administration for agreeing to the Wilson proposal. Edward Doyle, idealistic and incensed, had never accepted the fact that the strike was a lost cause. He told the convention that in setting forth his views, he had nothing to gain but everything to lose. Indeed he had. His long career in the UMWA was near its end.[81]

Shortly after the 1916 convention, White rewarded Lewis for his aid by appointing him union statistician. For the first time, Lewis formally entered the UMWA hierarchy to fill what had been a rather minor position.[82]

It was White, not Lewis, who administered the final blow to the dissenters in District 15. In January 1917, an executive board subcommittee met "to work out a policy of greater efficiency in administering the affairs of the organization." So that "no dual authority would intervene between national officers and coal miners," the committee proposed eliminating non–self-supporting districts, including District 15. In the future, organizing in these districts would be directed by a *Committee on Organization* made up of national officers, including the union president. The proposal went to the heart of the chronic conflict between District 15 and the national union by eliminating the district as an independent administrative unit. In proposing this drastic alteration in union structure, the committee realized that "we may incur the hostility and active opposition of some men who may be materially affected, and of others who may be influenced by reason of sentiment."[83]

The controversial proposal opened the door to radical changes in the structure of the union. It was put before the board in late January 1917, and it passed with eleven members of the board (who controlled 121 votes) in favor and nine (controlling 45 votes) opposed. Adamantly opposed, John Lawson, who sat as board member from Colorado, prepared to make an appeal to the national convention. Utilizing a procedural union rule, White blocked him from doing so.[84]

On 12 February 1917, District Secretary Doyle received a telegram informing him that three days hence the autonomy of District 15 would be suspended. White ordered him to turn over his office and all of its affairs to a representative of the national office.[85] Doyle's life in the union was over.

District 15 was the third district to lose autonomy; the following year, the executive board brought the total to five. Lawson accused the executive board of "placing the power of our great organization in the hands of a few men, which is directly in opposition to the principles upon which our union was founded, and concentration of power, without democratic safeguards is always dangerous."[86]

Mother Jones commiserated with Doyle, writing, "You certainly got a rotten deal."[87] A year and a half later, Doyle was still angry. "The greater the power centralized in the hands of a few at the head of any institution," he wrote to Mother Jones, "the less liberty of action there is . . . and from this follows a decay on one hand and an iron rule on the other."[88]

The following year, the national executive board sponsored a District 15 election, after which the district was to regain autonomy. Doyle and Lawson ran against the slate put up by the national office. The national office won. Lawson and Doyle charged massive election fraud and broke off to form a competing union. Like other such unions in other districts, it ultimately failed.[89]

The year 1917 was a momentous one both for the United States and for the United Mine Workers of America. On 6 April 1917, the United States declared war on Germany. The year saw John L. Lewis's effective rise to power in the UMWA. In the late summer, White appointed Lewis business manager of the *United Mine Workers Journal*, a unique opportunity to increase his visibility among the members. By autumn, Lewis's good fortune waxed further when President Wilson appointed John P. White as a permanent member of the wartime Federal Fuel Board.[90]

White's resignation left Vice President Frank Hayes, a severely debilitated alcoholic, holding the reins. Hayes appointed John L. Lewis, who had never been elected to any UMWA office, as vice president. White and Lewis were holding Hayes by the hand when he made the appointment. Still, it required ratification by the executive board. The meeting was held in October 1917. Four board members demurred. One member remarked that while he had no complaint against John L. Lewis, "We notice he is appointed statistician; we notice again he is appointed manager of the *Journal*, and now he is appointed Vice President. I want to say that democratic institutions do not want too many appointments. They prefer elections." Another board member protested, "I do not believe in elevating men to the high positions in our organization who have never been honored by an election to a high position by the rank and file of our organization." Despite these and other objections, the board confirmed Lewis's appointment, though not unanimously as his biographers state.[91] After Lewis's appointment, the executive board was not called into session again for more than a year. The UMWA was moving toward one man rule.[92]

With "a broken reed," Frank Hayes, in the presidency, it was only a matter of time before the union's governing body appointed Lewis as acting president. In July 1919, Hayes entered a hospital with alcohol-induced "brain fever." Under the circumstances, the executive board pressured him to take a leave of absence. He submitted his formal resignation on 1 January 1920.[93]

Lewis won his first election by coal miners at the riotous convention of 1921. He consolidated his power over the next ten years, forming alliances to eliminate his weakest opponents, and reforming them to eliminate new opponents. One by one, every district, whether self-supporting or not, came under his control. In March 1923, the UMWA vice president reported to Lewis (who was on a trip to Europe), "[E]verything is in exceedingly good shape and the affairs of the organization are running very smoothly. If any of these district officers peep their heads up, I will do just as you would do—kick them out of the organization, then revoke their charters."[94] Over the course of the decade, Lewis laundered every independent force out of the coal miners' union. In reaction, several competing coal miners' unions rose and declined during the 1920s and 1930s, but none could move John L. Lewis off center stage.

The legacy of John L. Lewis is an ambiguous one, fraught with riddles. He gained control of a declining union, in a declining industry. Part of the context was the rise of alternative fuels—natural gas and oil for heating houses, the diesel engine to run trains, the gasoline-fueled automobile to replace the old coal-fired steam locomotive. Another part of the context was the overextension of the industry in the profitable war period. In 1923, the U.S. Coal Commission issued an exhaustive study of the coal industry explaining that the work of mining coal in the United States was carried out in a grossly inefficient manner. For instance, miners spent half the day waiting for cars.[95]

Lewis epitomized the joining hands of capital and labor. He controlled the union with an iron hand and an impenetrable gaze. No more dissenting, anticorporate views could be heard there. In the 1920s, he began important efforts to save the coal industry itself. In 1928, the UMWA drew up legislation that incorporated ideas from two earlier bills that had failed to pass Congress. Lewis's version involved the federal government heavily in the coal industry to control both prices and wages. Government regulation of prices would prevent ruinous price cutting among competing firms. Obligatory collective bargaining would protect wages and prevent competition from non-union coal from overwhelming union employers. Lewis persuaded Indiana Senator James Watson to introduce the bill. It died in the Senate.[96]

But it was reborn in the New Deal. The main features of the Watson bill were incorporated in Franklin D. Roosevelt's National Industrial Recovery Act. For coal miners, the act was effected in 1933 in the Bituminous Coal Code. Coal miners got the eight-hour day, a minimum daily wage, and the right to live where they chose, to buy where they chose, to be paid in legal money, to employ a check-weighman, and to be represented by the United Mine Workers of America. This account was old and long overdue. It was settled by the New Deal. In 1937, when the Wagner Act replaced the earlier legislation, it gave workers the right to organize. The Supreme Court approved the Wagner Act, which stipulated that Rockefeller's Industrial Representation Plan and other company unions were unconstitutional.[97]

These were real reforms that brought real benefits to coal mine workers. Yet the John L. Lewis legacy can be read in two ways. In the 1930s, he emerged at the forefront of the sit-down strikes, as labor's militant hero. In the 1940s, this was to change. The story of Lewis's creation, in 1945, of the Welfare and Retirement Fund, and of his misuse of millions of dollars of that fund, cannot be told here. Neither is there space to elaborate his relationship with George Love, President of Pittsburgh Consolidation, their alliance within the Bituminous Coal Operators' Association, and their joint plan of pushing mechanization to squeeze out the small operator. Lewis's "organizing campaigns" were designed not to improve the lot of the miner but to run the small operator out of business. The new merger movement among the large coal operators was planned by George Love and aided by John L. Lewis. Under the circumstances, it is not surprising that Lewis increasingly failed to press his capitalist allies for concessions on behalf of working coal miners.[98]

In 1916, John D. Rockefeller, Jr., had published an article in the *Atlantic* proclaiming that capital and labor were partners. The burgeoning of wildcat strikes in the 1950s suggested that perhaps by then miners understood that John L. Lewis had carried this partnership one or two steps further than even Rockefeller had intended.[99]

The rise of corporate liberalism was, to use Gabriel Kolko's apt phrase, the triumph of conservatism. For reformers, it was the road toward making the system humanly tolerable, and many reforms did indeed improve the social system. But for those who held power, it was, above all, a system of control. That the Rockefeller-owned Standard Oil of New Jersey could form a cartel with the Nazi firm I. G. Farben, could continue this relationship throughout the time the German firm operated with slave labor, and could at the same time install the Industrial Representation Plan in its New Jersey plant seems less ironic when one sees that for Standard Oil liberalism was a tool, not a set of principles. This, in turn, made liberalism an ambiguous legacy for working people.

Within the Colorado Fuel and Iron Company, within the UMWA, and within the United States, liberalism often went along with the suppression of dissent. The Colorado Fuel and Iron Company could espouse democracy while also firing a man for "spreading propaganda disloyal to the company." The United States could fight a war "to make the world safe for democracy" while also turning to a systematic eradication of dissent. Reform went hand in hand with repression to secure a stable society.

On 15 June 1917, Congress passed the Espionage Act, which granted the attorney general the right to withhold from the mails matter urging "treason, insurrection, or forcible resistance to any law of the United States." Immediately, the postmaster general began removing socialist and prolabor publications from the mails. One of the first to go was the *Appeal to Reason.* Another was *The Masses.* That autumn, arrests began of antiwar socialists, IWW activists, and a wide variety of radicals—in all more than six thousand individuals.[100]

World War I did not make the world safe for democracy. Neither did patriotic hysteria end conflict between the classes. But the war marked a turning point in the terrain of class struggle. Previously, a significant body of opinion had held that the large corporations were un-American at best. When in 1871 a coal miner held that "the large corporations are riveting the chains of slavery around us," he was expressing a view that was neither original nor unusual.

By World War I, a profound change in the climate of opinion showed that the large corporations had gained the high ground. Mother Jones epitomized the change. In 1911, she told a convention of coal miners assembled to conduct the business of the largest trade union in the United States: "The industrial war is on in this country. Why? Because modern machinery plays a greater part in the production of wealth in this nation than it does in any other nation of the world. The class that owns the machine owns the government, it owns the governors, it owns the courts and it owns the public officials all along the line."[101]

By 1915, when she called John D. Rockefeller, Jr., "a much misunderstood young man," her tune had changed. In 1918, she stood before another UMWA convention, this time proclaiming, "We will stay with Uncle Sam. . . . There is no other uncle in the world like Uncle Sam, and the convention must express its deep appreciation of President Wilson. . . ."[102]

Mother Jones was a political chameleon who reflected the times. She would change again, more than once, before she died in 1930. The battle was not over, as the events of the 1930s would show. But from now on those fighting the causes of working people would have a new difficulty to contend with. The flags flying over Ludlow had burned to the ground. To the inhabitants of the coal camps, they had symbolized freedom: freedom to speak and read, freedom to choose, to assemble, and to organize. Previously, the people of the coalfields had claimed Americanism for themselves, and had accused the corporations of lacking the appropriate qualities. Now the flag flew over the large-scale corporation. There it symbolized something quite different: the triumph of American commerce, in the United States and around the world.

ABBREVIATIONS

AFL	*AMERICAN FEDERATION OF LABOR*
AT&SF	*ATCHISON, TOPEKA, AND SANTA FE RAILROAD*
CBLS	*COLORADO BUREAU OF LABOR STATISTICS*
CC&I	*COLORADO COAL AND IRON COMPANY*
CF&I	*COLORADO FUEL AND IRON COMPANY*
CICM	*COLORADO INSPECTOR OF COAL MINES*
CIR	*COMMISSION ON INDUSTRIAL RELATIONS*
CSHS	*STATE HISTORICAL SOCIETY OF COLORADO*
CUA	*CATHOLIC UNIVERSITY OF AMERICA*
D&RG	*DENVER & RIO GRANDE RAILWAY OR DENVER & RIO GRANDE RAILROAD*
EBM	*EXECUTIVE BOARD MINUTES*
HSP	*HISTORICAL SOCIETY OF PENNSYLVANIA*
HU	*MANUSCRIPTS AND ARCHIVES DEPARTMENT, BAKER LIBRARY, HARVARD UNIVERSITY*
LC	*LIBRARY OF CONGRESS*
MRUS	MINERAL RESOURCES OF THE UNITED STATES
NA	*NATIONAL ARCHIVES*
NMAH	*NATIONAL MUSEUM OF AMERICAN HISTORY (SMITHSONIAN)*
NSHS	*NEBRASKA STATE HISTORICAL SOCIETY*
RFA	*ROCKEFELLER FAMILY ARCHIVES*
RAC	*ROCKEFELLER ARCHIVE CENTER, TARRYTOWN, NEW YORK*
RME	*ROCKY MOUNTAIN ENERGY, BROOMFIELD, COLORADO*
RMN	ROCKY MOUNTAIN NEWS
SUNY	*STATE UNIVERSITY OF NEW YORK*
UMWA	*UNITED MINE WORKERS OF AMERICA*

UMWJ	UNITED MINE WORKERS JOURNAL
UPC	*UNION PACIFIC COAL COMPANY*
USGS	*UNITED STATES GEOLOGICAL SURVEY*
WA	THE WORKINGMAN'S ADVOCATE
WFM	*WESTERN FEDERATION OF MINERS*
WHD/DPL	*WESTERN HISTORY DEPARTMENT, DENVER PUBLIC LIBRARY*

NOTES

PREFACE
1. Merle Travis, "Folk Songs of the Hills," Capital 48001, released 9 June 1947. Archie Green, 281ff.
2. CBLS, *Seventh Biennial Report, 1899–1900* (1900), 99.

INTRODUCTION
1. George Kinghorn, "To the Miners of the West," *WA*, 3 May 1873, 2.
2. Thomas H. Huxley speech, 1875, reprinted in "On the Formation of Coal," *Earth Science* 35 (Spring 1982), 25 (4).
3. Before 1906 and after 1914 (between 1906 and 1914 no government agency collected strike statistics), coal mine workers struck more frequently than any other group except for construction workers. The number of workers in the building trades was smaller, however, and, although the building industry had more strikes, each strike involved fewer workers. For instance, from 1887 to 1894 the coal and coke industry had 709 strikes. At the same time, the building trades had 2,490 strikes. For the coal industry, this amounted to 675,128 strikers, or eighty-two percent of the coal mine work force, but for the building trades this involved 273,121 strikers, or fifty-six percent of the work force. U. S. Commissioner of Labor, *Tenth Annual Report*, 1564–65. This trend remained generally true until the 1920s, when the number of coal strikes dropped, while those in the building trades rose. Peterson (U. S. Bureau of Labor Statistics), 30, 38. For the contrary view that the building trades "set the record for industrial conflict" see Montgomery, "The Irish and the American Labor Movement," 49.

PART I: COAL IN AMERICA

PROLOGUE: COAL IS A ROCK THAT BURNS
1. Combs, 4.
2. Coal is described in Averitt (USGS), 133–34; Chamberlain (USGS), 16–17; Jensen and Bateman, 471–73; Moore, 130ff.; Jack A. Simon, "Coal," *McGraw Hill Encyclopedia of Science and Technology*, 1982, s.v.; G. A. Sparkham, "Coal," *Chamber's Encyclopedia*, 1973, s.v.; "Coal," and "Coals," *Encyclopaedia Britannica*, 1959, 1965, 1980, s.v.; Noyes, ed., 75–96.
3. Lignite contains fifty to seventy-five percent carbon. Bituminous coal contains sev-

enty-five to eighty-five percent carbon and has a heat value higher than lignite but lower than anthracite. Anthracite contains ninety to ninety-seven percent carbon.
4. Paul Averitt, "Coal," *Collier's Encyclopedia*, 1986, s.v.; Rhodes and Middleby, 6.
5. MacFarlane, 1.
6. Pratt, 206; Schurr and Netschert, 74ff.; Binder, 27–40; National Coal Board, 52; Richard Cowling Taylor, 264; U. S. Census (Eighth Census, 1860), *Manufactures of the United States in 1860* (1865), clxxii.
7. Finney and Mitchell.
8. Collier and Horowitz, 15–47; Jensen and Bateman, 488.
9. Noyes, ed., 14; U. S. Bureau of Mines, *Minerals Yearbook, 1976*, 364. See also Hershey.

CHAPTER 1: THE TURN FROM WOOD TO COAL
1. Thompson, 13.
2. "Reminiscences of an Old Timer," *UMWJ*, 5 May 1910, 6.
3. Wieck, 77.
4. Eavenson, 254; USGS, *Twenty-Second Annual Report*, 40.
5. U. S. Census (Eighth Census, 1860), *Manufactures of the United States in 1860* (1865), clxxi. For the British total, see Flinn, 26, Table 1.2.
6. Wieck, 77; Erickson, 107.
7. Berthoff, *British Immigrants in Industrial America*, 49.
8. John Hall Testimony, 13 Sept. 1871, *Report of the [Ohio] Mining Commission*, 148.
9. Works examining the causes of the first industrial revolution include Ashton, *The Industrial Revolution;* Dean; Dobb, 258–59; Cleland; Landes, *The Unbound Prometheus;* Pollard.
10. Ashton, *The Industrial Revolution*, 16; Berg, 5–6; Boyd, 28–29; Cleland; Flinn, 452; Nef, 77–81; Perelman, 55; Swank, 38; Wrigley.
11. Song of Solomon 8:6; Nef, 72–74; Nicolls, 50; Forbes, 6–7, 22.
12. Flinn, 18, 28; Landes, "The Industrial Revolution"; Martin, 180.
13. Te Brake, 337–59; Richard Cowling Taylor, 262.
14. Ashton and Sykes, 1; National Coal Board, 53; Pollard, 29; Flinn, 36–38.
15. Ashton and Sykes, 8, 10.
16. Ibid., 35; Boyd, 3, 5, 50–51; Cleland, 122.
17. Landes, "The Industrial Revolution," 230. See also Nef, 82–83.
18. Ashton, *Iron and Steel*, 60–64; Briggs, 6–48. Cardwell, 67–71; Derry and Williams, 312–20.
19. Some scholars believe the steam engine has been given too central a place in understanding the industrial revolution. Hunter, 160–210. Dolores Greenberg questions the significance of the steam engine in manufacturing, noting the increasing output of industries that continued to use traditional sources of power. However, much of Greenberg's argument rests on her questionable assertion that America's industrial development had caught up with Britain's by the 1840s. Greenberg, "Reassessing the Power Patterns of the Industrial Revolution," 1237–61; Greenberg, "Energy Flow in a Changing Economy, 1815–1880," 29–53.
20. Landes, "The Industrial Revolution"; Wrigley, 112–13.
21. Flinn, 454.
22. Boyd, 2, 55. Eight tons is a guess, based on the fact that a Newcastle chaldron measured about two and a half tons. Flinn, 461; Ashton and Sykes, 63. See also Wrigley, 108.
23. Richard Trevithick's first attempt at the steam locomotive was made on the coal-

fields of south Wales and the Tyne. George Stephenson and other leading contributors to the invention were closely associated with large collieries. Ashton and Sykes, 69.

24. Landes, 41.
25. It was thought that the scarcity of wood had caused a severe rise in the price of charcoal and a related decline of the British iron industry, but Flinn notes that wood prices didn't go up until after the discovery of coke smelting. Nevertheless, he observes, charcoal could not have sustained the iron industry during its rapid expansion in the seventeenth and eighteenth centuries. Flinn, 452.
26. Ashton, *Iron and Steel*, 26–38; Dean, 108; Derry and Williams, 474–75; Finney and Mitchell; Rosenburg, 300; Temin, 14–17.
27. National Coal Board, 53.
28. A fine description of preindustrial localism can be found in Chandler, *The Visible Hand*, 17–51.
29. Braverman provides a path-breaking discussion of the changes that affected the work process. See also Chandler, *The Visible Hand*, 17–51.
30. Pollard, 181–89.
31. See Note 9; Landes, *The Unbound Prometheus*, 7.
32. Handlin, *The Uprooted*, 3–12; Handlin, *A Pictorial History of Immigration*, 88–89, 127.
33. Deane, 20–36. See also James C. Riley, "Insects and the European Mortality Decline," *American Historical Review* 91, No. 4 (Oct. 1986), 833–58.
34. Deane persuasively argues that the standard of living for most people fell between 1780 and 1820, may or may not have risen between 1820 and 1840, and did rise in the decades following 1840. Deane, 268.
35. Handlin, *The Uprooted*, 34–36; Handlin, *A Pictorial History of Immigration*, 86–93.
36. Pollard, 207.
37. National Coal Board, 56; Pollard, 53 and 163; Flinn, 358.
38. Ashton and Sykes, 70–71; Boyd, 7; National Coal Board, 60; Raynes, 19.
39. The Children's Employment Commission (1842), quoted in Ashton and Sykes, 76.
40. Flinn notes that miners *favored* the bond. Flinn, 341–45, 356–58; see also Pollard, 53.
41. The crown lands included the Forest of Dean and the Alston Moor. Fisher; Ashton and Sykes, 100–114; Pollard, 41; Flinn, 12–13, 42.
42. A. J. Taylor, 216–17; Flinn, 55–57. See also Buttrick.
43. Ashton and Sykes, 20; Boyd, 87–88, 96.
44. The Royal Commission Report, 1842, quoted in Boyd, 88.
45. Boyd, 75. Regarding uncouth girls, Angela V. John writes that "immorality" below ground was exaggerated. Investigators who knew little about mining drew questionable conclusions from the fact that the women dressed in trousers and often worked stripped to the waist. John, 41, 45, 47.
46. Engels, 282.
47. Ashton and Sykes, 75. For a path-breaking study of women in nineteenth-century British mines that illuminates the dynamic between gender and class, see John. See also Flinn, 333–34.
48. Ashton and Sykes, 15, 18; Robert Bald, *General View of the Coal Trade in Scotland* (1818), quoted in Boyd, 91–92; Ritson, 79–98.
49. Robert Bald, *General View of the Coal Trade in Scotland* (1818), quoted in Ashton and Sykes, 24.
50. Nicolls, 161.

51. Ashton, *The Industrial Revolution,* 26; Nicolls, 189; John, 51–52.
52. John, 51–52.
53. Flinn, 412; Ashton and Sykes, 43.
54. Pollard, 205.
55. Ashton and Sykes, 51, 53; Flinn, 420.
56. Leifchild, 112; Ashton and Sykes, 20, 63.
57. Boyd, 54, 89.
58. Ibid., 17, 342.
59. National Coal Board, 59; Flinn, 359.
60. Boyd, 13; Berthoff, *British Immigrants in Industrial America,* 49; Flinn, 351; Raynes, 30; Wieck, 37, 39, 77; Yearly, *Britons in American Labor,* 18. For a study of the influence of Chartism on British immigrants, see Boston, xii–xiii.
61. Easterlin, 170.
62. Chandler, *The Railroads,* 7.
63. Chandler, "Anthracite Coal," 142–43, 146; Chandler, *The Railroads,* 3; Chandler, *The Visible Hand,* 47, 53. See also Myron W. Watkins, "Large Scale Production," *Encyclopedia of Social Sciences,* 1933, 176.
64. See Note 5. As late as 1870, British production (124 million tons) comprised 51.8 percent of world production, while U. S. production (some 37 million tons) made up about 15 percent, about the same as Germany. U. S. Census (Twelfth Census, 1900), *Mines and Quarries, 1902* (1905), 671.
65. Eavenson, 254.
66. In 1870, wood still supplied three-fourths of the country's energy supply. Coal overtook wood in the mid-1880s, with coal supplying fifty percent and wood forty-six percent of the energy in 1885. Schurr and Netschert, 34, 500; Pratt, 203.
67. For the history of conversion to coal burning locomotives, see Binder, 111–31. By 1862, all freight locomotives operated by the Pennsylvania Railroad were burning coal fuel. Burgess and Kennedy, 713. See also Wieck, 50–51; Schurr and Netschert, 52.
68. Temin, 51–80; Paul Paskoff adds to Temin's explanation that even after anthracite coal became available as a metallurgical fuel, the charcoal-burning furnace was cheaper to build than the anthracite furnace, and it was within the financial means of the individual- and partner-owned firm that predominated before the large-scale enterprise became common in the iron industry. Further, the larger anthracite furnace was economical only when run at capacity. Market conditions helped to keep iron makers loyal to the charcoal furnace. Paskoff, 128.
69. Chandler, "Anthracite Coal," 162. See also Stapleton, 148; Binder, 62. The excellent coking coal of the Connellsville region was not successfully demonstrated as a blast-furnace fuel until 1859. Finney and Mitchell, 12.
70. Bowen; USGS, *Twenty-Second Annual Report,* 76.
71. Chandler, *The Railroads,* 3, 21. Gordon, Edwards, and Reich convincingly argue that in the U. S. the 1840s began a "long swing" upwards that lasted until the depression of 1873. Gordon, Edwards, and Reich, 8–9.
72. George Kinghorn speech, *WA,* 10 Jan. 1874, 2.
73. In 1880, Pennsylvania supplied sixty-five percent of the national coal product. In 1901, it still supplied fifty-one percent. Not until the 1930s was Pennsylvania surpassed by West Virginia. USGS, *MRUS,* (1902), 418; USGS, *MRUS,* 1932. See also Billinger, 2, 33; Binder, 1–43; Eavenson, 171, 186–87; Pratt, 205; Rosenburg, 301.
74. In 1860, 25,000 miners (out of 36,500) worked in the anthracite region, which

produced seventy-three percent of the value of U.S. coal. U. S. Census (Eighth Census, 1860), *Manufactures of the United States in 1860* (1865), clxiii–clxiv. In 1870, bituminous production outstripped anthracite for the first time, but the anthracite region still contained the country's greatest concentration of miners (53,000 workers). U. S. Census (Ninth Census, 1870), *Vol. 3, Statistics of Wealth and Industry,* 764, 767.

75. Chandler, *The Visible Hand,* 91; Jenks, 82ff.; Pierce, 127; Temin, 20; Yearly, *Britons in American Labor,* 53; Wiman, 220ff.

CHAPTER 2: MY LAMP IS MY SUN

1. Braverman, 49.
2. John Siney, "From President Siney," *WA,* 31 Jan. 1874, 2.
3. This chapter is indebted to Braverman and to Dix. See also Yarrow; A. N. Humphries, "Mining Methods Practiced by the Westmoreland Coal Co.," *Annual Report of the Geological Survey of Pennsylvania for 1886* (Harrisburg, Pa., 1887).
4. Daddow and Bannan, 412–15 and 450–53; Dix, 1; Nicolls, 157; Schaefer. For the persistence of the primitive and wasteful drift method in Schuylkill Co, Pa., see Yearly, *Enterprise and Anthracite,* 110–11; Harvey, *The Best Dressed Miners,* 36.
5. Walter Sullivan, "Drills Bore Deeper Than Ever Into Earth," *New York Times,* 26 Mar. 1985, 17.
6. Nicolls, 163, 165; Dix, 77; Hudson Coal Co., 122; USGS, *Twenty-Second Annual Report,* 295.
7. *Report of the Inspector of Mines of the Anthracite Region of Pennsylvania, 1870* (Harrisburg, Pa., 1871), 110.
8. This describes the Pioneer mine in Ashland, Pa., now a demonstration mine. Franklin Platt, "A Special Report to the Legislature Upon the Causes, Kinds and Amount of Waste in Anthracite Mining," *Second Geological Survey of Pa., Vol. 2* (Harrisburg, Pa., 1881).
9. Hudson Coal Co., 122.
10. Brophy, 45.
11. Sinclair, 22.
12. Nicolls, 179.
13. Ibid., 164.
14. Simonin, 116.
15. Ibid., 130.
16. Alex Briggs to George L. Black, 13 Dec. 1904, p. 360, and 30 April 1904, p. 126, Letterbook: "Feb 1st, 1904 to," UPC/RME.
17. Brophy, 40.
18. Simonin, 108–9.
19. "Coal and Coal Mining," *Union Pacific Employees Magazine* 2 (1887–1888), 45. Unless otherwise stated, all modern information on gases is taken from Mosgrove, 197, 187–88.
20. Mosgrove, 187.
21. Nicolls, 179.
22. Ibid.; Patterson, "Reminiscences of John Maguire," 328.
23. CICM, *First Annual Report of the State Inspector of Coal Mines . . . for the Year Ending July 31, 1884* (1885), 8, 36; Nicolls, 179.
24. James Troughear (first quotation) and Martin Lennegan. *Report of the [Ohio] Mining Commission, 1871,* 123, 124.
25. Yearly, *Anthracite and Enterprise,* 116; Wieck, 33, 42; Wallace, *The Social Context of Innovation,* 125–33; Wallace, *St. Clair,* 39–48.

26. CICM, *First Annual Report . . . 1884* (1885), 8.
27. Roy, *The Coal Mines.*
28. Fay (Bureau of Mines), 69. Wallace, following contemporary newspaper accounts, gives the number of dead as 110. Wallace, *St. Clair,* 297; Korson, 180–81; Roy, *History of the Coal Miners,* 85–87.
29. Roy, *The Coal Mines,* 166–67.
30. Doyle Test., U. S. Commission on Industrial Relations, *Final Report and Testimony,* Vol. 7, 6925.
31. "Among the Coal Miners," *The Pueblo Courier,* 25 Jan. 1901, 3; White; Wallace, *St. Clair,* 52; DeKok.
32. Wallace, *St. Clair,* 13; Ill. Dept. of Mines and Minerals, *Compilation of the Reports,* 68.
33. Brophy, 36.
34. Ibid., 41–42.
35. CBLS, *First Annual Report, 1887–1888* (1888), 298.
36. See J. C. McNally (Consular Service USA, Liege, Belgium) to John Mitchell, 12 Nov. 1914, and enclosed report "Ankylosis," dated 12 Nov. 1904, Reel 8, Mitchell Papers (Microfilm Edition), CUA.
37. "Rats in Mines," *UMWJ,* 16 April 1891, 2; "Rats in the Coal Mines, *National Labor Tribune,* 25 June 1887, 3.
38. Appendix of "Statements Regarding the Delaware and Hudson Company Before the Anthracite Strike Commission," 11, Reel 5, Mitchell Papers (Microfilm Edition), CUA; *Rules and Regulations for the Government of All Employees of the Union Pacific Coal Co.,* (1 Oct. 1911), Box 1, UPC Papers, UPM, Omaha; Pa., *Industrial Statistics, Vol. 11, 1884,* 103a; *Rules and Conditions Governing the Hocking District, 1903–1904,* Reel 9, Mitchell Papers (Microfilm Edition), CUA.
39. Unless otherwise cited, descriptions of work are based on Brophy, 43–46; Dix, 8–12, 43–46; Goodrich, 23; Hudson Coal Co., 111–66; Nicolls, 165–68, 201. For a description of wedging, see Simonin, 116; Harvey, *The Best Dressed Miners,* 36–37.
40. Patterson, "Old W. B. A. Days," 374; Daddow and Bannan, 445. See also MacFarlane, 98; Wieck, 46.
41. *WA,* 20 Mar., 1875, 4.
42. Frank L. Gaddy, "Roof Control," in *Elements of Practical Coal Mining,* edited by Samuel Cassidy, New York: Society of Mining Engineers, 1973, 74; Pa., *Annual Report of the Geological Survey for 1886: Part I: Pittsburgh Coal Region* (Harrisburg, Pa., 1887), 377.
43. Fay (Bureau of Mines), 85.
44. Joseph S. Harris to John Veith, 20 Aug. 1875, Philadelphia and Reading Coal and Iron Co. Papers, NMAH; Pa., *Annual Report of the Geological Survey for 1886: Part I: Pittsburgh Coal Region* (Harrisburg, Pa., 1887), 392.
45. Dix, 77.
46. Samuel Daddow, quoted in Yearly, *Enterprise and Anthracite,* 121. See also *Annual Report of the Geological Survey of Pennsylvania for 1885* (Harrisburg, Pa., 1886), 297.
47. Flinn, 83–84, 90; Daddow and Bannon, 433; Fay (Bureau of Mines), 115, 139, 157, 171, 188, 196, 228, 244, 311.
48. I am indebted to Sam Witherspoon for clarifying this method of mining. Dix, 7; John L. Schroder, "Modern Mining Methods—Underground." *Elements of Practical Coal Mining* edited by Samuel Cassidy, New York: Society of Mining Engineers, 1973, 353–55.

49. Ill. Dept. of Mines and Minerals, *Compilation of the Reports,* 19; Drury, 119–20; U.S. Bureau of Mines, *Minerals Yearbook* 1 (1978), 344, and 390.
50. Roberts, *Anthracite Coal Communities,* 137.
51. U.S. Dept. of Labor, Women's Bureau. See also Maclean; Roberts, *Anthracite Coal Communities,* 57–150.
52. U. S. Dept. of Labor, Women's Bureau, 2.
53. Ibid., 17.
54. Schwieder, 81, 99. See also Corbin, 33–34.
55. Fraser, 249.
56. Pa., *Report of the Inspector of Mines of the Anthracite Region of Pennsylvania, 1870* (Harrisburg, Pa., 1871), 4; Korson, *Minstrels of the Mine Patch,* 188, 203.
57. Humphrey (Bureau of Mines), 29.
58. [?] Jones to Mr. H. K. Bennett, Hanna, Wyo., 19 Feb. 1909, File: "1908 Hanna Explosion," UPC/RME.
59. Fay (Bureau of Mines), 7, 14, 16, 115; Drury, 93, 175; Machisak, et al. (Bureau of Mines). See also Dix, 70–71.
60. "Every Member Dead," *UMWJ,* 18 Feb. 1904, 2.
61. Nicolls, 186; Simonin, 157, 160.
62. Chamberlain (USGS), 17, 43, 53; Drury, 91, 164; USGS, *The Prevention of Mine Explosions.* A summary of all explosions in the U. S. in which five or more men were killed is given in Humphrey (Bureau of Mines). See also White, 246.
63. This squeeze occurred on 12 January 1846. The writer was Andrew Bryden, and the foreman was his father, Alexander Bryden. Hudson Coal Co., 172–73.
64. Fay (Bureau of Mines), 318; Fraser, 252–54; Korson, *Coal Dust on the Fiddle,* 205–7; Korson, *Minstrels of the Mine Patch,* 139–43.
65. Hall and Snelling (USGS), 4, 8. See also Drury, 34, 39, 134; Machisak, et al. (Bureau of Mines), 56.
66. Kerr, 362.
67. Chamberlain (USGS), 17, 43, 53; Drury, 91, 164; USGS, *The Prevention of Mine Explosions.*
68. Trachtenberg, 32–49.
69. Testimony of Joseph H. Brown and A. B. Cornell, President and Secretary respectively of the Mahoning Coal Co., *Report of the [Ohio] Mining Commission, 1871,* 110–11; McAuliffe, 244.
70. Pa., *Reports of the Inspectors of Mines of the Anthracite Coal Regions of Pa., 1881* (Harrisburg, Pa., 1882), 160, 180.
71. Hall and Snelling (USGS), 18.
72. Andrew Roy, "Recollections of a Mine Inspector, No. 6," *UMWJ,* 4 July 1907, 4.
73. Dix, 73ff.
74. "Official Proceedings, Illinois Miners' Protective Association," *National Labor Tribune,* 20 Feb. 1886, 5; UMWA, *First Annual Convention,* 10–16 Feb. 1891, 5.
75. "Company Should Be Held Responsible for Deaths," (reprinted from Scranton *Times*), *UMWJ,* 26 Sept. 1907, 4.
76. *Union Pacific Employees Magazine* 2 (1887–1888), 338; *Report of the [Ohio] Mining Commission* (Columbus, Ohio, 1872), 153. Michael Flinn writes of Britain that in the "whole vast corpus of mining records of the eighteenth and early nineteenth centuries" there is exactly one reference to respiratory conditions. Flinn, 422–23.
77. Drury, 3, 48–50; U. S. Census, *Historical Statistics of the United States: Colonial Times to 1970* (1975), 182. See also Graebner, *Coal-Mining Safety.*
78. John L. Lewis is quoted in Corbin, 146.

CHAPTER 3: MY FATHER'S UNFAILING KINDNESS

1. Joint Conference of Miners and Operators, Pittsburgh, Pa., 7 Feb. 1888, Evans, Vol. 1, 286.
2. Walter C. Scott (Iowa), *UMWJ*, 14 May 1891, 4.
3. Pa., *Industrial Statistics, 1872–1873* (1874), 319.
4. Rosenblum, 69.
5. CBLS, *Eighth Biennial Report, 1901–1902* (1902), 127.
6. U. S. Census (Eighth Census, 1860), *Manufactures of the United States in 1860* (1865), clxiii–clxiv; U. S. Census (Twelfth Census, 1900), *Mines and Quarries, 1902* (1905), 666.
7. Thompson, 11.
8. Pa., *Industrial Statistics, 1872–1873*, 332.
9. "To the Miners of the West," *WA*, 3 May 1873, 2.
10. "From Illinois," *WA*, 28 June 1873, 2.
11. U. S. Census (Sixteenth Census, 1940), *Population: Comparative Occupational Statistics for the United States, 1870 to 1940*, (1943), 104.
12. U. S. Census (Eighth Census, 1860), *Manufactures of the United States in 1860* (1865), clxiii–clxiv.
13. The 1870 production figure was corrected to thirty-seven million tons in the 1902 Census. U. S. Census (Twelfth Census, 1900), *Mines and Quarries, 1902* (1905), 667, 668; U. S. Census (Ninth Census, 1870), *Vol. 3: Statistics of Wealth and Industry*, 764, 767. The 1940 census corrected the total number of coal mine operatives in 1870 (reported as 95,000) to 186,000. See Note 11.
14. U. S. Census (Ninth Census, 1870) *Vol. 3, Statistics of Wealth and Industry*, 840; Hartmann, 84; U. S. Industrial Commission, "The Foreign Born in the Coal Mines," 391.
15. U. S. Census (Ninth Census, 1870), *Vol. 3, Statistics of Wealth and Industry*, 840.
16. Barnum, 16–17; Gutman, "Black Coal Miners and the Greenback Labor Party." 506–35; Bailey, 144.
17. Roy, *History of the Coal Miners*, 72.
18. *WA*, 28 Feb. 1873, 2–3. See also Gutman, "Five Letters of Immigrant Workers," 384–99.
19. Welsh was the second president of the Workingmen's Benevolent Association. Killeen, 237–39; Patterson, "Old W. B. A. Days," 382. For patterns of Irish immigration to Scotland, see Laslett, *Nature's Noblemen*, 19; Flinn, 346.
20. Gudelunas and Shade, *Before the Molly Maguires*, 19.
21. Gibbons, 918.
22. Patterson, "Reminiscences of John Maguire," 315.
23. In Pennsylvania, 109 bituminous firms employed fewer than seventy-five persons; only 16 anthracite firms were that small. Pa., *Industrial Statistics, 1875–1876* (1877), 236–43.
24. U. S. Census (Tenth Census, 1880), *Report on the Mining Industries of the United States* (1886), xxv.
25. This mine had sixteen contract miners. Pa., *Industrial Statistics, 1877–1878*, 405.
26. Pa., *Industrial Statistics, 1875–1876*, 291.
27. Ibid.
28. Ibid.
29. Ibid.
30. Braverman, 59.
31. Patterson, "Reminiscences of John Maguire," 332.

32. "From Rees and Sarah Phillips," 20 Nov. 1869, Conway, ed., 185.
33. Pa., *Industrial Statistics, 1872–1873*, 380; Pa., *Industrial Statistics, 1875–1876*, 443.
34. Dix, 105.
35. Pa., *Industrial Statistics, 1872–1873*, 341; Brophy, 41. This discussion is based on Dix, 48–49.
36. Hudson Coal Co., 123–24.
37. " 'Befo de Wah' Mine Yields Curious Relics," *Coal Age* 25, No. 8 (21 Feb. 1924), 273–76.
38. Pa., *Industrial Statistics, 1877–1878*, 340.
39. Pa., *Industrial Statistics, 1875–1876*, 315.
40. *UMWJ*, 14 May 1891, 5; "Proceedings of the Convention," 14 Mar. 1863, Wieck, 75–76, 248.
41. "From . . . Jones, Oct., 1864," Conway, ed., 171.
42. George Archbold Test., "Evidence Taken at Pittsburgh," Pa., *Industrial Statistics, 1872–1873*, 472.
43. "An Appeal," *WA*, 7 Feb. 1874, 4.
44. Augustus Stimmer Test., "Evidence Taken at Pittsburgh," Pa., *Industrial Statistics, 1872–1873*, 477.
45. "Ohio's Convention," *UMWJ*, 28 Jan. 1892, 2.
46. Gibbons, 924.
47. Patterson, "Old W. B. A. Days," 373.
48. Brophy, 64, 76.
49. Korson, *Minstrels of the Mine Patch*, 10–11.
50. *WA*, 16 Aug. 1873, 3.
51. Quoted in Aurand, *From the Molly Maguires*, 192. The firm was probably the Pa. Coal Co. (a subsidiary of the Pa. Railroad). *WA*, 14 Apr. 1877.
52. "A Villainous Contract," *WA*, 16 Aug. 1873, 3.
53. Walter C. Scott (Iowa), *UMWJ*, 14 May 1891, 4.
54. Pa., *Industrial Statistics, 1872–1873*, 342; Pa., *Industrial Statistics, 1877–1878*, 403.
55. Yearly, *Enterprise and Anthracite*, 96–97.
56. Chandler, *The Visible Hand*.
57. Philip Conrad, Supt., to F. Mitchell, Secy.-Treas., 16 Nov. 1886; Philip Conrad to Seth Caldwell, Jr., Pres., 17 Jan. 1887, Superintendent's Reports (File XB-34), Buck Mountain Coal Co. Papers. The Buck Mountain Coal Co. began operation in 1840, not far from Mauch Chunk (now Jim Thorpe), in Luzerne Co. Hoffman, *Anthracite in the Lehigh Region*, 58.
58. Philip Conrad to Seth Caldwell, Jr., 6 Jan. 1887 and 17 Mar. 1887, Buck Mountain Coal Co., Papers.
59. Philip Conrad to Seth Caldwell, Jr., 13 Jan. 1887, Buck Mountain Coal Co. Papers.
60. Philip Conrad to Seth Caldwell, Jr., 15 Mar. 1887. See also reports of 21 Feb. 1887, 15 Mar. 1887, and 4 May 1887. Buck Mountain Coal Co. Papers; *Pennsylvania Coal Mine Inspectors' Reports, 1901* (Harrisburg, Pa., 1901), 42.
61. Gibbons, 920.
62. Patterson, "Old W. B. A. Days," 368; Corbin, 36–37.
63. Gibbons, 919.
64. Goodrich, 57; "A Villainous Contract," *WA*, 16 Aug. 1873, 3.
65. *UMWJ*, 16 July 1891, 4; Humphrey (Bureau of Mines), 27.
66. As late as 1909, the Census Bureau reported that 45.8 percent of all proprietors and firm members in the bituminous industry were still performing manual labor. U. S. Census (Thirteenth Census, 1910), *Vol. 11: Mines and Quarries, 1909* (1913), 191, 195.

67. Pa., *Industrial Statistics, 1877–1878*, 355.
68. Pa., *Industrial Statistics, 1875–1876*, 312.
69. Pa., *Industrial Statistics, 1875–1876*, 237.
70. "The Blind Spot," *Coal Age* 25, No. 4 (3 Apr. 1924), 483.
71. Pa., *Industrial Statistics, 1876–1877*, 162–63.
72. Pa., *Industrial Statistics, 1877–1878*, 408.
73. J. M. Tisdell to D. O. Clark, Supt., UPC, 11 May 1885, Letterbook, p. 142. UPC/RME.
74. "Evidence Taken at Pittsburgh, Nov. 8, 1873," Pa., *Industrial Statistics, 1872–1873*, 477.
75. Pa., *Industrial Statistics, 1872–1873*, 474.
76. Yearly, *Enterprise and Anthracite*, 51–54; Flinn, 48.
77. "From Illinois," *WA*, 9 Oct. 1875, 2.
78. The country's first was John Eltringham, appointed in May 1869 for Schuylkill Co. Patterson, "Reminiscences of John Maguire," 313.
79. Pa., *Industrial Statistics, 1876–1877*, 282–83.
80. Pa., *Industrial Statistics, 1877–1878*, 232–33. See also Pa., *Industrial Statistics, 1881–1882*, Vol. 10 (1883), 133a, 140a.
81. *Report of the [Ohio] Mining Commission, 1871.*
82. Pa., *Industrial Statistics, 1875–1876*, 252–53; John, 51–52.
83. Pa., *Industrial Statistics, 1875–1876*, 254.
84. George Archbold Test., "Evidence Taken at Pittsburgh," Pa., *Industrial Statistics, 1872–1873*, 471.
85. Pa., *Industrial Statistics, 1877–1878*, 102.
86. Pa., *Industrial Statistics, 1876–1877*, 209.
87. Pa., *Industrial Statistics, 1886*, Vol. 14 (1887), 46–47. After the turn of the century, the National Child Labor Committee estimated that nine to ten thousand boys under fourteen worked in the anthracite region and that a large number impossible to estimate also worked in the bituminous mines. Lovejoy, "Child Labor in the Soft Coal Mines," 26–34; Lovejoy, "The Extent of Child Labor in the Anthracite Coal Industry," 35–49.
88. Lovejoy, "Child Labor in the Soft Coal Mines," 26.
89. *Report of the [Ohio] Mining Commission, 1871*, 133, 138.
90. Brophy began his working life in a small bituminous mine in western Pennsylvania. Brophy, 36.
91. Ibid., 49–50.
92. Ibid., 60–61.
93. Second Geological Survey of Pa., *A Special Report . . . Upon the Causes, Kinds and Amount of Waste in Mining Anthracite*, Vol. A2 (Harrisburg, Pa., 1881), 23–29.
94. Pollard, 185; U. S. Census (Tenth Census, 1880) *Report of the Mining Industry of the U.S.*, 626, 638.
95. *Labor Standard*, quoted in Bimba, 30–31.
96. see ch. 2, note 47.
97. The lack of regular payment of wages is attested by numerous miners' complaints. See, for example, Pa., *Industrial Statistics, 1882–1883* (1884), 121–22.
98. *National Labor Tribune*, 26 Jan. 1884, 6; *National Labor Tribune*, 24 Jan. 1885, 8. In the 1890s, the low average for days worked per year was 178 (in 1894), about fourteen days per month. The high average was 216 (in 1890), still only eighteen days per month, about three-quarters time. USGS, "MRUS, 1899," in *Twenty-First Annual Report, 1899–1900*, 339.

99. George Archbold Test., "Evidence Taken at Pittsburgh," Pa., *Industrial Statistics, 1872–1873*, 470.
100. Pa., *Industrial Statistics, Vol. 9, 1882–1883*, 131; *UMWJ*, 6 Aug. 1891, 5.
101. Pa., *Industrial Statistics, Vol. 11, 1882–1883* (1884), 122, 131.
102. "Evidence Taken at Pittsburgh," Pa., *Industrial Statistics, 1872–1873*, 469–70 and 473–74.
103. Pa., *Industrial Statistics, 1882–1883*, Vol. 11, 146.
104. Pa., *Industrial Statistics, 1882–1883*, Vol. 11, 123.
105. "Evidence Taken at Pittsburgh," Pa., *Industrial Statistics, 1872–1873*, 469–70, 474–75.
106. The operator was W. H. Rend. Roy, *History of the Coal Miners*, 240.
107. These 1885 rules were printed in *UMWJ*, 6 Oct. 1910, 6.
108. D. O. Clark, Supt., UPC to H. G. Burt, Pres., UP Railroad, 2 Aug. 1900, File: July–Aug., 1900," Box 128, SG2-S1, Union Pacific Archives, NSHS; *Annual Report of the Union Pacific Coal Co. for the Year 1891*, 24, UPC/RME; George, 195.
109. Roy, *History of the Coal Miners*, 231; "A Mass Meeting," *National Labor Tribune*, 30 Jan. 1886, 8.
110. D. O. Clark, Supt., UPC, to Horace Burt, Pres., UP Railroad, 26 April 1899, File 33: "Mar.–Apr., 1899," Box 127, Union Pacific Archives, NSHS.
111. "Labor Troubles in the Coal Mines of Tioga County," Pa., *Industrial Statistics, 1872–1873*, 479.
112. Pa., *Industrial Statistics, 1872–1873*, 481.
113. Pa., *Industrial Statistics, 1872–1873*, 489.
114. "Proceedings of the Miners' National Association," *WA*, 25 Oct. 1873, 1.
115. Pa., *Industrial Statistics, 1872–1873*, 481–82.
116. "Proceedings of the Miners' National Association," *WA*, 25 Oct. 1873, 1.
117. Ibid.
118. Aurand, *From the Molly Maguires*, 50. Complaints abound on the subject of prices at the company store vs. those at independent stores. See for example, Gutman, "Black Coal Miners and the Greenback Labor Party," 519. The mass of miners' complaints on this issue is so great, encompassing such a wide geographical area and time span, that it adds up to incontrovertible evidence. Fishback's assertion that the company stores provided a service and did not charge higher prices was certainly true in some places, but as a generalization it just doesn't hold true. See Fishback, 317–22.
119. Pa., *Industrial Statistics, 1872–1873*, 495.
120. Ibid.
121. Pa., *Industrial Statistics, 1872–1873*, 496.
122. Ibid.
123. Roy, *History of the Coal Miners*, 107–8; Marcus, 416.
124. Fogel, 55.

CHAPTER 4: THE HAVEN OF JUSTICE AND RIGHT IS OUR DESTINATION
1. "Daniel Weaver's Letter . . . ," Apr. 1861, Wieck, 230–31.
2. "Letter From a Coal Miner, 1863," Wieck, 256.
3. U. S. Census (Eighth Census, 1860), *Manufactures of the United States in 1860* (1865), clxxiv.
4. Bruce Catton, *The Civil War* (New York: The Fairfax Press, 1980), 69–84.
5. U. S. Census (Eighth Census, 1860), *Manufactures of the United States in 1860* (1865), clxxii.
6. Montgomery, *Beyond Equality*, 4; Wieck, 20–21.

7. This account of the AMA is based on Wieck. His Appendix includes the major extant documents of the organization.
8. Wieck, 26.
9. Ibid., 204.
10. "Daniel Weaver's Letter . . . ," 30 Mar. 1861, Wieck, 229.
11. My view on this matter differs from Wieck's. Wieck, 85.
12. "From . . . Morgan," 14 Oct. 1865, Conway, ed., 178–79.
13. "From . . . Phillips," 20 Nov. 1869, Conway, ed., 185.
14. In 1863, the AMA modified its rules on "restriction on our labor," showing these rules to be fully in force. "Proceedings of the Convention of Coal Miners," 14 Mar. 1863, Wieck, 75–76, 248.
15. "From . . . Jones," Oct. 1864, Conway, ed., 171.
16. "Daniel Weaver's Address to the Miners . . . ," Wieck, 218–19.
17. "From . . . Jones," Oct. 1864, Conway, ed., 171.
18. "Daniel Weaver's Letter . . . ," 21 Mar. 1861, Wieck, 225.
19. Montgomery, Beyond Equality, x, 92.
20. "From . . . Watkins," 10 Mar. 1865, Conway, ed., 173.
21. "Daniel Weaver's Letter . . . ," 21 Mar. 1861, Wieck, 225.
22. "Letter from a Coal Miner, 1863," Wieck, 256.
23. Thompson is quoted in Boston, 10.
24. "Letter from a Coal Miner, 1863," Wieck, 256.
25. "Daniel Weaver's Letter . . . ," April 1861, Wieck, 230–31.
26. Harvey, "The Civil War . . . ," 363–64.
27. "From . . . Evans," 4 Oct. 1868, Conway, ed., 182–83.
28. The exchange included coal dealers as well as operators. Wieck, 107–10.
29. "From . . . Rees," 12 May 1865, Conway, ed., 180, 173.
30. "From . . . Powell," 12 May 1865, Conway, ed., 174.
31. The Fall Brook Coal Co. was organized by John Magee (1794–1868) in 1859. "John Magee to H. Brewer, Corning," 17 August 1864, in Wieck, 165. Andrew Roy describes this lockout as a strike, but Wieck convincingly argues that in fact it was a lockout. Roy, A History of the Coal Miners, 103–5; Wieck, 162–72, 206, 287–288.
32. "Rules and Regulations Adopted by the Fall Brook Coal Co., Tioga County, Pa. . . . ," Mar. 1865, Wieck, 289.
33. Wieck, 162.
34. U. S. Census (Eighth Census, 1860), Manufactures in the United States in 1860 (1865), clxiii; U.S. Census (Ninth Census, 1870), Vol. 3: Statistics of Wealth and Industry, Gudelunas and Shade, Before the Molly Maguires, 19.
35. "From . . . Roberts," 27 July 1864 [1865?], Conway, ed., 176.
36. See Berthoff, British Immigrants in Industrial America, 187; Berthoff, "The Social Order of the Anthracite Region."
37. Charlemagne Tower to Col. James B. Fry, 30 Sept. 1863, Pottsville, Pa., Entry 3050, Vol. 1, RG 110, NA. For a superb scholarly reconsideration of this widespread view of the Irish, see Eric Foner.
38. Montgomery, "Pennsylvania: An Eclipse of Ideology," 52; Gudelunas and Shade, Before the Molly Maguires, 84.
39. Gudelunas and Shade, Before the Molly Maguires, 104, 106.
40. Ibid., 38. For more on Bannan's character, see Palladino, 97, 103–6, 138–39; Gudelunas, "Nativism and the Demise of Schuylkill County Whiggery,"; Yearly, Enterprise and Anthracite, 134.
41. This notion is based on Palladino.

42. The attorney was Franklin B. Gowen. Schlegel, *Ruler of the Reading*, 10. For a laborer's wages, see Pa., *Industrial Statistics, 1872–1873*, 380.
43. Charlemagne Tower (1809–1899), a lawyer and coal operator, lived in Pottsville, Pa., from 1850 to 1875. He was heavily involved in litigation involving titles to coal lands. From April 1863 to May 1864, Tower served as U. S. Provost Marshal for the Tenth Congressional District of Pa. As a coal operator, he was one of the original proprietors of the Honeybrook Coal Co. and played a large role in transforming it into the Lehigh and Wilkesbarre Coal Co. *National Cyclopaedia of American Biography*, Vol. 5, 188–90. For the background of opposition to the draft and the Civil War, see Palladino, 10–31, 133; Wieck, 30. See also Wallace, *St. Clair*, 325–26.
44. C. Tower to Major Gilbert, 28 Nov. 1863, Pottsville; "Affidavit of William Burke of Cass Township," copied into C. Tower to Col. James B. Fry, 4 Dec. 1863, Pottsville; C. Tower to Major Gilbert, 18 Nov. 1863, Entry 3050, Vol. 1, RG 110, NA.
45. Aurand, *From the Molly Maguires*, 97.
46. Broehl, 91–92.
47. C. Tower to James B. Fry, 25 May 1863, Entry 3050, Vol. 1, RG 110, NA.
48. C. Tower to Col. J. V. Bomford, 11 Aug. 1863, Pottsville, Entry 3050, Vol. 1, RG 110, NA; Aurand, *From the Molly Maguires*, 97.
49. Leon F. Litwack, "Civil War," *Funk and Wagnalls New Encyclopedia*, 1983, q.v.; Shankman, 192, 195, 201; Montgomery, *Beyond Equality*, 104–7; Montgomery, "The Irish in the American Labor Movement," 209.
50. C. Tower to Col. James B. Fry, 7 Nov. 1863, Pottsville, Entry 3050, Vol. 1, RG 110, NA; Palladino, 119, 180–81.
51. "Irish-Americans and the late Civil War," *The Irish World*, 20 May 1871, 5.
52. C. Tower to Col. J. V. Bomford, 7 July 1863, Entry 3050, Vol. 1, RG 110, NA.
53. C. Tower to James B. Fry, 20 August 1863, Pottsville, Entry 3050, Vol. 1, RG 110, NA; Palladino, 212.
54. Palladino, 222–30; Entry 3050, Vol. 2, RG 110, NA.
55. Richard Heckscher to C. Tower, 1 Feb. 1864, Entry 3050, Vol. 2, RG 110, NA.
56. Aurand, *From the Molly Maguires*, 98.
57. Pa., *Industrial Statistics, 1872–1873*, 329–30.
58. Alonzo J. Snow Affidavit, 1 Feb. 1864, Entry 3050, Vol. 2, RG 110, NA.
59. "Labor Troubles in Pennsylvania," Pa., *Industrial Statistics, Vol. 11, 1880–1881*, 288.

CHAPTER 5: BEATEN ALL TO SMASH
1. Gowen Testimony, Pa. Senate (Doc. No. 39), 1534.
2. Quoted in Aurand, *From the Molly Maguires*, 65.
3. Gordon, Edwards, and Reich, 108–9.
4. Also quite large was the Knights of St. Crispin, a shoe workers' union. Montgomery, "The Irish in the American Labor Movement," 210.
5. "Labor Troubles in Pennsylvania," Pa., *Industrial Statistics, 1880–1881*, 288; Schlegel, *Ruler of the Reading*, 96.
6. In 1870, the WBA officially changed its name to the Miners and Laborers Benevolent Association, though people continued calling it the WBA. For histories of the WBA, see Aurand, *From the Molly Maguires;* Aurand, "The Workingmen's Benevolent Association"; Montgomery, *Beyond Equality;* Killeen; Roy, *History of the Coal Miners*, 76–83, 87–102; Schlegel, "The Workingmen's Benevolent Association"; Schlegel, *Ruler of the Reading*, 16–31, 62–76; Wallace, *St. Clair*, 388–402.
7. Pa., *Industrial Statistics, Vol. 11, 1882–1883*, 125–26; Trachtenberg, 18–19. Mont-

gomery points out that despite the loophole, the law was progressive for its time. The eight-hour law was repealed by a 1913 law that merely established a ten-hour day for women. Montgomery, "Pennsylvania: An Eclipse of Ideology," 58.

8. Pa., *Industrial Statistics, 1872–1873*, 332.
9. Aurand, *From the Molly Maguires*, 68; Pa., *Industrial Statistics, 1872–1873*, 333; Killeen, 106–112.
10. The mix of ethnic cultures within the WBA is contrary evidence for the view that ethnic conflict subverted class solidarity in the anthracite region, a view argued in Gudelunas, and Shade, *Before the Molly Maguires*, and in Walker, 361–76. Walker (386) errs in describing the WBA as "essentially an extension of the Welsh Community."
11. *Biographical Dictionary of American Labor Leaders*, 332–33; "American Labor Portraits—No. 3," *WA*, 22 Nov. 1873, 1; Boston, 95; Killeen, 33–39, 57; Roy, *History of the Coal Miners*, 77; Yearly, *Britons in American Labor*, 123; Siney Test., Pa. Senate (Doc. No. 39), 1694–95.
12. James Kealey Test., Pa. Senate (Doc. No. 39), 1769. Richard Williams Testimony, Pa. Senate (Doc. No. 39), 1721. Roy, *History of the Coal Miners*, 100–01.
13. Aurand, *From the Molly Maguires*, 69; Roy, *History of the Coal Miners*, 78–81.
14. Quoted in Aurand, *From the Molly Maguires*, 69.
15. Aurand, *From the Molly Maguires*, 69.
16. Aurand, *From the Molly Maguires*, 70, 72; Schlegel, "Workingmen's Benevolent Association," 25; U. S. Commissioner of Labor, *Third Annual Report*, 1053; WBA General Council Meeting, 11 May 1869, "Labor Troubles in Pennsylvania," Pa., *Industrial Statistics, Vol. 11, 1880–1881*, 287.
17. "From Benjamin James," 8 Feb. 1871, Conway, ed., 189.
18. *The Irish World*, 29 April 1871, 6.
19. Aurand, *From the Molly Maguires*, 71.
20. Aurand made an important contribution in bringing out the relations between miners and laborers in *From the Molly Maguires*. The omission of this conflict-ridden relationship is glaring in several of the works cited in this chapter that argue that the conflict in the region was due to ethnic antagonism.
21. "From T. Thomas," [probably 1870s], Conway, ed., 194.
22. Quoted in Aurand, *From the Molly Maguires*, 76.
23. Ibid., 78–79.
24. Quoted in Ibid., 82.
25. Ibid.
26. Quoted in Ibid., 65.
27. "Labor Troubles in Pennsylvania," Pa., *Industrial Statistics, Vol. 11, 1880–1881*, 317.
28. Aurand, *From the Molly Maguires*, 86.
29. Quoted in Ibid., 81.
30. Quoted in Ibid., 82.
31. Pa. Senate (Doc. No. 39), 1538.
32. Pa., *Industrial Statistics, 1872–1873*, 335.
33. Patterson, "Old W. B. A. Days," 356.
34. "From E. Jones," 21 Mar. 1871, Conway, ed., 190.
35. This was Thomas Dickson, Pres., Delaware and Hudson Canal Co., quoted in Aurand, *From the Molly Maguires*, 71.
36. "Coal Strike," *The Irish World*, 22 April 1871, 5.
37. The small enterprise in the Schuylkill County industry could predominate because

the coal was near the surface, and the field was closer than other anthracite fields to the market (of Philadelphia). This and the refusal of the state legislature to grant incorporation privileges to Schuylkill County canals and railroads allowed individuals with little capital to get involved. Yearly, *Enterprise and Anthracite*, 16, 57–59. In 1869, Gowen became acting president because of Pres. Charles E. Smith's poor health. In 1870, he became president in his own right. Schlegel, *Ruler of the Reading*, 10, 11, 18; Schlegel, "The Workingmen's Benevolent Association," 250, 253. See also Davies.

38. Pratt, 206.
39. Davies, 64; Yearly, *Enterprise and Anthracite*, 208. The railroad acquired the coal lands in the name of the Laurel Run Improvement Co. Moody's *Manual of Railroads and Corporation Securities*, 1910, 918.
40. Gowen Test., Pa. Senate (Doc. No. 39), 1534.
41. Packer Test., Pa. Senate (Doc. No. 39), 1538–40. See also Gowen Test. Pa. Senate (Doc. No. 39), 1526, 1532–33.
42. Killeen, 189–94; Schlegel, "The Workingman's Benevolent Association," 250, 253; Schlegel, *Ruler of the Reading*, 19, 153; Virtue (Dept. of Labor), 737; Pa. Senate (Doc. No. 39), 1527, 1697.
43. For an overview of the depression of 1870s, see Albro Martin, "Economy from Reconstruction to 1914," in *Encyclopedia of American Economic History* Vol. 1, edited by Glen Porter (New York, 1980). In positing "ethnic group identity as the most enduring source of identity among Scranton workingmen," Walker uses as evidence, the "speed with which working-class militancy dissipated," neglecting to note that, rather than dissipating, working-class militancy was smashed. Walker, 370.
44. Aurand, *From the Molly Maguires*, 88–95; Roy, *History of the Coal Miners*, 99; Schlegel, *Ruler of the Reading*, 62–76.
45. Roy, *History of the Coal Miners*, 99; Joseph S. Harris to John Reese, 14 June 1875, Letterbook of Joseph S. Harris (Vol. 3, Feb.–Dec. 1875), Philadelphia and Reading Coal and Iron Co. Papers, NMAH.
46. Patterson, "After the W. B. A.," 172.
47. Patterson, "Old W. B. A. Days," 383.
48. "Rules and Regulations for the Government of Workmen . . . ," in Pa., *Industrial Statistics, 1875–1876* (1877), 379.
49. *The Labor Standard*, 26 Aug. 1876, quoted in Bimba, 68.
50. Bimba, 69–70. For weekly accounts of crimes, see contemporary issues of newspapers such as the *Shenandoah Herald*.
51. Aurand and Gudelunas, "The Mythical Qualities . . . ," 91–105; Aurand, *From the Molly Maguires*, 96–114; Bimba; Broehl; Coleman; John R. Commons et al., *History of Labour in the United States*, Vol. 2 (New York, 1918), 181–85; Lane; Schlegel, *Ruler of the Reading*, 87–152.
52. *Miners' Journal*, 10 Oct. 1857, quoted in Gudelunas and Shade, *Before the Molly Maguires*, 68. See also Wallace, *St. Clair*, 320–23.
53. Pa. Senate, (Doc. No. 39), 1531, 1538, 1545, 1548.
54. For a description of Gowen's ambivalent stance toward violence, see Schlegel, *Ruler of the Reading*, 9–10. Schlegel's account of the Molly Maguires is one of the older and most balanced.
55. Cummings's and McParlan[d]'s reports are extant in summaries provided by their superintendent (of the Pinkerton agency), Benjamin Franklin. Molly Maguire Papers, HSP.
56. Schlegel, *Ruler of the Reading*, 139.

57. The most objective account of the Molly Maguires is still Coleman. Broehl's account is based on sources unavailable to Schlegel, including Alan Pinkerton letterbooks and financial records of payment from the Philadelphia & Reading Railroad, as well as summaries of McParlan[d] reports that were written in the Pinkerton office. Broehl's narrative history is largely based on McParlan[d]'s reports, which the author treats as "documentation" of what happened. His enthusiasm for his newfound primary sources blinds him to the fact that there is still no corroboration whatever of McParlan[d]'s version. Ann Lane notes this in her excellent review of the literature. Melvyn Dubofsky's suggestion that McParlan[d] may have acted as an *agent provocateur* appears in *Industrialism and the American Worker, 1865–1820* (Arlington Heights, Ill., 1975), 34–38. I read Wallace's account of the Mollies after completing my own. It seems to add evidence to support a skeptical view. However Wallace suggests that there may have been an inner circle of the AOH that exacted retributive justice in a society felt to dispense only injustice to the Irish. Wallace, *St. Clair*, 314–61.
58. Darrow is quoted in Broehl, 355.
59. The *Irish World* version is quoted in Schlegel, *Ruler of the Reading*, 151.
60. Erickson, 44.
61. John Siney, "Tioga Again," *WA*, 21 Feb. 1874, 1.
62. *WA*, 28 Feb. 1874, 2, 3. Erickson notes that Swedes were used as strikebreakers at Blossburg, Pa. (1873), Mahoning Valley in Ohio (1873), Shenango (1875), and Harmony Clay, Indiana (1875). Erickson, 110.
63. Interview with William S. Griffen, reprinted in *WA*, 3–10 Oct. 1874, quoted in Gutman, "The Buena Vista Affair," 198.
64. Gutman, "The Buena Vista Affair," 197.
65. "Labor Troubles in Pennsylvania," Pa., *Industrial Statistics, Vol. 9, 1880–1881*, 314.

CHAPTER 6: "PENNSYLVANIA IS SWARMING WITH FOREIGNERS"
1. "Unions and Strikes," *UMWJ*, 25 June 1891, 3.
2. "From Illinois," *WA*, 28 June 1873, 2.
3. U. S. Census (Twelfth Census, 1900), *Mines and Quarries, 1902* (1905), 669.
4. The 1940 Census corrected gross errors in earlier labor force statistics. U. S. Census (Sixteenth Census, 1940), *Population: Comparative Occupational Statistics for the United States, 1870–1940* (1943), 104. The 1900 Census corrected 1870 production figures. U. S. Census (Twelfth Census, 1900), *Mines and Quarries, 1902* (1905), 666–68, 671.
5. USGS, *MRUS, 1890*, (1891), 85, 148, 172. By 1900, West Virginia had replaced Ohio in third place. USGS, *MRUS, 1902* (1903), 680.
6. Wood supplied seventy-five percent of the country's energy in 1870, and forty-six percent in 1885. Schurr and Netschert, 34–35, 500; Pratt, 203–4.
7. Temin, 3–4, 125–38; Edward T. McClellen, "Skyscraper," *Academic American Encyclopedia*, 1985, q. v.
8. For these insights, I am indebted to Dr. Darwin Stapleton of the Rockefeller Archive Center.
9. Pa., *Industrial Statistics, Vol. 15, 1887*, p. 2F.
10. U. S. Census (Twelfth Census, 1900), *Mines and Quarries, 1902* (1905), 669.
11. Veblen, 24.
12. Klein, *The Life and Legend of Jay Gould;* LeKachman, 183–84.
13. Sereno S. Pratt, "Our Financial Oligarchy," *The World's Work* 10 (Oct. 1905), 6704–14.
14. USGS, *MRUS, 1901* (1902), 280.

15. U. S. Industrial Commission (1901), "The Organization of the Pittsburgh Coal Co.," 99.
16. Johnson, 25.
17. In contrast, 665 operators did eighty-four percent of the coal business. U. S. Census (Twelfth Census, 1900), *Mines and Quarries, 1902* (1905), 62, 64.
18. Bunting, 4, 63, 76, 94, 108, 124, 152, 158, 168, 188, 240, 245; Pittsburgh-Buffalo Co.
19. U. S. Census (Twelfth Census, 1900), *Mines and Quarries, 1902* (1905), 73.
20. Chandler, *The Visible Hand,* 503–13.
21. U. S. Census (Thirteenth Census, 1910), *Volume 11, Mines and Quarries, 1909* (1913), 192; Interstate Commerce Commission, *Special Report No. 1,* 48.
22. Beachley; "Coal: The 'Pitt-Consol Adventure,'" *Fortune,* July 1947, 99; *Commercial and Financial Chronicle,* Oct. 1903, 1472, 28 July 1906, 161, 5 Nov. 1910, 1256; Harvey, *The Best-Dressed Miners,* 38, 340–41, 360, 369; Love, 12, 15; *Moody's Manual of Railroad and Corporation Securities, 1910,* 2628–29, *1915,* 2383, *1916,* 2432–37; Williams, 154–55.
23. Harvey, *The Best-Dressed Miners,* 342.
24. Lambie, 65, 190–91, 324–25, 362–63; U. S. Industrial Commission, "The Organization of the Pittsburgh Coal Co.," 99.
25. Following the United States Coal Commission of the 1920s, a number of scholars have described a sick industry. They fail to distinguish the earlier period from that which came into its own in the 1920s. Arthur Donovan blames the industry's woes on its inability to adopt modern management practices, without taking into account the profitability of inefficient practices. William Graebner asserts, "The lack of uniform accounting practices would make an accounting determination of profits virtually impossible before 1917 or 1918. This essay accepts the contemporary view that profits were low, a logical consequence of the industry's industrial and market structure." Graebner, "Great Expectations . . . ," 50. Richard Simon's account is more complex and interesting but still relies on rather late (beginning 1917) statistics.
26. U. S. Census (Thirteenth Census, 1910), *Mines and Quarries, 1909* (1913), 192, 205.
27. Ibid., 205.
28. Anthracite was also mined in Rhode Island, Colorado, and New Mexico, where production statistics were combined with those of bituminous. (See various issues of *MRUS.*) Bituminous and anthracite were each capable of making inroads on the other's markets. A number of firms mined both types, and many miners had experience with both types.
29. USGS, *Twenty-Second Annual Report,* 107; Whitten, 151–67.
30. Jones; Nearing; *Moody's Manual of Railroads and Corporation Securities,* 1910, 913–24.
31. Jones, 72–73.
32. "Consolidations in 1905," USGS, *MRUS, 1905* (1906), 505.
33. Lambie, 33–35, 49, 65.
34. Interstate Commerce Commission, *Report of Investigation.*
35. U. S. Census, *Statistical Abstract of the United States: 1950* (Washington, D.C., 1950), 97.
36. U. S. Census chart reprinted in Rosenblum, 71.
37. Pa., *Industrial Statistics, Vol. 10, 1881–1882* (1883), 52a.
38. In a typical anthracite community, 65.8 percent of 562 foreign-born males employed

in mining had worked on farms before coming to the United States. U.S. Commission on Immigration, "Representative Community E," 376.

39. "The Bituminous Coal Region," (Leg. Doc. No. 6) in Pa., *Industrial Statistics, Vol. 13, 1885* (1886), n.p.
40. Bodnar, "Immigration and Modernization," 333–60.
41. *UMWJ,* 4 Feb. 1892, 2.
42. U. S. Industrial Commission, "The Foreign Born in the Coal Mines," 391.
43. Bailey, 144; Sheridan (U. S. Bureau of Labor), 435.
44. Fishback, 65; U. S. Census (Sixteenth Census, 1940), *Population: Comparative Occupation Statistics for the United States, 1870–1940* (1943), 159; U.S. Industrial Commission, "The Foreign Born in the Coal Mines," 390. In 1900, blacks constituted 54.3 percent of Alabama's coal mine work force, 25.9 percent of Virginia's 28.4 percent of Tennessee's, 23.7 percent of Kentucky's, and 22.2 percent of West Virginia's. Blacks in the work force slowly declined, until in 1930 they comprised 22.5 percent of the coal mine work force in the southern Appalachian field. After 1930, the decline continued, partly due to discrimination against blacks within the industry (they were blocked from many skilled jobs). In the 1930s and 1940s, loading was mechanized, with the resulting job loss falling disproportionately on blacks. Further, the fact that the industry has turned increasingly to stripping, with employees drawn from the traditionally white (and racist) construction industry, precipitated a further decline of blacks in coal mining—this despite the UMWA's integrationist racial policy. Barnum, Table 7, p. 18 and passim. See also Corbin, 79.
45. This view was put forward by Warne, and propagated further by the U. S. Commission on Immigration. See also Jenks and Lauck; Leiserson. Virtually the lone voice in pointing out the gross error of this view was Hourwich, 12. For a review of the biases of *scholars* toward the new immigrants, see Barendse.
46. Pa., *Industrial Statistics, Vol. 15, 1887* (1888), p. 4F.
47. Erickson, 136, note 80.
48. Pa., *Industrial Statistics, Vol. 10, 1881–1882* (1883), 167–68.
49. U. S. Industrial Commission, "The Foreign Born in the Coal Mines," 394–95; Roberts, *The Anthracite Coal Industry* 394–95; Ill. Bureau of Labor Statistics, *Annual Report for 1892,* 468–69.
50. "From . . . Williams," 10 Nov. 1895, Conway, ed., 205.
51. Roberts, *The Anthracite Coal Industry,* 104, 127.
52. Berthoff, *British Immigrants in Industrial America,* 28.
53. USGS, *Twenty-Second Annual Report,* 353.
54. Pa., *Industrial Statistics, Vol. 14, 1886* (1887), 197a.
55. Berthoff, *British Immigrants in Industrial America,* 28; Wieck, 77; Yearly, *Britons in American Labor,* 20; Yearly, *Enterprise and Anthracite,* 66; Hartmann, 85.
56. Ill. Bureau of Labor Statistics, *Twenty-Second Annual Coal Report,* 134–40.
57. Wilson mined coal in Tioga County, Pa., and was active in the Knights of Labor in the 1880s. In 1899, he became president of UMWA District 2. *Biographical Dictionary of American Labor,* 379; Brophy, 23–24.
58. Hourwich, 422.
59. Ginger, 156.
60. Rosenblum, 124, 132.
61. Pa., *Industrial Statistics, Vol. 15, 1887,* p. F5.
62. Sheridan (U. S. Bureau of Labor), 479–81.
63. Ibid., 404.
64. *UMWA,* 25 June 1891, 1.

65. LaGumina, 34; Bailey, 153.
66. Warne, 70.
67. U. S. Commission on Immigration, "Representative Community E," 387–89; Schwieder, "Italian Americans," 273–74.
68. Erickson, 117, note 82.
69. Roberts, *Anthracite Coal Communities*, 37.
70. Pa., *Industrial Statistics, Vol. 12, 1884* (1885), 64, 66, 68.
71. Roberts, *Anthracite Coal Communities*, 39.
72. Richard L. Davis, quoted in Gutman, "The Negro and the United Mine Workers of America," 345.
73. John Chessa oral interview in LaGumina, 30.
74. Combs, 1.
75. Goodrich, 33–34.
76. *National Labor Tribune*, 23 Jan. 1886, 8; *National Labor Tribune*, 2 Jan. 1886, 2.
77. I had worked out the significance of blowing off the solid, (which was well understood at the time), before reading the excellent analysis of it in Alexander MacKenzie Thompson, III, 91–94. Shooting off the solid had long been standard procedure in mining anthracite coal, where the hardness of the coal counteracted some of the problems. Fay (Bureau of Mines), 284.
78. Pa., *Industrial Statistics, Vol. 10, 1881–1882* (1883), p. 15a.
79. USGS, *Twenty-Second Annual Report*, 353.
80. Alexander MacKenzie Thompson III, 43–44; USGS, *MRUS, 1914*, (1916), 625; Fay (Bureau of Mines), q.v. the states.
81. The Harrison Machine was manufactured by the Whitcomb Co. of Chicago. The Lechner Machine, first sold to operators in Ohio's Hocking Valley, was manufactured by the Lechner Mining Co., (later the Jeffrey Manufacturing Co.). Dix, 17–18; Roy, *History of the Coal Miners*, 148–53; Geological Survey of Pennsylvania, "Mining Methods of the Pittsburgh Coal Region," *Annual Report for 1886* (Harrisburg, Pa., 1887). The mining machine was not successfully adopted in anthracite mining until around 1915. USGS, *Twenty Second Annual Report*, 87; Fay (Bureau of Mines), 284.
82. USGS, *MRUS, 1914*, (1916), 58.
83. Ibid., 587, 616, 623; Dix, 27–28.
84. U. S. Industrial Commission, "Foreign Born in the Coal Mines," 400; *UMWJ*, 19 May 1904, 3.
85. Francis Lechner (1903), quoted in Dix, 25.
86. D. O. Clark, "Report for the Coal Department for the Year 1886," p. 212, MS3761. SG 2, S. 1, Box 43, Union Pacific Archives, NSHS.
87. The preliminary warning of roof noises offered considerable protection under the older methods of mining, which involved little or no noise at the face. In mines where the machine replaced blowing off the solid, however, the machines improved safety because less blasting powder was used. Fay (Bureau of Mines), 109.
88. Drury, 11, 152; Dix, 25.
89. Pa., *Industrial Statistics, Vol. 10, 1881–1882*, 86a.
90. From "Miner," *UMWJ*, 6 Aug. 1891, 1.
91. Quoted in U. S. Industrial Commission, "The Foreign Born in the Coal Mines," 393.
92. Quoted in Ibid., 393.
93. Pa., *Industrial Statistics, Vol. 10, 1881–1882*, p. 86a.
94. Dix, 33; Res. No. 14, UMWA, *Minutes of the Sixteenth Annual Convention*, 16–23 Jan. 1905, 172; Schwieder, *Black Diamonds*, 138–39.

95. Pa., *Industrial Statistics, Vol. 12, 1884,* (1885), p. 27a.
96. Pa., *Industrial Statistics, Vol. 13, 1885,* (1886), p. 38b.
97. *Report of the Ill. Bureau of Labor Statistics for 1888,* quoted in U. S. Industrial Commission, "The Foreign Born in the Coal Mines," 393.

CHAPTER 7: A CLOSING UP OF OUR RANKS

1. Ware, *Labor in Modern Industrial Society,* 188.
2. "Unions and Strikes," *UMWJ,* 25 June 1891, 3.
3. Gordon, Edwards, and Reich, 230–31.
4. For histories of the Knights of Labor, see John L. Butler, "History of the Knights of Labor Organization in Pennsylvania," Pa., *Industrial Statistics, 1887,* pp. G33–G39; Fink, "The Uses of Political Power"; Fink, *Workingmen's Democracy;* Grob; Rosenblum, 111–15; Ware, *Labor in Modern Industrial Society,* 185–88; Wright.
5. U.S. Industrial Commission, "National Labor Organizations in the United States," 3, 5; Wright, 142.
6. Aurand, *From the Molly Maguires,* 117.
7. The Knights also excluded professional gamblers and those who lived by the manufacture or sale of intoxicating liquors. Grob, 34–39.
8. Wright, 157.
9. Quoted in Grob, 38; "Preamble . . . of the Knights of Labor," Pa. *Industrial Statistics, Vol. 15, 1887,* p. 34G.
10. The Knights allotted substantial space in their *Journal of United Labor* to the advocacy of cooperatives. See issues from 1880 to 1885. See also U. S. Industrial Commission, "National Labor Organizations in the United States," 21; *National Labor Tribune,* 3 Jan. 1885, 2.
11. *Journal of United Labor,* 15 Nov. 1880, 69.
12. *Journal of United Labor,* 15 May 1880, 9.
13. Grob, 48–49; Brier, "Interracial Organizing," 20. But see also Campbell.
14. Terrence V. Powderly, "Among the Huns: How They Live and Sleep," *Journal of United Labor,* 25 July 1885, 1040–41. An exception to this was a Colorado Assembly that praised its Italian members as early as 1884. Henry W. Reyes (Coal Creek), *Journal of United Labor,* 10 Nov. 1884, 832–33.
15. Erickson, 115.
16. Erickson, 3. Higham points out that the Knights couldn't very well oppose immigration itself, due to the large numbers of immigrants in the membership. Higham, 49.
17. U. S. Industrial Commission, "National Labor Organizations in the United States," 16.
18. Grob, 50–51.
19. Erickson, 154.
20. Lee, 23, 32.
21. U. S. Industrial Commission, "National Labor Organizations in the United States," 5.
22. Grob, 42.
23. U. S. Industrial Commission, "National Labor Organizations in the United States," 20.
24. U. S. Commissioner of Labor, *Third Annual Report,* 830.
25. U. S. Commissioner of Labor, *Third Annual Report,* 1010.
26. During the same period, the building trades, the second most militant sector, struck 380 times. Pa., *Industrial Statistics, Vol. 15, 1887,* (1888), p. 46F.
27. USGS, *MRUS, 1889 and 1890* (1891), 169.
28. Harvey, "The Knights of Labor," 573.

29. Nicholas Stack, State Master Workman of Ala. superintended a Birmingham coal company. McLaurin.
30. Pa., *Industrial Statistics, Vol. 10, 1881–1882,* 149, 147–63.
31. Harvey, "The Knights of Labor," 570.
32. Pa., *Industrial Statistics, Vol. 15, 1887* (1888), p. F7.
33. Ibid., p. F9.
34. Ibid., p. F11.
35. *National Labor Tribune,* 27 Feb. 1886, 5.
36. Ware, *Labor in Modern Industrial Society,* 188. See also Grob, 12.
37. Quoted in Grob, 38.
38. "Dist. No. 3 Coke," *National Labor Tribune,* 1 Jan. 1887, 5.
39. Evans, Vol. 1, 87.
40. Ibid., 387.
41. Ibid; Roy, *A History of the Coal Miners,* 263.
42. Throughout I have omitted the names of local trade unions that joined the National Federation. Articles in *National Labor Tribune* show that the union was formed out of state and local organizations as well as out of a split in the Knights of Labor. See also Roy, *A History of the Coal Miners,* 262–63.
43. Evans, Vol. 1, 225.
44. Ibid., 139.
45. Ibid., 140–43.
46. "A Truthful Picture," *National Labor Tribune,* 24 Jan. 1885, 3.
47. In 1886, the paper represented local trade unions that later formed the National Federation. *National Labor Tribune,* 13 Feb. 1886, 2. See also Greene, *The Slavic Community on Strike,* 86–87.
48. This statement came from an Ohio Miners' Convention, one of the founding bodies of the National Federation. *National Labor Tribune,* 2 Feb. 1884, 5.
49. "Ohio State Miners' Convention," *National Labor Tribune,* 17 Jan. 1885, 5.
50. John Picket, *National Labor Tribune,* 4 April 1885, 5.
51. Tyr Connell, New Straitsville, Ohio, *National Labor Tribune,* 8 Jan. 1887, 5.
52. Evans, Vol. 1, 417ff.
53. Roy, *A History of the Coal Miners,* 271.
54. Evans, Vol. 1, 204–05; Aurand, "The Anthracite Strike of 1887–1888," 175; Greene, *The Slavic Community on Strike,* 81–85.
55. Greene, *The Slavic Community on Strike,* 108–10. See also Barendse, 1–8.
56. The local trade union was called the Miners and Laborers Amalgamated Association. Pa., *Industrial Statistics, Vol. 15, 1887,* pp. 14f–15f; Roy, *History of the Coal Miners,* 265–72.
57. Pa., *Industrial Statistics, Vol. 15, 1887* (1888), p. F3; *National Labor Tribune,* 1 Jan. 1887, 5.
58. In 1889, a split from the Knights had merged with the National Federation; this new formation took the name the National Progressive Union. The following year, it was actually the National Progressive Union that merged with NTA 135 to become the UMWA. Grob, 124–27. See also Ware, *The Labor Movement in the United States,* 212–18.
59. UMWA, "Joint Convention, 1890"; UMWA, "Constitution and Laws, 1890."
60. In July 1903, the UMWA had 253,000 members. The next largest union was the United Brotherhood of Carpenters and Joiners, with 167,000 members. The 109 other AFL unions had fewer than 100,000 members. "Membership of National and International Unions," Reel 6, Mitchell Papers, Microfilm Edition, CUA.

61. UMWA, "Joint Convention, 1890"; UMWA, "Constitution and Laws, 1890."
62. UMWA "Proceedings, First Annual Convention, 1891," in UMWA Papers, Washington, D.C.
63. UMWA, "Constitution and Laws, 1890," 21.
64. Jones speech, "Report of Second Annual Convention of District 6 [Ohio]," *UMWJ*, 28 Jan. 1892, 2.
65. John B. Rae, "Eight Hours," *UMWJ*, 23 April 1891, 4.
66. Jones speech, "Report of Second Annual Convention of District 6. [Ohio]," *UMWJ*, 28 Jan. 1892, 2.
67. UMWA, "Proceedings, First Annual Convention, 1891," 1, 4; "The Convention," *UMWJ*, 18 Feb. 1892, 2; Laslett, *Labor and the Left*, 200.
68. Roy, *History of the Coal Miners*, 313–15.
69. Ibid., 314–15.
70. *UMWJ*, 25 July 1891, 4.
71. "Report of Secretary McBryde," UMWA, *Proceedings of the Sixth Annual Convention, 1895.*
72. Penna address, UMWA, *Proceedings of the Seventh Annual Convention, 1896*, 5. At its low point, the UMWA membership sank to eleven thousand members. U. S. Industrial Commission, "The Foreign Born in the Coal Mines," 405; "Report of Secretary Pearce, UMWA, *Proceedings of the Ninth Annual Convention, 1898*, 11; U. S. Industrial Commission "Labor Organization of Mine Workers," 185.
73. Brophy, 74.
74. George, 186–94.
75. "Pres. Ratchford's Address," UMWA, *Proceedings of the Ninth Annual Convention, 1898*, 5; U. S. Industrial Commission, "The Coal Mine Industry," 326.
76. Greene, *The Slavic Community on Strike*, 129–30.
77. Greene, "A Study in Slavs, Strikes, and Unions"; Greene, *The Slavic Community on Strike*, 129–51; Pinkowski; Novak.
78. "Ratchford's Address," UMWA *Proceedings of the Ninth Annual Convention, 1898*, 3–4, 10, and Secy. Pearce Report, 11.
79. "Ratchford's Address," UMWA, *Proceedings of the Ninth Annual Convention, 1898*, 6.
80. Ibid., 10; UMWA *Proceedings of the Tenth Annual Convention, 1899*, 16, 38.
81. Greene, *The Slavic Community on Strike*, 158–59.
82. The strike succeeded partly due to the intervention of Mark Hanna, an Ohio coal operator and industrialist who was president of the Republican National Committee. Believing a strike would damage McKinley's election campaign, Hanna pressured the operators to settle, which they did. However, the agreement lasted only until April 1901. Greene, *The Slavic Community on Strike*, 158–62.
83. Parton, ed., 90.
84. Mother Jones speech, UMWA, "Proceedings of the Annual Convention, 1901," UMWA Papers, Washington, D.C.
85. McDonald and Lynch, 43.
86. Roy, *History of the Coal Miners*, 348; McDonald and Lynch, 49; *Biographical Dictionary of American Labor Leaders*, 300–301.
87. Bailey, 149–50.
88. John Campbell, "A Vigorous Letter," *UMWJ*, 8 Oct. 1891, 2.
89. Walter S. Scott, "From the Districts," *UMWJ*, 14 May 1891, 4.
90. UMWA, "Joint Convention, 1890," 9.
91. Gutman, "The Negro and the UMWA."

92. UMWA, *First Annual Convention, Minutes of Feb. 16, 1891*, 8.
93. Brier, "Interracial Organizing."
94. *UMWJ*, 30 April 1891, 5.
95. F. M. Moran, *UMWJ*, 30 April 1891, 5.
96. *UMWJ*, 10 March 1892, 5; "Riley Writes," *UMWJ*, 7 April 1892, 8.
97. "From . . . Williams," 10 Nov. 1885, Conway, ed., 209.
98. *UMWJ*, 30 April 1891, 5.
99. The vice president of the Alabama district in 1903, B. L. Greer, was black. "Forced him to Kiss Negro," *UMWJ*, 13 Aug. 1903, 4. Three consecutive vice presidents of District 17 (W. Va.), Horace Smith, John L. Edmonds, and J. J. Wren, respectively, were all black. Stephen Brier, "Interracial Organizing," 29, 31. Other black officials include W. J. Campbell, Secy.-Treas. (in 1902) of District 23 (Kentucky). *UMWJ*, 24 April 1902, 1.
100. "Willing Hands." *UMWJ*, 14 April 1892, 8; "Riley Writes," *UMWJ*, 7 April 1892, 8.
101. UMWA, *Proceedings of the Fifteenth Convention, 1904*, 163; UMWA, *Proceedings of the Seventeenth Annual Convention, 1906*, 171, 196; UMWA, *Proceedings of the Eighteenth Convention, 1907*, 255–60.
102. UMWA, *Proceedings of the Seventeenth Convention, 1906*, 168, 230.
103. Nyden.
104. Gutman, "The Negro and the UMWA," 361.
105. Ibid., 342.
106. "Hallier Tells Story Horse Creek Affair," *UMWJ*, 20 Aug. 1903, 4; "Forced Him to Kiss Negro," *UMWJ*, 13 Aug. 1903, 4.
107. Quoted in Gutman, "The Negro and the UMWA," 359; Ward and Rogers, 21.
108. UMWA, "Joint Convention, 1890"; UMWA, "Constitution and Laws, 1890," 15.
109. *UMWJ*, 25 June 1891, 5.
110. From an Illinois miner. *UMWJ*, 7 April 1892, 3.
111. U. S. Industrial Commission, "The Foreign Born in the Coal Mines" 407; Weitz.
112. *UMWJ*, 23 July 1891, 1.
113. U. S. Industrial Commission, "The Foreign Born in the Coal Mines," 409; Bailey, 143.
114. John P. Jones, Pres., District 6 (Ohio) *UMWJ*, 7 May 1891, 5.
115. "Forms of Socialism," *UMWJ*, 25 June 1891, 8.
116. UMWA, *Proceedings of the Fourth Annual Convention, 1903*, 12.
117. Quoted in Gutman, "The Negro and the UMWA," 347.
118. "Convention Proceedings," *UMWJ*, 18 Feb. 1892, 2.
119. UMWA, *Proceedings of the Twelfth Convention (1901)*, 98.
120. Thomas Duffy, (Pres., District 7), "The Proper Method," *UMWJ*, 21 Nov. 1901, 2.
121. UMWA, *Proceedings of the Thirteenth Convention (1902)*, 143.
122. *UMWJ*, 28 April 1904, 3.
123. UMWA, *Proceedings of the Fifteenth Convention (1904)*, 151.
124. *UMWJ*, 17 Oct. 1901, 2; *UMWJ*, 30 Jan. 1902, 4.
125. *UMWJ*, 3 March 1904, 3.
126. UMWA, *Proceedings of the Seventeenth Annual Convention (1906)*, 187.
127. In 1910, 1,039 coal mine workers (out of a total of 613,500) were classified as "Indian, Chinese, Japanese, and all others." U. S. Census (Sixteenth, 1940), *Population: Comparative Occupational Statistics for the United States 1870 to 1914* (1943), 159.

PART II: COAL IN THE AMERICAN WEST

PROLOGUE: REFLECTIONS AT MIDPOINT

1. U. S. Census (Twelfth Census, 1900), *Mines and Quarries, 1902* (1905), 680. CBLS, *Seventh Biennial 1899–1900* (1900), 330; USGS, *Coal Resources of Colorado*, 131.
2. George P. West, "Report on Ludlow," File: "West Report," Box 10, Commission on Industrial Relations Papers, RG 174, NA.

CHAPTER 8: CAPITALISTS, RAILROADS, AND COAL

1. McClurg, 23.
2. This was the death song of White Antelope, a Cheyenne chief killed in the Sand Creek Massacre. Stan Hoig, *The Peace Chiefs of the Cheyenne* (Norman, Okla., 1980), 66.
3. Bancroft; Fritz; Hafen; Ubbelohde et al.
4. L. S. Storrs, "The Rocky Mountain Coalfields," USGS, *Twenty-Second Annual Report*, 22, 421; Jensen and Bateman, 477.
5. See Note 3. Forbes Parkhill, "Colorado's Earliest Settlements," *The Colorado Magazine* 34 (Oct. 1957), 241–53.
6. See Note 3.
7. U. S. Census (Twelfth Census, 1900), *Mines and Quarries, 1902* (1905), 46, and 52.
8. Scamehorn, *Pioneer Steelmaker*, 5.
9. Loretta Fowler, *Arapahoe Politics, 1851–1928* (Lincoln, Nebr., 1982), 15–17; Hafen, 34–38; Lillian B. Shields, "Relations With the Cheyennes and Arapahoes in Colorado to 1861," *The Colorado Magazine* 4 (Aug. 1927), 145–54; Virginia Cole Trenholm, *The Arapahoes, Our People* (Norman, Okla., 1970), 1–12; E. B. Renaud, "The Indians of Colorado," *Colorado: Short Studies of Its Past and Present,* by Junius Henderson et al., (New York, 1927, 1969), 23ff.
10. Hafen, 34–38.
11. Bancroft, 355–56, 363; Fritz, 86; Trenholm, *The Arapahoes*, 83, 88–89.
12. Joe Mills, "Early Range Days," *Colorado: Short Studies of Its Past and Present*, by Junius Henderson et al., (New York, 1927, 1969), 92; Chauncey Thomas, "Butchering Buffalo," *The Colorado Magazine* 5 (Apr. 1928), 42ff.
13. Fowler, *Arapahoe Politics*, 34; Trenholm, *The Arapahoes*, 139–40.
14. Fritz, 104; Henderson (USGS), 5–8; Rodman Wilson Paul, *Mining Frontiers of the Far West, 1848–1880* (New York, 1963), 40.
15. Medicine man, quoted in Fowler, *Arapahoe Politics*, 38.
16. Eavenson, 344; McClurg, 9.
17. Frank Fossett, *Colorado: Its Gold and Silver Mines* (Crawford, N.Y., 1880), 58.
18. Hafen, 124.
19. Ovando J. Hollister, *The Mines of Colorado* (Springfield, Mass., 1867), 11.
20. Raymond G. Carey, "The Puzzle of Sand Creek," *The Colorado Magazine* 41 (Fall 1964), 279–98; Margaret Coel, *Chief Left Hand: Southern Arapahoe* (Norman, Okla., 1981), 237–91; Fritz, *Colorado*, 204; Hafen, 165; Fowler, *Arapahoe Politics*, 39; Harry E. Kelsey, Jr., "Background to Sand Creek," *The Colorado Magazine* 45 (Fall 1968), 279–300; Janet LeCompte, "Sand Creek," *The Colorado Magazine* 41 (Fall 1964), 314–35; Trenholm, *The Arapahoes*, 190–96; Ubbelohde et al., 108–12; William E. Unrau, "A Prelude to War," *The Colorado Magazine* 41 (Fall 1964), 299–313.
21. Fritz, 44, 268; Hafen, 214.
22. Hafen, 188.

23. MacFarlane, 536, 539; Richard C. Overton, *Gulf to Rockies* (Austin, Tex.: 1953), 13.

24. USGS, *MRUS, 1882* (1883), 86. For the first few years, the UP leased the coal land at Rock Springs to Thomas Wardell, a Missouri coal dealer. Klein, *The Life and Legend of Jay Gould*, 151. The description of Rock Springs is based on a letter from "Coxwain," Rock Springs, Wyo., 22 Sept. 1889, *Union Pacific Employees Magazine* 1 (1886), 285. A history of the Hanna mines is given in I. N. Bayless, "Hanna Mines Suspends Operations," *The Explosives Engineer*, July-August 1954, 118–19; Pryde. One of several typescript histories of UPC is "Historical Document No. 97," Box 3, UPC Papers, UPM, Omaha. For histories of UP coal operations in Utah, see UP Coal Department, *Annual Report . . . 1891*, UPC Papers, RME; USGS, *MRUS, 1885* (1886), 69; USGS, *MRUS, 1886*, (1887), 351.

25. "Statement of Mining . . . 1884," Ms. No. 3761, File "Colorado Coal," Box 4, SG2-S1, UP Archives, NSHS; Bancroft, 578; McClurg, 21, 25; Eavenson, 345–47; USGS, *MRUS, 1883–1884* (1885); CBLS, "Report of the Colorado Joint Legislative Committee," *Eighth Biennial Report, 1901–1902* (1902), 136.

26. Robert G. Athearn, *Rebel of the Rockies: A History of the Denver and Rio Grande Railroad* (New Haven, Conn. 1962), 3–5; Bancroft, 603; Scamehorn, *Pioneer Steelmaker*, 7; Robert Edgar Riegel, *The Story of Western Railroads* (Lincoln, Nebr., 1926), 113–15; O. Meredith Wilson, *The Denver and Rio Grande Project* (Salt Lake City and Chicago, 1982), 4–5.

27. Klein, *The Life and Legend of Jay Gould*, 180–85.

28. The road reached Trinidad in 1879. L. L. Waters, *Steel Trails to Sante Fe* (Lawrence, Kans. 1950), 51, 55; Merle Armitage, *Operations Sante Fe* (New York: 1948), 8, 204; Carl Ubbelohde et al., 125. In 1883 and 1884, the Starkville mine was operated for the AT&SF by the Trinidad Coal and Coke Co. USGS, *MRUS, 1883 and 1884* (1885), 29. See also USGS, *MRUS, 1882*, 62.

29. Eavenson, 346–47; USGS, *MRUS, 1883 and 1884*, 12.

30. Bancroft, 554; Paul S. Logan, "Building the Narrow Gauge From Denver to Pueblo," *Colorado Magazine* 8 (1931), 201–8; Ubbelohde et al., 123–25; USGS, *MRUS, 1882*, (1883), 41.

31. The D&RG companies in Utah were the Pleasant Valley Coal Co., operating mines in Winter Quarters and Castlegate, near Scofield, and the Utah Fuel Co., organized in 1887 to operate mines in Sunnyside and Clear Creek, as well as the Sommerset mine in Colorado. The two coal companies merged in 1899. Powell, *The Next Time We Strike*, 19–23. See also Helen Z. Papanikolas, "Utah's Coal Lands: A Vital Example of How America Became a Great Nation," *Utah Historical Quarterly* 43 (1975), 104–24.

32. USGS, *MRUS, 1882* (1883), 41; "Predecessor Companies of the Colorado Coal and Iron Co." File 202, CF&I Papers, CSHS. In 1880, all sixty-eight Colorado silver-lead smelters were fueled with coke manufactured by the Colorado Coal and Iron Co. CC&I *Second Annual Report*, 31 Dec. 1880 (1881), 8.

33. George L. Anderson, *Kansas West* (San Marino, Ca., 1963), 148–79; Athearn, *Rebel of the Rockies.* 53–65; Bancroft, 606–8; Riegel, *The Story of Western Railroads*, 185–88; Overton, *Gulf to Rockies*, 21.

34. Thomas Kimball to C. F. Adams, Jr., 3 August 1885, Ms. No. 3761, File: "D&RG Railway," Box 5, SG2-S1, UP Archives, NSHS.

35. USGS, *MRUS, 1886*, 247–49; Overton, *Gulf to Rockies*, 22–23.

36. Frances W. Gregory and Irene D. Neu, "The American Industrial Elite in the 1870's: Their Social Origin," and William Miller, "American Historians and the

Business Elite," in *Men in Business: Essays on the Historical Role of the Entrepreneur*, edited by William Miller, (New York and Evanston, Ill.: 1952, 1962). Drawing similar conclusions is an extensive study of the business classes of the anthracite region. Davies.

37. *Biographical Dictionary of American Business Leaders*, edited by John N. Ingham (Westport, Conn.; 1983), q. v.; John S. Fisher, *A Builder of the West: The Life of General William Jackson Palmer* (Caldwell, Idaho, 1939); Judge Wilson McCarthy, *General William Jackson Palmer (1836–1909) and the Denver and Rio Grande Railroad!* (New York, 1954); *Dictionary of American Biography*, 195; *National Cyclopaedia of American Biography*, Vol. 22, 398–400; Scamehorn, *Pioneer Steelmaker*, 7–8; *Who's Who in America, 1908–09*, 1436.

38. Frank Jackson to William J. Palmer, 2 Oct. 1885, File: "Frank J. Jackson," Box 11, Palmer Papers, CSHS.

39. Ibid., 11 April 1860; Palmer Diary, Box 6, Palmer Papers, CSHS. For an account of Palmer's role in the the development of the coal burning locomotive, see Binder, 127–29.

40. F. H. Jackson to William Palmer, 11 Aug. 1865, File: "Frank J. Jackson," Box 11, Palmer Papers, CSHS.

41. Judge Wilson McCarthy, *General William Jackson Palmer*, 15; F. H. Jackson to William J. Palmer, 15 Aug. 1866 and 27 Sept. 1866, Palmer Papers, File: "Frank J. Jackson," Palmer Papers, CSHS.

42. "Predecessor Companies of the Colorado Coal and Iron Co.," File 202, CF&I Papers, CSHS; "General Palmer Dies, Leaving 15,000,000," *New York Times*, 14 Mar. 1909, 11.

43. *National Cyclopaedia of American Biography*, Vol. 24, 404; Scamehorn, *Pioneer Steelmaker*, 8; Athearn, *Rebel of the Rockies*, 11–13.

44. Jess Augustin Castro, "Alexander Cameron Hunt: Colorado Territorial Governor, 1867–1869," Master's Thesis, Univ. of Denver, 1957; *National Cyclopaedia of American Biography* 6, 447; Morris F. Taylor, "El Moro: Failure of a Company Town," *Colorado Magazine* 48 (Spring 1971), 131.

45. Chandler, *The Visible Hand*, 80.

46. Quoted in Fisher, *A Builder of the West*, 61.

47. Hafen, Appendix.

48. D&RG, *First Annual Report* (1873), 8.

49. Ibid., 6.

50. Ibid., 29.

51. William J. Palmer to "Queen," [Mary Lincoln Mellen], 11 Dec. 1869, Box 15, Palmer Papers, CSHS.

52. Scamehorn, *Pioneer Steelmaker*, 5–10; "Predecessor Companies of the Colorado Coal and Iron Co," File 202, CF&I Papers, CSHS.

53. D&RG, *First Annual Report*, 41, 83.

54. *Report of the Secretary of the Interior*, vol. 1 (Washington, D.C., 1887) 50th Cong., 1st Sess., H. R. Exec. Doc. 1, Part 5, 136–37; Scamehorn, *Pioneer Steelmaker*, 32–34.

55. D. O. Clark to H. G. Burt, 24 Aug. 1899, File: "Burt-Coal-Oct-Dec., 1899," Box 127, SG-2, S-1, UP Archives, NSHS.

56. Scamehorn, *Pioneer Steelmaker*, 9.

57. O. Meredith Wilson, *The Denver and Rio Grande Project*, 15.

58. Colin B. Goodykoontz, "The Exploration and Settlement of Colorado," in *Colorado: Short Studies of Its Past and Present* by Junius Henderson et al. (New

York, 1927, 1969), 53–54; Wilson, *The Denver and Rio Grande Project,* 11; and Harold H. Dunham, "Coloradans and the Maxwell Grant," *Colorado Magazine* 32 (Apr. 1955), 131, 133.

59. Athearn, *Rebel of the Rockies,* 11; Scamehorn, *Pioneer Steelmaker,* 9; Wilson, *The Denver and Rio Grande Project,* 11.
60. Dorothy R. Adler, *British Investments in American Railways, 1834–1898* (Charlottesville, Va., 1970), 27, 136–37.
61. D&RG, *First Annual Report,* 12; Athearn, *Rebel of the Rockies,* 2–3, 18; *Colorado Magazine* 24, (Nov. 1947), 244.
62. D&RG, *First Annual Report,* 9, 15.
63. Wilson, *The Denver and Rio Grande Project,* 20–22; Athearn, *Rebel of the Rockies,* 22–23; Ubbelohde et al., 126; Paul S. Logan, "Building the Narrow Gauge From Denver to Pueblo," *Colorado Magazine* 8 (1931), 206.
64. Ubbelohde et al., 126; Athearn, *Rebel of the Rockies,* 25.
65. Athearn, *Rebel of the Rockies,* 33–37; Wilson, *The Denver and Rio Grande Project,* 26.
66. D&RG, *First Annual Report,* 36.
67. D&RG, *Report to the Stockholders, 1880,* 27; "Report of T. E. Sickels, Consulting Engineer, Dec. 30, 1882, N.Y." in D&RG, *Annual Report, 1882,* 49.
68. D&RG, *Report to the Stockholders, 1880,* 27.
69. Ibid., 28.
70. Scamehorn, *Pioneer Steelmaker,* 5; CC&I, *Third Annual Report, 1881,* 6–7, 16–19.
71. George Engle, Supt., El Moro, to J. R. Cameron, Genl. Supt., CC&I, 18 May 1881, File 30 (CC&I), Box 2, CF&I Papers, CSHS.
72. George Engle, "Report for Week Ending April 30, 1881," File 30 (CC&I), Box 2, CF&I Papers, CSHS.
73. E. S. McKinlay, (So. Pueblo Machinery Co.) to Chas. Lamborn, 28 Mar. 1881, File 30, Box 2, CF&I Papers, CSHS.
74. George Engle, "Report for Week Ending June 7, 1881," File 30, Box 2, CF&I Papers, CSHS.
75. George Engle to Chas. Lamborn, 11 May 1881, File 30, Box 2 CF&I Papers, CSHS.
76. CC&I, *Second Annual Report, 1880,* 5, 8; CC&I, *Third Annual Report, 1881.*
77. Joseph Simons to H. E. Sprague, [1886?], File 16, CF&I Papers, CSHS.
78. In six out of eleven years prior to 1893, the steel works reported a loss. Scamehorn, "John C. Osgood and the Western Steel Industry," 136. See also Scamehorn, *Pioneer Steelmaker,* 64. In 1901, an engineer surveying the firm for a Chicago investment firm wrote: "The fuel department of the Colorado Fuel and Iron Co. has in the past been the principal source of profit." (Julian Kennedy, "Engineer's Report to Blair and Co., Aug. 6, 1901," CF&I Papers, Corporate Reports Dept., Baker Library, Harvard Univ.).
79. Joseph Simons to H. E. Sprague, [1886?], File 16, CF&I Papers, CSHS.
80. J. Imbrie Miller, "Report on Coal Mines," 24 Jan. 1889, File 28, CF&I Papers, CSHS.
81. George S. Ramsey, Genl. Supt., to M. E. M. Steck, Acting General Manager, CC&I, 1 Jan. 1890 and 1 Jan. 1891, File 28, Box 1, CF&I Papers, CSHS.
82. CICM, *First Annual Report . . . 1884,* 86; U.S. Census (Tenth Census, 1880), *Report of the Mining Industries* (1886), 686. U.S. Census (Eleventh Census, 1890), *Report on Mineral Industries* (1892), 347–49.
83. Colorado Fuel Co., "Minutes of the . . . Meetings of the Stockholders and Directors," Bound Items, CF&I Papers, CSHS. See also Sylvia Ruland, *The Lion of Redstone* (Boulder, Colo., 1981); Scamehorn, "John C. Osgood . . ."

84. J. Imbrie Miller, "Report on Coal Mines," 24 Jan. 1889, File 28, CF&I Papers, CSHS.
85. Bunting, 57.

CHAPTER 9: NO TIME TO IMPROVE HIS INTELLECT

1. CBLS, *First Biennial Report, 1887–1888* (1888), 301.
2. CBLS, *Eighth Biennial Report, 1901–1902* (1902), 383.
3. CICM, *First Annual Report, 1884* (1885), 12–13.
4. Ibid., 42.
5. Ibid., 13–15.
6. Ibid., 42.
7. Ibid., 23. See also McClurg, 64–65.
8. "A Coal Mine Horror," *National Labor Tribune*, 2 Feb. 1884, 8.
9. From "Coxwain," Rock Springs, 22 Sept. 1889, *Union Pacific Employees Magazine* 4 (1889–1890), 285.
10. U.S. Industrial Commission, "The Foreign Born in the Coal Mines," 396. See also Ethelbert Stewart, "The Place of Labor Turnover in Government Statistics," Reel 1, Series 1, Part 1, Stewart Papers. In 1923, the Children's Bureau found that in five company towns in W. Va. more than half of the families had lived there for less than a year. Corbin, 40.
11. From "Errinoch," Rock Springs, 24 April 1899, *Union Pacific Employees Magazine* 4 (1889–1890), 126.
12. Grogan, 325; George R. [Mitchell?], "Report Upon the Boulder Valley Coal Lands at Erie, Colo., Aug. 1873," UPC Papers, UPM.
13. CICM, *First Annual Report, 1884*, 4; Grogan, 338; Knight, 34, 38.
14. CBLS, *First Biennial Report*, 118–23; Marlatt, 89.
15. Ibid., 89–102; Scamehorn, *Pioneer Steelmaker*, 62–63; CBLS, *First Biennial Report*, 139–40.
16. "The Strike in Fremont," *Labor Enquirer*, 16 Aug. 1884, 4.
17. "Miners Organize," *Labor Enquirer*, 13 Aug. 1884, 2.
18. "The Proper Ring," *Labor Enquirer*, 30 Aug. 1884, 4.
19. *Labor Enquirer*, 13 Sept. 1884, 4.
20. Marlatt, 92–93; Brundage, 184.
21. Thomas L. Kimball to S. R. Callaway, File: "1884, Coal Operators Association of Colorado," Box 1, SG2-S1, UP Archives, NSHS.
22. D. O. Clark to Pres. Adams, Jr., 6 Nov. 1884, File: "1884 Coal Operators Association of Colorado," Box 1, SG2-S1, UP Archives, NSHS.
23. S. R. Callaway to Charles F. Adams, Jr., 16 Jan. 1885, File: "Labor Disputes," Box 8, SG2-S1, UP Archives, NSHS.
24. Marlatt, 93–95.
25. S. R. Callaway to Charles Francis Adams, Jr., 29 Jan. 1885, File: "Labor Disputes, Coal Strike," Box 8, SG2-S1, UP Archives, NSHS.
26. CBLS, *First Biennial Report*, 139–40.
27. McClurg, 64. In 1910, the Immigration Commission enumerated the employees of six mines in the northern lignite field and found that, of 528 employees, 136 were Italians; 110 were English, Scottish, or Welsh; and 53 were the sons of Britons. Germans constituted another significant minority. U.S. Commission on Immigration, *Reports . . . Part 25: Japanese and Other Immigrant Races in the . . . Rocky Mountain States, Vol. 3*, 241.
28. USGS, *MRUS, 1883 and 1884* (1885), 27.
29. "John Cameron's Supt.'s Report," CC&I, *Seventh Annual Report, 1885*, 15.

30. John Cameron, Supt., "No. 15 General Reports Ending Dec. 31, 1885," File 28, Box 1, CF&I Papers, CSHS.
31. "John Cameron's Supt's Report," CC&I, *Seventh Annual Report, 1885,* 18.
32. CC&I, *Seventh Annual Report, 1885,* 5.
33. S. R. Callaway to C. F. Adams, Jr., 11 Mar. 1885, File: "Colorado Coal," Box 4, SG2-S1, UP Archives, NSHS.
34. Gomer [?] Thomas, (Supt., Grass Creek) to J. M. Tisdell, Esq., 7 Mar. 1885, Letterbook, UPC Papers, RME.
35. Klein, *The Life and Legend of Jay Gould,* 141–54.
36. Quoted in Ibid., 152.
37. P. J. Quealy to Frank A. Manley, 16 Mar. 1917, Hist. Doc. 97, Box 3, UPC Papers, UPM, Omaha.
38. *History of the Union Pacific Coal Mines,* 80.
39. Black, 68; Crane and Larson; Grob, 50–51; *History of the Union Pacific Coal Mines,* 75–91; Klein, *Union Pacific,* 481–88.
40. S. R. Callaway to Charles F. Adams, Jr., 20 Sept. 1885, File: "Labor Disputes/Coal Strike," Ms. No. 3761, SG-2,S-1, Box 8, UP Archives, NSHS.
41. Isaac H. Bromely to Charles F. Adams, Jr., 4 Oct. 1885, File: "Labor Disputes/Coal Strike," Ms. No. 3761, SG-2,S-1, Box 8, UP Archives, NSHS.
42. S. R. Callaway to C. F. Adams, Jr., 3 Sept. 1885, File: "Labor Disputes/Coal Strike," Ms. No. 3761, Box 8, SG-2,S-1, UP Archives, NSHS.
43. Thomas Neasham, Chmn., and J. N. Corbin, Secy., Denver Committee, to General Manager, UP Railroad, 26 Sept. 1885, File: "Labor Disputes/Coal Strike." Ms. No. 3761, SG-2,S-1, Box 8, UP Archives, NSHS.
44. Black, 68; Grob, 50–51.
45. C. F. Adams, Jr. to S. R. Callaway, 21 Sept. 1885, File: "Labor Disputes/Coal Strike," Ms. No. 3761, SG-2, S-1, Box 8, UP Archives, NSHS.
46. D. O. Clark to S. R. Callaway, 11 Dec. 1885, File: "UPRR/Coal Dept.," SG-2, S-1, Box 15, UP Archives, NSHS.
47. "Memorandum of Understanding," 27 Dec. 1889, Hist. Doc. 38, Box 2, UPC Co. Papers, UPM, Omaha; Thomas Neasham and J. S. Corbin, Exec. Board, UP Employees Association, to S. H. H. Clark, 5 Nov. 1891; and S. H. H. Clark to Sidney Dillon, 6 Nov. 1891, Ms. no. 3761, File "S. H. H. Clark," Box 92, UP Archives, NSHS.
48. George S. Ramsay to E. M. Steck, Genl. manager CC&I, 1 July 1890, File 37, Box 2, CF&I Papers, CSHS.
49. "Circular Issued to the Stockholders of the CC&I Co., Sept. 1, 1892," and "Letter to Stockholders, Nov. 25, 1892," CF&I Papers, Corporate Reports Dept., Baker Library, Harvard Univ.
50. R. C. Hills, "Coal and Iron Lands of the CF&I Co., Sept. 1, 1893," File: "CF&I Co., Misc. I," CF&I Papers, Corporate Reports Dept., Baker Library, Harvard Univ.
51. CF&I, *Fourth Annual Report, 1896,* 8; USGS, *MRUS, 1896–1897* (1897), 474.
52. Scamehorn, *Pioneer Steelmaker,* 70; CBLS, *Third Biennial Report, 1891–1892* (1892), 174.
53. CBLS, *Third Biennial Report, 1891–1892,* 177–79.
54. "In Colorado," *UMWJ,* 17 Dec. 1891, 5.
55. CBLS, *Third Biennial Report, 1891–1892,* 178.
56. Ibid., 179.
57. "Report of Coal (Tons) Mined for Week of Feb. 6., 1886," File: "Incoming, S. R.

Callaway, Vol. 55, 1886," MS no. 3761, Box 27; and File: "Incoming, Jan.–Mar. 1887," SG-2, S-1, Box 38, UP Archives, NSHS.

58. "Black Diamond," Omaha *Weekly Republican,* 3 June 1881.

59. McClurg, 58; Evans, Vol. 2, 331.

60. CBLS, *Fifth Biennial Report, 1895–1896* (1896), 8–9, 12–16.

61. The speaker is Edward Boyce. U. S. Commissioner of Labor, *A Report on Labor Disturbances,* 41; Melvyn Dubofsky, *We Shall Be All: A History of the IWW* (New York, 1969), 56; Laslett, *Labor and the Left,* 208; Suggs, "Catalyst for Industrial Change," 322–37.

62. *UMWJ,* 26 Nov. 1903, 3.

63. Laslett, *Labor and the Left,* 192–240; Weinstein, *The Decline of Socialism,* 36; James R. Green, 193–204, 279–86.

64. WFM, EBM, 11 June 1902 and 5 Dec. 1903; UMWA, EBM, April 10, 1903.

65. CBLS, *Ninth Biennial Report, 1903–1904* (1904), 193–94.

66. CBLS, *Seventh Biennial Report, 1899–1900* (1900), 236–38; CBLS, *Eighth Biennial Report, 1901–1902* (1902), 384; "The Coal Mines of Colorado," *Pueblo Courier,* 8 Feb. 1901, 2; "The World of Labor and Its Workers," Denver *Times,* 25 Oct. 1900, 5.

67. CBLS, *Eighth Biennial Report, 1901–1902,* 131.

68. D. O. Clark to Horace G. Burt, Omaha, 10 Aug. 1900, File: "July–Aug., 1900," Box 128, SG2-S1, UP Archives, NSHS.

69. D. O. Clark to Horace G. Burt, 12 Mar. 1900, Ms. No. 3761, File 334, Box 128, SG2-S1, UP Archives, NSHS.

70. See "Struggle for Bread," *RMN,* 20 Jan. 1901, 3. In 1893, Populist Governor Waite refused to appoint David J. Griffiths as inspector despite exemplary examination scores, probably because he was perceived as a "company man." For his dual career as mine inspector and mine superintendent, see "D. J. Griffiths," *Portrait and Biographical Record of the State of Colorado* (Chicago, 1899), 81–82; Frank Hall, *History of the State of Colorado* 4, (Chicago, 1895), 459; *The Colorado Fuel and Iron Co. Reference Book No. 2,* Sept. 1901, 38; "Circular No. 591," Box 12, CF&I Papers, CSHS.

71. Bowers to Rockefeller, Jr., 4 Nov. 1912, Item 99, Box 28, Bowers Papers. See also Memorandum of S. J. Donleavy to Rockefeller, 10 Mar. 1915, File 17, Jesse F. Welborn Papers, CSHS.

72. "Strike in the Northern Coal Field," *Pueblo Courier,* 18 Jan. 1901, 2; CBLS, *Eighth Biennial Report, 1901–1902,* 130–58; McClurg, 89–97; Powell, *The Next Time We Strike,* 37–50.

73. McClurg, 92.

74. "Denver Sends to Iowa for Coal," *RMN,* 16 Jan. 1901, 5.

75. "Alarmed Over Famine," *RMN,* 17 Jan. 1901, 1–2.

76. "Southern Miners Are Involved in Strike," *RMN,* 20 Jan. 1901, 1.

77. "Legislature Will Take Up Coal Strike Today," *RMN,* 18 Jan. 1901, 1–2; "Intimidation Will Not Be Permitted," *RMN,* 22 Jan. 1901, 1–2. See also *RMN,* 2 Jan. 1901, 3, 8 Jan. 1901.

78. The organizer was Charles Duncan. McClurg, 92.

79. "Strike in the Northern Coal Fields," *Pueblo Courier,* 18 Jan. 1901, 2. "Colorado Legislative Committee Report," CBLS, *Eighth Annual Report, 1901–1902,* 137, 140. This reprints Colorado General Assembly, 536–73.

80. *Pueblo Courier,* 25 Jan. 1901, 3.

81. "Among the Coal Miners," *Pueblo Courier,* 15 Jan. 1901, 3. In the mid-1890s,

CF&I employees worked about half-time. CBLS, *Fifth Biennial Report, 1895–1896* (1896), 26.

82. "Colorado Legislative Committee Report," CBLS, *Eighth Annual Report, 1901–1902,* 137, 140.
83. CBLS, *Seventh Biennial Report, 1899–1900* (1900), 274, and chart on 94.
84. *Pueblo Courier,* 11 Jan. 1901, 1; "An Open Letter to Senator Tanquary," *Pueblo Courier,* 25 Jan. 1901, 1.
85. "The Miners' Side," *Pueblo Courier,* 21 Dec. 1900, 1, 25 Jan. 1901, 1.
86. Charles Duncan, "To the People of Denver and Colorado," *Pueblo Courier,* 15 Mar. 1901, 3.
87. Colorado General Assembly, 541.
88. "Lafayette, Colo.," *UMWJ,* 19 Sept. 1901, 7.
89. Ibid.
90. Colorado General Assembly, 559–60; McClurg, 165.
91. "Among the Coal Miners," *Pueblo Courier,* 25 Jan. 1901, 3.
92. CBLS, *Eighth Biennial Report, 1901–1902,* 144.
93. "Among the Coal Miners," *Pueblo Courier,* 25 Jan. 1901, 3.
94. "Intimidation Will Not Be Permitted," *RMN,* 22 Jan. 1901, 1–2.
95. Colorado General Assembly, 564, 570. See also, *UMWJ,* 6 Feb. 1902, 6.
96. "The Coal Miners of Colorado," *Pueblo Courier,* 8 Feb. 1901, 2.
97. Ibid.; CBLS, *Eighth Biennial Report, 1901–1902,* 156–57.
98. John Gehr, "Coal Mining at Gallup," *Pueblo Courier,* 11 Jan. 1901, 4.
99. "From New Mexico," *UMWJ,* 3 Oct. 1901, 7; "New Mexico," *UMWJ,* 23 Jan. 1902, 11; CBLS, *Eighth Biennial Report, 1901–1902,* 130–31, 148, 155, 157–58.
100. "Western Mining Company's Sociological Work," *Outlook* 72, (20 Sept. 1902), 149–50.
101. "Pictou," *Camp and Plant* 1, No. 2 (20 Dec. 1901), 16.
102. John Simpson, "Colorado," *UMWJ,* 6 Feb. 1902, 6.
103. Ibid.
104. "Threaten to Kill John Gehr," *UMWJ,* 20 Mar. 1902, 2.
105. CBLS, *Eighth Biennial Report, 1901–1902,* 156.
106. CBLS, *Ninth Biennial Report, 1903–1904* (1904), 193–94.
107. Scamehorn, *Pioneer Steelmaker,* 177; Bunting, 57.
108. *RMN,* 7 Aug. 1903, 1, 2; "Memorandum of Agreement," CF&I Papers, File: "Misc. I," Corporate Reports Dept., Baker Library, Harvard Univ.; Scamehorn, *Pioneer Steelmaker,* 157.
109. J. C. Osgood, "To the Stockholders . . ." 11 Nov. 1902, CF&I Papers, File: "Misc. I," Corporate Reports Dept., Baker Library, Harvard Univ. See also CF&I, *Tenth Annual Report, 1902* (Dec. 1902), 15.
110. CF&I, *Tenth Annual Report, 1902* (1902) 2.
111. See the *RMN,* 2 July 1903, 1; 21 July 1903, 14; 1 Aug. 1903, 2; 2 Aug. 1910, 9; 4 Aug. 1903, 1; 5 Aug. 1903, 1, 2; 7 Aug. 1903, 1. For the financial details of the takeover, see File: "CF&I, Gates, 1903," Box 22, RFA, RG 2, Business Interests, RAC.
112. In 1909, the two Osgood companies merged to become the Victor-American Fuel Co. Scamehorn, "John C. Osgood . . . ," 133–48; "Osgood's Men Will Resign," *RMN,* 7 Aug. 1903, 1, 2; *RMN,* 5 Aug. 1903, 1.
113. John Simpson, *UMWJ,* 9 July 1903, 5.
114. *RMN,* 5 Aug. 1903, 1; 18 Aug. 1903, 1; Scamehorn, *Pioneer Steelmaker,* 72.
115. "Pres. F. J. Hearne Proud," *RMN,* 25 Aug. 1903, 3.

116. Ibid.
117. Quoted in U. S. Commissioner of Labor, *A Report on Labor Disturbances*, 47–48.
118. Melvyn Dubofsky, *We Shall Be All*, 47. The employer's anti-union offensive across the nation is described in Selig Perlman and Philip Taft, *History of Labor in the United States, 1896–1932* (New York, 1935), 129–37.
119. UMWA, EBM, 10 April 1903; WFM, EBM, 10 April 1903.
120. CBLS, *Ninth Biennial Report, 1903–1904*, 194.
121. "Colorado Tramp," Lafayette, Colo. *UMWJ*, 12 Feb. 1903; *UMWJ*, 19 Feb. 1903, 3.
122. J. W. Trul, Rock Springs, 31 Aug. 1903, *UMWJ*, 10 Sept. 1903; "Strike Is Threatened," *UMWJ*, 24 Sept. 1903, 1.
123. Anderson F. Barnett ("Broncho Bill"), Rock Springs, *UMWJ*, 4 June 1903, 8; A. F. Barnett ("Broncho Bill"), Rock Springs, 18 April 1903, *UMWJ*, 30 April 1903, 6.
124. *UMWJ*, 24 Sept. 1903, 1.
125. Anderson F. Barnett, *UMWJ*, 4 June 1903, 8.
126. *RMN*, 1 July 1903, 1, 5–6; and 2 July 1903, 1, 2. The disaster occurred in the Hanna No. 1 mine on 30 June 1903. It took until 16 December to remove all the bodies but one, which was never recovered. Also killed were seven horses and twenty-eight mules. "Explosion," UPC Papers, RME; Humphrey (Bureau of Mines), 24.

CHAPTER 10: WE HAVE GRIEVOUS WRONGS

1. Duncan McDonald, "The Situation in District 15," *UMWJ*, 12 Nov. 1903, 2.
2. "A Coal Miner's Complaint," *RMN*, 31 Oct. 1903, 16.
3. "Manifesto," enclosed in William Howells to Gov. James Peabody, 13 Aug. 1903, File: "Strikes, Miners, Cripple Creek, etc.," Peabody Papers.
4. I believe it unlikely that the union endorsed the strike by mistake, as Suggs suggests. For the controversy preceding the vote, see Suggs, "The Colorado Coal Miners' Strike," 40; "Coal Miners to Meet in Pueblo," *RMN*, 23 Sept. 1903, 8; "Urged to Arbitrate," *RMN*, 24 Sept. 1903, 6; CBLS, *Ninth Biennial Report, 1903–1904* (1904), 185–86; UMWA, *Minutes of the Sixteenth Convention, 1905*, 186–87, 204; "Operators and Miners Confer," *RMN*, 3 Nov. 1903, 9; "Strike Is Deferred," *RMN*, 8 Oct. 1903, 16.
5. U. S. Commissioner of Labor, *A Report on Labor Disturbances*, 333.
6. The speaker was John Gehr. "Coal Miners to Meet in Pueblo," *RMN*, 23 Sept. 1903, 8; "Miners to Vote on Strike," *RMN*, 18 Sept. 1903, 2. "Urged to Arbitrate," *RMN*, 24 Sept. 1903, 6; "Miners to Convene Here," Pueblo *Labor Advocate*, 25 Sept. 1903.
7. Quoted in CBLS, *Ninth Biennial Report, 1903–1904*, 185.
8. Knight, 42–48; Lonsdale; U. S. Commissioner of Labor, *A Report on Labor Disturbances*, 51–61; CBLS, *Seventh Biennial Report, 1899–1900*, 125–53.
9. "Crisis in Colorado Centers at Polls," *UMWJ*, 19 May 1904, 7.
10. *RMN*, 25 Sept. 1903, 2; 26 Sept. 1903, 12. For a summary of UMWA relations with the WFM, see Laslett, *Labor and the Left*, 208–9.
11. "Confidential Report," to Samuel Gompers, 28 Sept. 1903, carbon copy in Reel 6, Mitchell Papers.
12. "Miners Condemn Gehr's Attitude," *RMN*, 27 Sept. 1903, 9.
13. Ibid.; U. S. Commissioner of Labor, *A Report on Labor Disturbances*, 332.
14. UMWA, EBM, 6 Oct. 1903.
15. Ibid.
16. Ibid.

17. CBLS, *Ninth Biennial Report, 1903–1904*, 185–86.
18. Suggs, "The Colorado Coal Miners' Strike," 40. See also Suggs, *Colorado Conservatives Versus Organized Labor*, 468.
19. *RMN*, 31 Oct. 1903, 6.
20. Clip. from Denver *Republican* in "Mother Jones Clippings," WHD/DPL. Mother Jones arrived in Denver on 19 Oct. 1903. *RMN* 20 Oct. 1903, 4. She writes in her autobiography that just before the 1903 strike she went to Colorado and disguised herself as a peddler, clandestinely visiting all the coal camps. Mother Jones was so well known at the time, and her arrival was so widely announced in the press, that this account seems improbable. Jones, 95.
21. "Great Power of Mother Jones . . ." (reprinted from Denver *Post*), *UMWJ*, 29 Oct. 1903, 7.
22. "Mother Jones Here," *UMWJ*, 29 Oct. 1903, 1; "Great Power of Mother Jones Who Comes Here," *UMWJ*, 29 Oct. 1903, 7.
23. Long, *Mother Jones*; Featherling.
24. "Mother Jones in Colorado," *UMWJ*, 12 Nov. 1903, 3.
25. CBLS, *Ninth Biennial Report, 1903–1904*, 195; Zeph Hill to James H. Peabody, 18 Nov. 1903, File: "Strikes, Miners, Cripple Creek," Box: "Office of the Governor," Peabody Papers.
26. *UMWJ*, 19 Nov. 1903, 5.
27. "Vote on Coal Strike," *RMN*, 16 Nov. 1903, 3; "Reports From the Coal Districts . . . ," *RMN*, 10 Nov. 1903, 1, 9; "Northern Mines Are Rapidly Resuming," *RMN*, 12 Nov. 1903, 1, 8; "Miners Quit State," *RMN*, 17 Nov. 1903, 1, 4; U. S. Commissioner of Labor, *A Report on Labor Disturbances*, 338, 340.
28. *UMWJ*, 7 Jan. 1904, 6.
29. All four states were included in the strike call. See "Mitchell's Letter," *UMWJ*, 12 Nov. 1903, 2.
30. For more on the Rockefeller/Harriman/Gould coal-railroad interests in the Rocky Mountain West, see *Commercial and Financial Chronicle* 76 (Apr. 1903), 51; *Commercial and Financial Chronicle* 77 (Oct. 1903), 1471, 1455; *Commercial and Financial Chronicle* 78 (19 Mar. 1904), 1171. John C. Osgood, still a minority stockholder in the CF&I, controlled the Victor Fuel Co. in southern Colorado and the American Fuel Co. at Gallup, N. Mex. "CF&I is Solid," *RMN*, 21 Dec. 1903, 1, 6; "Only a Bit of Economy," *RMN*, 8 April 1904, 2.
31. For the relationship between the Northern and the UP, see File: "Northern Coal and Coke Co.," Box 4, UPC Papers, RME.
32. "Reports From the Coal Districts . . ." *RMN*, 10 Nov. 1903, 1, 9. John Mitchell told the 1905 UMWA convention that the miners of Rock Springs "would not have struck if we had offered them ten dollars a day." UMWA, *Minutes of the Sixteenth Convention, 1905*, 190.
33. "Miners Quit State," *RMN*, 17 Nov. 1903, 1, 4; "The Coal Miners Strike," *UMWJ*, 19 Nov. 1903, 5; "Union Organizers Arrested in Hastings," *RMN*, 19 Nov. 1903, 11; UMWA, *Minutes of the Sixteenth Convention, 1905*, 196.
34. "Northern Mines Are Rapidly Resuming," *RMN*, 12 Nov. 1903, 1, 8; Powell, *The Next Time We Strike*, 51–80; Powell, "The 'Foreign Element,'" 127.
35. "Trying to Prevent Coal Strike in Utah," *RMN*, 16 Nov. 1903, 10; "Coal Situation Is Improving," *RMN*, 14 Nov. 1903, 5; Powell, "The 'Foreign Element,'" 127, 129.
36. "Resolution No. 34," UMWA, *Minutes of the Fifteenth Annual Convention, 1904*, 154; UMWA, EBM, 4 Jan. 1904.
37. U. S. Commissioner of Labor, *A Report on Labor Disturbances*, 338–40.

38. "Mines to Resume," *RMN*, 15 Nov. 1903, 1, 4.
39. "Northern Coal Miners . . . Again at Work," *RMN*, 1 Dec. 1903, 1.
40. U. S. Commissioner of Labor, *A Report on Labor Disturbances*, 338, 340; UMWA, *Minutes of the Sixteenth Convention, 1905*, 192.
41. Suggs, "The Colorado Coal Miners' Strike," 50; UMWA, *Minutes of the Sixteenth Convention, 1905*, 11. See also "Vote on Coal Strike," *RMN*, 16 Nov. 1903, 3; UMWA, EBM, 24 Jan. 1904.
42. The manager was Julian Kebler. "Getting Ready for the Strike," *RMN*, 6 Nov. 1903, 5.
43. Ibid.
44. "Order Eastern Coal," *RMN*, 16 Nov. 1903, 3.
45. "Vote on Coal Strike," *RMN*, 16 Nov. 1903, 3.
46. "Coal Men in Southern Colorado Walk Out . . . ," *UMWJ*, 12 Nov. 1903, 1.
47. This was Sheriff Clark. "Reports From the Coal Districts . . . ," *RMN*, 10 Nov. 1903, 1, 9; "Getting Ready for the Strike," *RMN*, 6 Nov. 1903, 5; "Coal Men in Southern Colorado Walk Out . . . ," *UMWJ*, 12 Nov. 1903, 1; "Coal Miners Ready to Lay Down Tools," *RMN*, 9 Nov. 1903, 1, 2.
48. "To Open Up Leyden," *RMN*, 26 Nov. 1903, 3; "Miners to Take Action Towards Settlement," *RMN*, 2 Dec. 1903, 1.
49. *UMWJ*, 12 Nov. 1903, 1.
50. "Vote on Coal Strike," *RMN*, 16 Nov. 1903, 3.
51. Reprinted in *UMWJ*, 3 Dec. 1903, 4. For Wardjon's biography, see "Case of Miners Presented . . . ," Pueblo *Labor Advocate*, 27 Nov. 1903, 1.
52. William Beard to *UMWJ*, 28 April 1904, 6.
53. Steel, xlv; "Colorado Situation," *UMWJ*, 10 Dec. 1903, 8.
54. Notarianni, 48.
55. "Bituminous Mining," *UMWJ*, 3 Dec. 1903, 8; Notarianni, 55.
56. Quoted in Notarianni, 53. For Demolli's biography, see Powell, "The 'Foreign Element,'" 133.
57. For example, John Gehr accused the Italians of disloyalty. UMWA, EBM, 6 Oct. 1903. Demolli proved Gehr's point by forming a dual union, the International Prosperity League, for which the UMWA fired him. "Rival to UMW," *RMN*, 18 April 1904, 2.
58. Notorianni, 51.
59. This was the Amerigo Vespucci Society. *RMN*, 8 Nov. 1903, 1, 2; "Coal Miners Ready to Lay Down Tools," *RMN*, 9 Nov. 1903, 1, 2; "Coup in the Strike," *RMN*, 18 Nov. 1903, 2.
60. "Coal Situation Is Improved," *RMN*, 14 Nov. 1903, 3.
61. For a discussion of anticlerical tendencies among Italian miners, see Schwieder, *Black Diamonds*, 75–79.
62. "Priest Is Assaulted," *RMN*, 20 Nov. 1903, 9.
63. "Struck by Meat Ax," *RMN*, 8 Dec. 1903, 9; U. S. Commissioner of Labor, *A Report on Labor Disturbances*, 341; "The Coal Miners' Strike," Pueblo *Labor Advocate*, 13 Nov. 1903, 1; *UMWJ*, 19 Nov. 1903, 5.
64. "Getting Ready for Resumption . . . ," *RMN*, 21 Nov. 1903, 16.
65. "Confidential Report" to S. Gompers, 28 Sept. 1903, carbon copy in Reel 6, Mitchell Papers.
66. CBLS, *Ninth Biennial Report, 1903–1904*, 189; "Negotiations Fail," *RMN*, 16 Nov. 1903, 1; U. S. Commissioner of Labor, *A Report on Labor Disturbances*, 335.
67. "John Mitchell is Needed . . . ," *RMN*, 22 Nov. 1903, Sect. 2, 5.

68. "Getting Ready for Resumption . . . ," *RMN*, 21 Nov. 1903, 16.
69. CBLS, *Ninth Biennial Report, 1903–1904*, 190; Clip., "Mother Jones Bars Settlement Plans," Governor Peabody Scrapbook No. 1, Western History Collection, Univ. of Colorado Library, Boulder.
70. Denver *Post*, 22 Nov. 1903, reprinted in Foner, 104.
71. Italians constituted a significant minority of the northern Colorado miners. Marshall had an Italian-speaking local. "Negotiations in Colorado Fail . . . ," *RMN*, 16 Nov. 1903, 1; "Mitchell May Come . . . ," *RMN*, 23 Nov. 1903, 2.
72. "Order Eastern Coal," *RMN*, 16 Nov. 1903, 3.
73. *RMN*, 23 Nov. 1903, 2; W.B. Wilson speech, AFL, *Report of Proceedings of the Twenty-Fifth Annual Convention, 1905*, 255;
74. For an important discussion of Mitchell's relations with operators, see Gowaskie.
75. Weinstein, *The Corporate Ideal*, 3–39; Scheinberg, 23–42.
76. "Case of the Miners Presented . . . ," Pueblo *Labor Advocate*, 27 Nov. 1903, 1. For evidence of Wardjon's socialism, see UMWA, *Minutes of the Sixteenth Convention, 1905*, 232. On Howell's socialism, see "Aid Strike in Colorado," *RMN*, 22 Jan. 1904, 3.
77. "Case of the Miners Presented . . . ," Pueblo *Labor Advocate*, 27 Nov. 1903, 1.
78. "Cause of Bitterness . . . ," *RMN*, 23 Nov. 1903, 2.
79. CBLS, *Ninth Biennial Report, 1903–1904*, 190; U. S. Commissioner of Labor, *A Report on Labor Disturbances*, 336.
80. "They Have Coal," *RMN*, 19 Nov. 1903, 11; "More Coal From South," *RMN*, 26 Nov. 1903, 1; "More Coal Now Being Received . . . ," *RMN*, 28 Nov. 1903, 4.
81. "His Views of Refusal," *RMN*, 22 Nov. 1903, 15; Randall speech, UMWA, *Minutes of the Sixteenth Convention, 1905*, 179.
82. "Cause of Bitterness . . . ," *RMN*, 23 Nov. 1903, 2.
83. W. R. Fairley speaking. UMWA, *Minutes of the Sixteenth Annual Convention, 1905*, 197.
84. "Eighty-five Laborers Go to Coal Fields," *RMN*, 5 Dec. 1903, 16; "Coup in the Strike," *RMN*, 18 Nov. 1903, 2; "Will Import Italians . . . ," *RMN*, 11 Nov. 1903, 9; "Best of Order," *RMN*, 11 Nov. 1903, 1, 4; "Northern Mines Are Rapidly Resuming," *RMN*, 12 Nov. 1903, 1, 8; "Every Independent Mine Working in Fremont," *RMN*, 27 Nov. 1903, 3; "Secretary Simpson Writes," *UMWJ*, 10 Dec. 1903, 5; "Chris Evans Writes," *UMWJ*, 31 Dec. 1903, 1; "Iron Miners Arriving . . . ," *RMN*, 7 Jan. 1904, 7.
85. "Mitchell's Letter," *UMWJ*, 12 Nov. 1903, 2.
86. *UMWJ*, 5 Nov. 1903, 4.
87. This occurred on 7 Dec. The dead were Luciano De Santos and Joseph Vilano. U. S. Commissioner of Labor, *A Report on Labor Disturbances*, 342; "Big Battle at Segundo," *RMN*, 8 Dec. 1903, 1; "Under Guard," *RMN*, 9 Dec. 1903, 1.
88. "Coal Situation Is Improved," *RMN*, 14 Nov. 1903, 3; "More Coal From South," *RMN*, 26 Nov. 1903, 1.
89. "Getting Ready for Resumption . . . ," *RMN*, 21 Nov. 1903, 16.
90. "Best of Order," *RMN*, 11 Nov. 1903, 1, 4.
91. Quoted in Suggs, *Colorado Conservatives Versus Organized Labor*, 473.
92. *Polly Pry* 1, No. 18 (2 Jan. 1904), 1.
93. That McParlan[d] had become the superintendent of the Denver Pinkerton office was reported in CBLS, *First Biennial Report, 1887–1888* (1888), 60; Pinkerton's National Detective Agency, *Second Annual Report, 1896*, File 314, CF&I Papers,

CSHS; Suggs, *Colorado Conservatives Versus Organized Labor*, 490; Mrs. Leonel Ross O'Bryan (Polly Pry) Papers, WHD/DPL.

94. In a legal dispute involving the Socialist Party in 1910, Mother Jones asserted in a written statement:

> That his [Barnes of the SP] still more infamous charge that I have led an immoral life is a more infamous lie. I am now 77 years old, have been a Widow 43 years. During the last 30 years I have been active in the labor movement . . . required to travel all over the United States and live with and under the observation of thousands of Fathers, Mothers, and Husbands and Wives, and but once has my honor as a woman been questioned prior to the attack of Barnes . . . and that was by the Mine Owners, Citizens Alliance, and Pinkerton Detective Agency in Denver Colo. . . .

Mother Jones statement, State of Illinois, County of Cook, in File 64 (Barnes Trial by SP), Thomas J. Morgan Papers, Univ. of Illinois, Champaign. For accounts contradicting this, see Featherling, 137. The UMWA attorney advised against a slander suit, noting the careful language in which the accusation was framed in order to avoid one. Henry Cohen (Denver) to John Mitchell, 19 January 1904, Reel 7, Mitchell Papers.

95. In 1914, the accusation was read into the *Congressional Record*. Hon. George J. Kindel, "Extension of Remarks," *Congressional Record*, 63d Congress, 2d Session, 13 June 1914, 3. It became the subject of correspondance by the President of the United States. Louis F. Post (Asst. Secy., Dept. of Labor) to Mr. Tumulty (Secy. to the President), 15 Nov. 1913, Series 4, File: "x902 Colorado Coal Problems Oct.–Nov. 15, 1913," Woodrow Wilson Papers. The story also appears in Mother Jones's F.B.I. file. File No. OG 8128, "F. B. I. Files Dated 1909–1924," RG 60, NA.

96. CBLS, *Ninth Biennial Report, 1903–1904*, 195–98; U. S. Commissioner of Labor, *A Report on Labor Disturbances*, 342.

97. "Coup in the Strike," *RMN*, 18 Nov. 1903, 2.

98. Clipping, "Awful Deed," New Castle *Nonpareil*, 18 Dec. 1903, John Lawson Scrapbooks, WHD/DPL; *RMN*, 19 Dec. 1903, 5.

99. For accounts of various beatings, see "John Simpson Report," CBLS, *Ninth Biennial Report, 1903–1904*, 195–98.

100. "Attempt to Murder Coal Leader," *UMWJ*, 17 Mar. 1904, 1; "Killing and an Assault . . . ," *RMN*, 5 Mar. 1904, 5.

101. UMWA, EBM, 4 Jan. 1904. UMWA, EBM, 24 Jan. 1904.

102. *Polly Pry* 1, No. 23 (6 Feb. 1904); UMWA, *Minutes of the Sixteenth Annual Convention, Jan. 16–23, 1905*, T. L. Lewis, 187–88; W. R. Fairley, 197–99; Chris Evans, 212; "Coal Miners Vote to Stay on Strike," *RMN*, 9 Feb. 1904, 12; Suggs, *Colorado Conservatives Versus Organized Labor*, 473.

103. "John Mitchell Address," UMWA, *Minutes of the Sixteenth Convention, 1905*, 11–12; "Special Convention, District No. 15," *UMWJ*, 7 April 1904, 7.

104. "Militia at Trinidad," *RMN*, 24 Mar. 1904, 5; Suggs, "The Colorado Coal Miners' Strike," 46; Suggs, *Colorado Conservatives Versus Organized Labor*, 468.

105. The paper was *Il Lavatore Italiano*. U. S. Commissioner of Labor, *A Report on Labor Disturbances*, 349–50; Suggs, "The Colorado Coal Miners' Strike," 46, 48; *RMN*, 23 Mar. 1904, 1.

106. Jones, 103; The deportation occurred on 26 Mar. 1904. U. S. Commissioner of Labor, *A Report on Labor Disturbances*, 351. See "Special Orders # 21, Mar. 26,

1904, Headquarters, First Prov. Battalion, NGC, Las Animas County, Military Campaign," Report, Trinidad Campaign, Colorado State Archives. See also Zeph T. Hill (Major Commanding) to Hon. James Peabody, 27 Mar. 1904 (Trinidad), Colorado National Guard Record, Roll 3, WHD/DPL.

107. "Union Leaders Arrested . . . ," *RMN*, 27 Mar. 1904, 1.

108. U. S. Commissioner of Labor, *A Report on Labor Disturbances*, 354; CBLS, *Ninth Biennial Report, 1903–1904*, 197; "Another Dastardly Assault," *UMWJ*, 5 May 1904, 3.

109. Wardjon speech, UMWA, *Minutes of the Sixteenth Annual Convention, 1905*, 231; Elizabeth Morris to John Mitchell, 7 May 1904, Reel 8, Mitchell Papers.

110. CBLS, *Ninth Biennial Report, 1903–1904*, 197.

111. *UMWJ*, 26 May 1904, 8.

112. "Serious Charges Against Soldiers," *RMN*, 1 April 1904, 7.

113. Smedley, 92–93.

114. Suggs, "The Colorado Coal Miners' Strike," 49; "Killing and an Assault . . . ," *RMN*, 15 Mar. 1904, 5.

115. G. H. Wilson to *UMWJ*, 24 Mar. 1904, 6.

116. UMWA, EBM, 30 April 1904; U. S. Commissioner of Labor, *A Report on Labor Disturbances*, 356–57; Secy.-Treas. Wilson Report, *Minutes of the 16th Annual Convention, 1905*, 43.

117. The convention lasted from 20 to 24 June 1904. "Chris Evans Writes," *UMWJ*, 14 July 1904, 6; UMWA, EBM, 24–25 Aug. 1904; U. S. Commissioner of Labor, *A Report on Labor Disturbances*, 357.

118. UMWA, EBM, 24 Aug. 1904; See Elizabeth Morris to John Mitchell, 24 and 28 June 1904, 9 July 1904, Reel 8, Mitchell Papers.

119. Printed in UMWA, EBM, 25 Aug. 1904.

120. William Beard to *UMWJ*, 18 Aug. 1904, 4.

121. U. S. Commissioner of Labor, *A Report on Labor Disturbances*, 359.

122. Robert Randall speech, UMWA, *Minutes of the Sixteenth Annual Convention, 1905*, 185.

123. "Chris Evans Writes," *UMWJ*, 8 Dec. 1904, 6; *UMWJ*, 8 Dec. 1904, 4; Randall speech, UMWA, *Minutes of the Sixteenth Annual Convention, 1905*, 182.

124. Wm. Beard to *UMWJ*, 18 Aug. 1904, 4.

125. Delegate Robert Randall speech, UMWA, *Minutes of the Sixteenth Annual Convention, 1905*, 176–78, 184–85.

126. John Mitchell speech, UMWA, *Minutes of the Sixteenth Annual Convention, 1905*, 14.

127. UMWA, *Minutes of the Sixteenth Annual Convention, 1905*, 229.

128. Ibid., 213.

129. See Laslett, *Labor and the Left*, 193–240.

130. Gluck, 221.

131. UMWA, *Minutes of the Sixteenth Annual Convention, 1905*, 230.

132. The new officers were John McLennan and Harry Douthwaite. "Chris Evans Writes," *UMWJ*, 8 Dec. 1904, 6.

133. UMWA, EBM, 28 April 1904; UMWA, *Minutes of the Sixteenth Annual Convention, 1905*, 179.

134. UMWA, *Minutes of the Sixteenth Annual Convention, 1905*, 281–82.

135. This was a headline. "Chris Evans Writes," *UMWJ*, 16 June 1904, 8.

136. Randall speech, UMWA, *Minutes of the Sixteenth Annual Convention, 1905*, 180.

CHAPTER 11: WHO LIVE LIKE RATS

1. L. M. Bowers to John D. Rockefeller, Jr., Denver, 12 Apr. 1909, Item 96, Box 28, Bowers Papers.
2. Mike Sekoria Test., U. S. Congress, House Subcommittee on Mines and Mining, 692.
3. UMWA, *Minutes of the Sixteenth Annual Convention, 1905,* 44, 176.
4. Adding to the troubles of the steel department was a recent ambitious and untimely expansion of the plant. CF&I, *Thirteenth Annual Report, 1905,* 1, 4; CF&I, *Twelfth Annual Report, 1904,* 8.
5. F. Gates to L. M. Bowers, 12 Nov. 1907, Item 81, Box 27, Bowers Papers.
6. Frederick Gates to L. M. Bowers, 19 Nov. 1907, Item 81, Box 27, Bowers Papers; *The National Cyclopaedia of American Biography* Vol. 31, 202–3. See also "Jesse Floyd Welborn," File 52, Welborn Papers.
7. Undated clip., Item 82, Box 27, Bowers Papers.
8. L. M. Bowers to Frederick Gates, Denver, 5 Oct. 1911, Item 98, Box 28, Bowers Papers.
9. L. M. Bowers to Rockefeller, Denver, 1 Mar. 1909, Item 96, Box 28, Bowers Papers.
10. Frederick Gates to L. M. Bowers, 10 June 1907, Item 81, Box 27, Bowers Papers; L. M. Bowers to F. T. Gates, Denver, 14 May 1909, Item 96, Box 28, Bowers Papers.
11. L. M. Bowers to Frederick Gates, Denver, 30 May 1907, Item 81, Box 27, Bowers Papers.
12. L. M. Bowers to Rockefeller, Denver, 2 Oct. 1909, Item 96, Box 28, Bowers Papers.
13. L. M. Bowers to Rockefeller, Denver, 1 Mar. 1909, Item 96, Box 28, Bowers Papers; CF&I, *Annual Report, 1909,* 3.
14. See Note 1.
15. "President's Remarks to Board of Directors," 26 June 1907, File 59, Welborn Papers.
16. *UMWJ,* 10 Mar. 1910, 2; CBLS, *Twelfth Biennial Report, 1909–1910* (1911), 19, 21.
17. CBLS, *Twelfth Biennial Report, 1909–1910,* 23.
18. Ibid.
19. Ibid., 28
20. *UMWJ,* 10 Feb. 1910, 1.
21. CICM, *Fourteenth Biennial Report, 1909–1910* (1911), 145; "More than 100 may be dead," *New York Times,* 1 Feb. 1910, 1; "Blow Up Kills Many in Kentucky Mine," *New York Times,* 2 Feb. 1910, 5; "New Mine Disaster Claims 68 Victims," *New York Times,* 3 Feb. 1910, 1; Humphrey, 31. See also *UMWJ,* 10 Feb. 1910, 1; L. M. Bowers to Frederick Gates, Denver, 1 Feb. 1910, Item 97, Box 28, Bowers Papers.
22. CBLS, *Twelfth Biennial Report, 1909–1910,* 25–26, 29.
23. "Facts About the Primero Explosion," *UMWJ,* 24 Feb. 1910, 1, 8.
24. L. M. Bowers to Frederick Gates, Denver, 1 Feb. 1910, Item 97, Box 28, Bowers Papers.
25. CICM, *Fourteenth Biennial Report, 1909–1910,* 7, 136.
26. See Note 66, Chapter 9.
27. CBLS, *Twelfth Biennial Report, 1909–1910,* 28.
28. "Lindsey Exhibit No. 5," U. S. Commission on Industrial Relations, Vol. 8, 7388.
29. "Las Animas Plans Home for Orphans," *RMN,* 12 Jan. 1913, 6.
30. CBLS, *Twelfth Biennial Report, 1909–1910,* 30. The CBLS had jurisdiction over the surface of the mines and the scales; the State Mine Inspector had jurisdiction

underground. Edwin V. Brake testimony, U. S. Commission on Industrial Relations, Vol. 8, 7234.
31. Kornbluh, 7.
32. WFM, EBM, 13 Dec. 1905.
33. WFM, EBM, 18 May 1905; 10 June 1905; 12 and 13 Dec. 1905; 22 and 24 May 1906; 14 June 1906; 10 Dec. 1906.
34. UMWA, EBM, 27 Mar. 1907; Thomas Kerby to *UMWJ*, 9 May 1907, 1.
35. John Lawson, John McLennan, and E. L. Doyle to John P. White, Denver, Colo., 10 July 1913, File: "D 15, 1913," UMWA Papers, Washington, D.C.
36. Alex Oberinsky was killed. Ibid; John Lawson Test., U. S. Congress, House Committee on Mines and Mining, 286.
37. Thos Kerby to *UMWJ*, 9 May 1907, 1.
38. John Lawson, UMWA, EBM, 28 Mar. 1911.
39. Harvey O'Higgins, "John Lawson's Story," *Metropolitan Magazine*, Dec. 1916; John Lawson Test., U. S. Congress, House Committee on Mines and Mining, 233.
40. WFM, Report of the Executive Board, 17 Dec. 1907.
41. UMWA, EBM, 21 Dec. 1907, 1 July 1908.
42. Frederick Gates to L. M. Bowers, 4 Dec. 1907, Item 81, Box 27; L. M. Bowers to Rockefeller, 2 Dec. 1912, Item 99, Box 28, Bowers Papers.
43. Brundage, 246.
44. UMWA, *Proceedings of the Seventeenth Annual Convention (1901)*, 180.
45. UMWA, EBM, 28 June 1907, 43.
46. "Agreement Between the UMWA and the Coal Operator's Association of Wyoming," 30 Aug. 1910, UPC Papers, RME.
47. "Corporate History of the Union Pacific Coal Co.," File: "Coal," Public Relations Department, UP Railroad, Omaha.
48. "Mitchell Comes to Denver . . . ," *RMN*, 14 July 1907, 2.
49. "Two Japs Are Delegates," *RMN*, 11 July 1907, 3.
50. McGovern, 124; John Lawson, UMWA, EBM, 30 June 1911.
51. U. S. Commission on Immigration, *Reports . . . Part 25: Japanese and Other Immigrant Races in the . . . Rocky Mountain States, Vol. 3.*
52. Application to the Social Security Board, Env. 6, Doyle Papers. See also E. L. Doyle Test., U. S. Congress, House Committee on Mines and Mining, 2185, 2187; Clip., "New Sec'y," Env. 15, Doyle Papers; Denver *Post*, 19 July 1954.
53. Doyle Diary, 1909–Mar. 1912, Env. 2, Doyle Papers.
54. Doyle Diary, 3 and 10 Feb. 1910, Env. 2, Doyle Papers.
55. Doyle Diary, 16, 17, 18, 19, and 27 Aug. 1909, and 3 Sept. 1909, Env. 2, Doyle Papers.
56. Doyle Diary, 14 Jan. 1910, Env. 2, Doyle Papers.
57. UMWA, EBM, 21 July 1910; McGovern, 125.
58. Geo Bisler, "Colorado Miners Stand Firm," *UMWJ*, 21 July 1910; McGovern, 125.
59. McDonald and Lynch, 86. The Cincinnati Wage Convention held in March 1910 was one of three raucous conventions that set this policy. *RMN*, 17 Jan. 1910, 2; "Laugh Stops Riot in Miners' Meeting," *RMN*, 23 Jan. 1910, 3; *RMN*, 1 April 1910, 1; "Report of President Lewis," and "Some Convention Happenings," *UMWJ*, 24 Mar. 1910, 1, 2; "Joint Convention at Cincinnati Fails to Agree," *UMWJ*, 31 Mar. 1910, 1.
60. McGovern, 125; Beshoar, 21. See also Doyle Test., U. S. Congress, House Committee on Mines and Mining, 2237.
61. Mother Jones speech, UMWA, *Convention Proceedings of the Twenty-Second Annual Convention*, 1911.

62. Doyle Test., U. S. Commission on Industrial Relations, 7023; Frank Garnier Test., U. S. Congress, House Committee on Mines and Mining, 2156, 2159; CBLS, *Twelfth Biennial Report, 1909–1910*, 267–68.
63. Doyle Diary, 2 and 3 Oct. 1910, Env. 2, Doyle Papers.
64. D. 15, UMWA, *Proceedings of the Tenth Annual Convention [of District 15] . . . 1913*, 48, found in Pamphlet File, Western History Collection, Univ. of Colorado, Boulder.
65. *UMWJ*, 29 Sept. 1910, 7; *UMWJ*, 13 Oct. 1910.
66. Injunction Decree, *Northern Coke Co. v. United Mine Workers* (District Court, Denver, 1910), Env. 3, Doyle Papers. See also McGovern, 128; and Doyle Test., U. S. Congress, House Committee on Mines and Mining, 2210.
67. Mother Mary Jones to Thomas J. Morgan, 27 Mar. 1911, Denver, Folder 7, Box 1, Thomas J. Morgan Papers, Univ. of Chicago; UMWA, EBM, 30 June 1911; *The Miners' Magazine*, 9 Feb. 1911, 9, 16 Feb. 1911, 3.
68. "Money Spent During Colorado Strike," Env. 7, Doyle Papers; Doyle Test., U. S. Congress, House Committee on Mines and Mining, 2190.
69. UMWA, EBM, 2 June 1911, 8 July 1912.
70. See E. L. Doyle to John P. White (President, UMWA), 30 April 1912, Denver, File: "D 15/1912," UMWA Papers, Washington, D.C.
71. CICM, *Fifteenth Biennial Report, 1911–1912* (Denver, 1913), 8–9; "Mortgage," Box 57, Rocky Mountain Fuel Co. Papers.
72. "Miners Will Win Inside a Month . . . ," *RMN*, 2 April 1910, 2.
73. "Reports of the Globe Inspection Company to the Rocky Mountain Fuel Co.," Box 6, Roche Papers (hereafter, Globe Inspection Co.).
74. Inspector D-85, "Reports for Trinidad Under Date of 7th, Denver Colo.," 9 Apr. 1910, Globe Inspection Co.
75. Inspector D-30 Piedmont, "Report," 19 and 20 Apr. 1910, Globe Inspection Co.
76. Inspector D-30, Piedmont, Denver, 27 Apr. 1910, Globe Inspection Co.
77. Inspector D-3, Trinidad, "Reports," 27 Apr. 1910, Globe Inspection Co.
78. Inspector D-3, Trinidad, "Reports," 20 and 21 May 1910, Denver, Globe Inspection Co.
79. Inspector D-30, Piedmont, "Reports," 30 Apr., Globe Inspection Company.
80. Inspector D-85, "Under Date of 6th, Denver, April 8th, 1910," Globe Inspection Co.
81. Inspector D-3, Trinidad, "Reports," 17 and 18 May 1910, Globe Inspection Co.
82. Inspector D-85, "Under Date of 7th, Denver, April 9, 1910," Globe Inspection Co.
83. Inspector D-3, "Reports," 13 Apr. 1910, "Trinidad," 14 Apr. 1910; E. P. Weaver, Pres. Globe Inspection Co. to E. E. Shumway, Pres., Rocky Mountain Fuel Co. 21 Apr. 1910, Globe Inspection Co.
84. Lawson Test., U. S. Congress, House Committee on Mines and Mining, 233.
85. "Special Report," 14 Apr. 1910; Inspector D-3, "Reports from Trinidad Under date of 12th," Globe Inspection Co.; UMWA, EBM, 5 Feb. 1913, 3 Feb. 1916; author interview with Mike Livoda, Denver, 1977.
86. Beshoar, 5.
87. "Inspector D-85 Under Date of the 6th," 8 Apr. 1910; Inspector D-3, "Reports from Trinidad Under Date of 13th," Denver, 14 Apr. 1910; "Reports Under date of May 2," Denver, 3 May 1910. Globe Inspection Co.
88. Robert Young Test., U. S. Commission on Industrial Relations, 6914.
89. James Fyler Test., U. S. Congress, House Committee on Mines and Mining, 1095.
90. Dave Williams Test., Ibid., 1116.

91. Mike Sekoria Test., Ibid., 692.
92. Chas. P. Gilday, "Strike in the West . . . ," *UMWJ*, 3 Nov. 1910.
93. District 15, UMWA, *Proceedings of the Tenth Annual Convention [of District 15] . . . 1913*, 7, found in Pamphlet File, Western History Collection, Univ. of Colorado Library, Boulder.
94. Inspector D-87, "Florence Reports Under date of May 11, Denver, May 12, and under May 19 (delayed)," Trinidad, 13 May 1910, Globe Inspection Co.
95. Chas. Moyer, Pres. WFM, to John Lawson in UMWA, EBM, 21 Oct. 1910; "Notes From the Convention of the W.F.M.," *The Miners' Magazine*, 30 July 1908, 5; Dubofsky, 106, 115–19.
96. *UMWJ*, 26 Sept. 1907, 5; UMWA, "Minutes of the Eighteenth Annual Convention, 1907," *UMWJ*, 24 Jan. 1905, 2; UMWA, EBM, 18 July 1913.
97. *UMWJ*, 20 Oct. 1910, 1; Robert Uhlich, "Colorado Explosion Discussed," *UMWJ*, 3 Nov. 1910, 8; Humphrey, 31. CBLS, *Twelfth Biennial Report, 1909–1910* (1910), 37. Again the mine inspector had found the mine in "reasonably safe condition." CICM, *Fourteenth Biennial Report, 1909–1910* (1911), 151.
98. "CF&I Executive Committee Meeting, Jan. 23, 1911 and Mar. 27, 1911," File 58, Welborn Papers.
99. The Victor American Fuel Co. explosion occurred on 8 Nov. *UMWJ*, 17 Nov. 1910, 1; CICM, *Fourteenth Biennial Report, 1909–1910*, 6; Humphrey, 31; CBLS, *Twelfth Biennial Report, 1909–1910*, 13; CICM, *Fifteenth Biennial Report, 1911–1912* (1913), 9, 10, 68; Fay (Bureau of Mines), 20, 246.
100. CICM, *Fourteenth Biennial Report, 1909–1910* (1911), 6.
101. UMWA, EBM, 21, 27, and 28 Oct. 1910.
102. Ibid., 18 July 1913.
103. "Edwin Perry to the Executive Board," Ibid., 1 May 1913.
104. In 1911, the UMWA was $310,026 in debt. UMWA, EBM, 24 May 1911; Secy.-Treas. Report, UMWA, *Proceedings of the Twenty-Second Convention, 1911*, 82.
105. T. L. Lewis speech, UMWA. *Proceedings of the Twenty-Second Annual Convention, 1911*, 68.
106. In April 1911, President White reported to the executive board that "the correspondence seems to be missing." UMWA, EBM, 4 Apr. 1911; Gluck, 243; McDonald and Lynch, 93–94. In his resignation speech to the executive board, White said of the beginning of his administration, "I was not left any records in this office to guide its destiny. . . ." UMWA, EBM, 25 Oct. 1917.
107. McDonald and Lynch, 92. For another incident, see UMWA, EBM, 30 June 1911, 318, 323–24.
108. *UMWJ*, 27 Jan. 1910, 15; *UMWJ*, 17 Feb. 1910.
109. Doyle to Wm. Green, 13 July 1913; Doyle to Edwin Perry, 7 June 1912, 10 Oct. 1912; Doyle to John P. White, 11 Mar. 1913, Env. 4, Doyle Papers.
110. John P. White, an opponent of T. L. Lewis, served as UMWA President from 1911–1917. Dubofsky and Van Tine, 24; McDonald and Lynch, 92.
111. E. L. Doyle to John P. White, 30 Apr. 1912, Env. 4, Doyle Papers.
112. UMWA, EBM, 16 Dec. 1911.
113. McDonald Test., U. S. Congress, House Subcommittee on Mines and Mining, 2020; John McLennan, E. L. Doyle, and John Lawson to John P. White, 10 July 1913, Denver, File: "D 15/1913," UMWA Papers, Washington, D.C.; McLennan Test., U.S. Congress, House Subcommittee on Mines and Mining, 2513–14.
114. L. M. Bowers to Frederick Gates, 15 Apr. 1912, Item 99, Box 28, Bowers Papers.
115. "Minutes of Joint Ways and Means Committee," 25 June 1912, Env. 10, Doyle Papers.

116. John P. White to E. L. Doyle, 23 June 1913, Env. 3, Doyle Papers.
117. Mother Jones speech, Charleston, W. Va. "Mother Jones on the Lawn of the Y.M.C.A.," Folder 2, Box 1, 1912–1913 Coal Strike Papers, W. Va. Collection, W. Va. U., Morgantown.
118. CICM, *Fifteenth Biennial Report, 1911–1912* (1913), 5; "Nationalities Employed in the Mines of Colorado During . . . 1912 . . . ," File: "D 15/1912," UMWA Papers, Washington, D.C.
119. Pete Hanraty to John P. White, 30 July 1912; Moran, Fitzgibben, and Paulsen to John P. White, 14 Aug. 1912, Denver, File: "D 15/1912," UMWA Papers, Washington, D.C.
120. Pete Hanraty et al. to John P. White, 30 July 1912, File: "D 15/1912," UMWA Papers, Washington D.C.
121. J. F. Welborn to L. M. Bowers, 8 Aug. 1912, Item 86, Box 27, Bowers Papers.
122. J. F. Welborn to L. M. Bowers, 24 Aug. 1912, Item 86, Box 27, Bowers Papers.
123. Doyle, McLennan and Lawson to John P. White, 10 July 1913, Denver, File: "D15/1913," UMWA Papers, Washington, D.C.
124. James McDonald Test., U. S. Congress, House Subcommittee on Mines and Mining, 2019–20.
125. Moran, Fitzgibbon, and Paulsen to John P. White, 14 Aug. 1912, File: "D 15/1912," UMWA Papers, UMWA, Washington, D.C.
126. L. M. Bowers to J. B. McKennan, Denver, 19 May 1911, Item 98, Box 28, Bowers Papers.
127. Militello Test., U. S. Congress, House Subcommittee on Mines and Mining, 2074, 2077.
128. McLennan Test., Ibid., 2513–14.
129. CBLS, *Twelfth Biennial Report, 1909–1910*, 32.
130. CBLS, *Twelfth Biennial Report, 1909–1910*, 27; Waleryon Korda to Woodrow Wilson, 29 Apr. 1914, Chicago, File 168733-a, RG 60, NA.
131. John McQuarrie Test., U. S. Commission on Industrial Relations, 6782; Inis Weed, "The Miner as Citizen," File: Commission on Industrial Relations, Box 10, RG 174, NA.
132. Author interview Livodas, Denver, Jan. 1977.
133. Tikas was born Illias Spantidakis. Brewster Test. and Doyle Test., U. S. Commission on Industrial Relations, 6664–65, 7023; Zeese Papanikolas.
134. L. M. Bowers to Rockefeller, 30 Sept. 1912, Item 99, Box 28, Bowers Papers.
135. James McDonald Test., U. S. Commission on Industrial Relations, 6770, 6772.
136. Author interview with Livodas, Denver, Jan. 1977; "Affidavit of Mike Livoda," U. S. Commission on Industrial Relations, 6950; CBLS, *Thirteenth Biennial Report, 1911–12* (1913), 60; Doyle Test., U. S. Commission on Industrial Relations, 6952.
137. "President's Report. CF&I, Executive Committee Meeting," 21 Apr. 1913, Denver, File 58, Welborn Papers.
138. Louis Tikas to Secy.-Treas. D.15, Denver, 28 Aug. 1913, File: "D 15/1913," UMWA Papers, Washington, D.C.
139. E. L. Doyle to Wm. Green, 30 July 1913, and E. L. Doyle to John P. White, May 19, 1914, Denver, Env. 4, Doyle Papers.
140. No. 52 report from Walsenburg; 24 Aug. 1913, U. S. Commission on Industrial Relations, 7062.
141. No. 94 from Superior, Colo., 18 Sept. 1913, Ibid., 7063.
142. "To Local Unions, UMWA," from John P. White, Frank J. Hayes, and Edwin Perry, 23 July 1913, Env. 6, Doyle Papers; UMWA, EBM, 16–19 July 1913.

143. No. 94, 18 Sept. 1913, Superior, Colorado, U. S. Commission on Industrial Relations, 7063.
144. Louis Tikas to Secy.-Treas., District 15, Denver, Colorado, 28 Aug. 1913, File: "D 15/1913," UMWA Papers, Washington D.C.
145. Adolph Germer to John P. White, 2 Sept. 1913, Walsenburg, File: 1911–1913, Box 1, Germer Papers.
146. Clip., from *The Miner's Magazine*, File: "Colo 2," Box 2, Mother Jones Papers. The detectives were G. W. Belcher (who shot Lippiatti) and Walter Belk. "Minutes Doyle Kept While Secretary for the Policy Committee, Aug. 4, 1913–Nov. 5, 1914," 16 Aug. 1913, Env. 11, Doyle Papers; Harry C. Farber Test., U. S. Congress, House Subcommittee on Mines and Mining, 1565; *UMWJ*, 21 Aug. 1913; "Town in Turmoil," *RMN*, 17 Aug. 1913, 1, 7; "Mine Organizer Killed in Fight," Denver *Times*, 17 Aug. 1913, 1, 3; "Miner and Detective in Pistol Duel," *The [Trinidad] Chronicle News*, 19 Aug. 1913, 1, 2.
147. "Minutes Doyle Kept . . . ," Env. 11, Doyle Papers. Mother Jones to T. V. Powderly, 20 Sept. 1913, Trinidad, File: "Personal Correspondence, I-J," Box A1–118, Powderly Papers.
148. Author interview with Livodas, Denver, Jan. 1977.
149. E. L. Doyle, "Minutes of the Policy Committee, Sept. 12, 1913," Env. 11, and Doyle to Frank Hayes, 15 Oct. 1912, Env. 4, Doyle Papers; *UMWJ*, 25 Sept. 1913, 2.
150. O'Neal, 77, 85.
151. "An Appeal to All Mine Workers in District 15 . . . ," Doyle Test., U. S. Commission on Industrial Relations, 7067. The Policy Committee consisted of UMWA Vice President Frank Hayes and District officers Doyle, McLennan, and Lawson. *UMWJ*, 4 Sept. 1913, 1.
152. McGovern, 145; Louis Tikas to the Policy Committee, 7 Oct. 1913, Denver, File: "D 15/1913," UMWA Papers, Washington, D.C.; *RMN*, 9 Sept. 1913, 4, 12 Sept. 1913, 3; "Ammons Meets Miners' Leader," *RMN*, 13 Sept. 1913, 12; "Expect Miners' Strike Tuesday," *RMN*, 14 Sept. 1913, 4.
153. "Hayes Predicts Strike Victory," *RMN*, 15 Sept. 1913, 3; E. L. Doyle, "Minutes of the Policy Committee," 14 Sept. 1913, Env. 11, Doyle Papers; *UMWJ*, 25 Sept. 1913, 2.
154. O'Neal, 87–88.
155. *UMWJ*, 25 Sept. 1913, 2; "Hayes Predicts Strike Vote," *RMN*, 15 Sept. 1913, 3.
156. "Strike Crisis Near," *RMN*, 16 Sept. 1913, 1, 5. "Hayes Predicts Strike Victory," *RMN*, 15 Sept. 1913, 3.
157. The coal miners grievances are all taken from "Read The Grievances of the Colorado Coal Miners," Trinidad, 15–16 Sept. 1913, File: "D 15/1913," UMWA Papers, Washington D.C.; *Proceedings, Special Convention of District 15, . . . Sept 16, 1913*, 4, Env. 10, Doyle Papers.
158. The demands of the Colorado Fuel and Iron strike were as follows:

> First, We demand recognition of the union.
> Second, We demand a 10 percent advance in wages on the tonnage rates and the following day-wage scale, which is practically in accord with the Wyoming day-wage scale.
> Third, We demand an eight hour day for all classes of labor in or around the coal mines and at coke ovens.
> Fourth, We demand pay for all narrow work and dead work, which includes brushing, timbering, removing falls, handling impurities, etc.

Fifth, We demand checkweighmen at all times, to be elected by the miners, without any interference by company officials in said election.

Sixth, We demand the right to trade in any store we please, and the right to choose our own boarding place and our own doctor.

Seventh, We demand the enforcement of the Colorado Mining laws and the abolition of the notorious and criminal guard system which has prevailed in the mining camps of Colorado for many years.

E. L. Doyle Test., U. S. Commission on Industrial Relations, 7017. The four demands already part of state law were the third, fifth, sixth, and seventh. McClurg, 206.

159. Louis Tikas to policy committee, Denver, 7 Oct. 1913, File: "D 15/1913," UMWA Papers, Washington D.C.

CHAPTER 12: THE VOICE OF THE GUN

1. Ethelbert Stewart to Louis F. Post, in Post to Pres. Wilson, 22 Nov. 1913, Woodrow Wilson Papers.
2. Clip., Lucy Huffaker interview with Mary Petrucci, File: "Colorado, 4," Box 2, Mother Jones Papers.
3. E. L. Doyle to Wm. Green, 30 Sept. 1913, Env. 4, Doyle Papers.
4. This was Don MacGregor. McGovern and Guttridge, 105.
5. Mother Jones speech, Starkville, Colo., 24 Sept. 1913, CIR Report, 7253.
6. E. L. Doyle to William Green, 30 Sept. 1913; E. L. Doyle to John P. White, 6 Oct. 1913, Env. 4, Doyle Papers.
7. L. M. Bowers to Rockefeller, 29 Sept. 1913, Item 100, Box 28, Bowers Papers. See also CF&I, *Twenty-Second Annual Report . . . 1914*, 5; W. B. Wilson, "Memorandum for the President," 10 Dec. 1913, File x902: "Colorado Coal Problems, Nov. 17–Jan. 1914," Series 4, Woodrow Wilson Papers.
8. The union set up tent colonies at Walsenburg, Aguilar, Rugby, Ludlow, Sopris, Piedmont, Segundo, and Tercio. "CF&I Marshall Shot . . . ," *RMN*, 25 Sept. 1913, 2; Boughton, Danks, VanCise, "Report of Military Commander," CIR Report, 7313; O'Neal, 105.
9. O'Neal, 100–101.
10. Ibid., 102.
11. Hayes Test., CIR Report, 7196.
12. McGovern, 163; *UMWJ*, 26 Feb. 1914; O'Neal, 100, 109–10, 138.
13. McLennan Test., CIR Report, 6526, 6980; Env. 18, Doyle Papers; "Mrs. Thomas Tells About Ludlow," *UMWJ*, 1 July 1964, 6.
14. Denver *Post*, 12 May 1945, 12; *UMWJ*, 15 Sept. 1910, 2; *RMN*, 15 May 1945, 25; Lawson Test., CIR Report, 7187; Benson; Beshoar; O'Neal, 104.
15. O'Neal, 108–9; author interview with Livodas, Denver, 1977; "U.S. Officers Came Back on It," *RMN*, 21 Oct. 1913, 4.
16. Petrucci Test., CIR Report, 8192.
17. "Mrs. Thomas Tells About Ludlow," *UMWJ*, 1 July 1964, 6.
18. *RMN*, 19 Feb. 1914, 2.
19. Author interview with Livodas, Denver, 1977.
20. O'Neal, 115.
21. Author interview with Livodas, Denver, 1977.
22. *The Miner's Advocate* (Denver), 17 Feb. 1914.
23. O'Neal, 112.
24. McGovern, 171; "Women and Children Beat Injure Nonunion Man," *RMN*, 25

Sept. 1913, 2; "Strikers at Walsen Attack Coal Miner," Trinidad *Chronicle-News*, 24 Sept. 1913, 1.

25. O'Neal, 114.
26. "Extracts From the Press . . . ," *The Miner's Magazine*, 23 Oct. 1913, 6.
27. O'Neal, 112.
28. Advertisement, *RMN*, 19 Oct. 1913, 11.
29. Mother Jones speech at Starkville, 24 Sept. 1913, CIR Report, 7253; UMWA, *Answer to the Military Occupation of the Coal Strike Zone, 1913–1914*, WHD/ DPL.
30. L. M. Bowers to Rockefeller, 29 Sept. 1913, Item 100, Box 28, Bowers Papers.
31. In autumn 1913, Ethelbert Stewart (Dept. of Labor) attempted to arbitrate the strike and failed. In late November, Secretary of Labor W. B. Wilson formulated an arbitration proposal that both sides rejected. In late autumn, President Woodrow Wilson corresponded with CF&I officials, to no avail. In February 1914, a congressional committee investigated the strike. On 1 May, the U. S. Army entered the strike region. On 29 April 1914, President Wilson ordered the Dept. of Labor to make another attempt at mediation. Federal mediators Hywel Davies and W. R. Fairley spent two months in the strike district. Their plan became the "Wilson Proposal." Finally, in late 1914 President Wilson appointed a three-man commission, ostensibly to arbitrate the strike, although by this time the strikers were defeated and there was nothing to arbitrate. Berman, 76–99.
32. L. M. Bowers to Frances [Bowers], 24 Sept. 1913, File: "Outgoing, 1913," Box 2, Bowers Papers.
33. Frame 000265–000281, Reel 1, Stewart Papers. For Stewart's reports, see File 41/10, Boxes 89, 90, RG 280, NA.
34. Ethelbert Stewart to Louis F. Post, in Post to Pres. Wilson, 22 Nov. 1913, Woodrow Wilson Papers.
35. McGovern, 187.
36. Ralph to Taylor, 24 Oct. 1913, Rocky Mountain Fuel Co. Papers.
37. "First Strike Death. . . . ," *RMN*, 25 Sept. 1913, 1; McGovern, 169–71.
38. *UMWJ*, 16 Oct. 1913; McGovern, 174.
39. McGovern, 190; "One Killed . . . ," *RMN*, 18 Oct. 1913, 1, 3; Zamboni Test., U. S. Congress, House Subcommittee on Mines and Mining, 790.
40. "Statement of Mr. Don MacGregor," File: "Colorado Strike, CIR Report," CIR Papers, RG 174, NA. See also McGovern and Guttridge, 105.
41. L. M. Bowers to Rockefeller, 18 Nov. 1913, Item 100, Box 28, Bowers Papers; McGovern, 195.
42. "Terror Reigns: Claim of Union," *RMN*, 20 Oct. 1913, 3; "Send No Troops: Plea of Union," *RMN*, 19 Oct. 1913, 10.
43. "Statement of Mr. Don MacGregor," File: "Colorado Strike, CIR Report," CIR Papers, RG 174, NA.
44. Clip. stamped 20 Oct. 1913, John Chase Scrapbook, WHD/DPL; "Ammons Spends Day in Strike Zone," *RMN*, 22 Oct. 1913 3; *UMWJ*, 30 Oct. 1913, 1; "Doyle Minutes of the Policy Committee," Env. 11, Doyle Papers.
45. O'Neal, 113.
46. The Seventh Street incident occurred on 24 Oct. McGovern, 191; CIR Report, 7334–43.
47. McGovern, 193; CIR Report, 7334–7343.
48. "Statement of Mr. Don MacGregor," File: "Colorado Strike, CIR Report," CIR Papers, RG 174, NA.

49. Pres. Woodrow Wilson to J. F. Welborn, 30 Oct. 1913; L. M. Bowers to Pres. Woodrow Wilson, 8 Nov. 1913, Item 100, Box 28, Bowers Papers.
50. Rockefeller to Pres. Woodrow Wilson, 21 Nov. 1913, Item 87, Box 27, Bowers Papers.
51. L. M. Bowers to Mr. Streeter, 23 Dec. 1913, Item 100, Box 28, Bowers Papers.
52. *UMWJ*, 6 Nov. 1913, 1.
53. McGovern and Guttridge, 139.
54. "Statement of Mr. Don MacGregor," File: "Colorado Strike, CIR Report," CIR Papers, RG 174, NA.
55. "Crooked Work Laid Bare," *UMWJ*, 20 Nov. 1913, 2.
56. McGovern, 239.
57. "Statement of Mr. Don MacGregor," File: "Colorado Strike, CIR Report," CIR Papers, RG 174, NA.
58. Karl Linderfelt Test., CIR Report, 6883.
59. Helen Ring Robinson Test., CIR Report 7212–13.
60. McGovern, 212.
61. The Trinidad *Chronicle-News*, 21 Nov. 1913, 1.
62. Van Cise Exhibit No. 2, CIR Report, 7324–26. See also p. 6809.
63. "Transcripts of Statements of Witnesses . . . ," Vol. 1, Box 1, Farrar Papers.
64. "Lifelines," 3rd Draft, 11 Feb. 1969, Minot Papers. One soldier voiced objections to being a "scab herder," for which he was court-martialed. Court-Martial Case No. 27: "Record of the General Court Martial," Military District of Colorado, Government Records, Special Series CNG, Colorado State Archives.
65. Mother Jones was deported on 4 Jan. 1914. On 12 Jan., she returned, was rearrested, and incarcerated in the San Rafael Hospital. She was released on 16 Mar. "The Military Occupation of the Coal Strike Zone of Colorado . . . ," WHD/DPL.
66. "Troops Deport Mother Jones," *RMN*, 5 Jan. 1914, 2; Clip., "Mother Jones Deported by Militia," 5 Jan. 1914, John Chase Scrapbooks, WHD/DPL.
67. "Troops Deport Mother Jones," *RMN*, 5 Jan. 1914, 2.
68. "Mother Jones Deported," Pueblo *Star Journal*, 7 Jan. 1914.
69. "Denver Suffragists Raise Cry . . . ," *RMN*, 7 Jan. 1914, 1; "Suffragists Score Ammons' Failure to Appoint Board," Denver *Post*, 7 Jan. 1914, 5; "Mother Jones Divides Women," *RMN*, 12 Jan. 1914, 4.
70. *RMN*, 7 Jan. 1914, 1; *RMN*, 13 Jan. 1914, 8; "MJ Divides Women," *RMN*, 12 July 1914; 4; *UMWJ*, 22 Jan. 1914, 17.
71. The telegrams can be found in File 168733-a, RG 60, NA.
72. *RMN*, 16 Jan. 1914, 4.
73. Report of the Commanding General, "The Military Occupation of the Coal Strike Zone . . . ," WHD/DPL.
74. The women's demonstration occurred on 22 Jan. "Great Czar Fell," *UMWJ*, 29 Jan. 1914, 4; Clip., Trinidad *Free Press*, George Minot Papers; *RMN*, 23 Jan. 1914, 1; Denver *Post*, 23 Jan. 1914, 9; J. W. Brown, "Military Despotism in Colorado," *UMWJ*, 26 Mar. 1914, 2.
75. Report of the Commanding General, "The Military Occupation of the Coal Strike Zone . . . ," WHD/DPL; Boughton Test., CIR Report, 6375.
76. The scheduled speaker was Mrs. Katherine Williamson. *RMN*, 23 Jan. 1914, 1; "Women Angry . . . ," Trinidad *Chronicle-News*, 23 Jan. 1914, 1, 3.
77. "Eyewitness Tells Story . . . ," Trinidad *Chronicle-News*, 23 Jan. 1914, 4.
78. Hayes, Lawson, Frampton, and Wilkinson to E. L. Doyle, 17 Dec. 1913, Doyle Papers.

79. Wm. Diamond to E. L. Doyle, 10 Jan. 1914, Doyle Papers.
80. E. L. Doyle to William Diamond, 31 Jan. 1914, Env. 4, Doyle Papers.
81. This letter gives the most coherent account in existence of Tikas's part in the strike. Tikas was killed on 20 Apr. 1914. Louis Tikas to UMWA, 10 Feb. 1914, File: "D 15, 1914," UMWA Papers, Washington D.C.
82. Davis Robb (Branch Office No. 3, Florence) to E. L. Doyle, 12 Dec. 1913, 21 Mar. 1914, 26 Mar. 1914, Env. 3, Doyle Papers.
83. William Green to Paul J. Paulson, 21 Mar. 1914, Env. 6, Doyle Papers.
84. E. L. Doyle to Wm. Green, 23 Mar. 1914, Env. 4, Doyle Papers.
85. UMWA, EBM, 2 Oct. 1913, 4 Feb. 1914, 8 May 1914; Edwin Perry, Secy.-Treas., UMWA to Doyle, 11 Apr. 1912, 19 Dec. 1912, 20 Jan. 1913, 10 Feb. 1913, Env. 2, Doyle Papers.
86. "Report Shows How Union Organizers Spend Thousands . . . ," Trinidad *Chronicle-News*, 24 Jan. 1914, 1.
87. Untitled typescript, Env. 18, Doyle Papers, WHD/DPL.
88. "Strike Probers Arrive Sunday," *RMN*, 6 Feb. 1914, 3.
89. *UMWJ*, 18 Feb. 1914, 1; "Probers Visit Strike Camps," *RMN*, 19 Feb. 1914, 2.
90. The committee's report and testimony are contained in U: S. Congress, House Subcommittee on Mines and Mining.
91. John P. White, Frank J. Hayes, and William Green to Pres. Woodrow Wilson, File: 168733-a, RG 60, NA.
92. E. L. Doyle to William Green, 11 Mar. 1914, Env. 4, Doyle Papers.
93. *UMWJ*, 9 Apr. 1914, 1.
94. Lt. C. A. Conners to Capt. Frost, Starkville, 18 Apr. 1914, Frost Papers.
95. "Mother Jones," *UMWJ*, 23 Apr. 1914, 2; "The Military Occupation of the Coal Strike Zone of Colorado . . . ," WHD/DPL.
96. CBLS, *Fourteenth Biennial Report*, 197; "G. P. West Report on Ludlow," 9, File: "West Report on Ludlow," Box: "Colorado Strike," CIR Papers, RG 174, NA.
97. This letter was obviously written before the massacre. "Military Rule Ended," *UMWJ*, 23 Apr. 1914, 1.
98. L. M. Bowers to Rockefeller, 18 Apr. 1914, Item 101, Box 28, Bowers Papers.
99. Lt. C. A. Conners to Capt. Frost, 14 Apr. 1914, Frost Papers.
100. Robinson Test., CIR Report, 7211.
101. Ibid.; Low Test., CIR Report, 6853.
102. Robinson Test., Ibid., 7211.
103. Mary Thomas O'Neal to Arthur Biggs, Pres., D 15, UMWA, 5 June 1970, [in author's possession].
104. "Pearl Jolly Speech . . . ," 21 May 1914, Meredith Papers.
105. Mary Thomas O'Neal to Arthur Biggs, Pres., D. 15, UMWA, 5 June 1970, [in author's possession].
106. CIR Report: Mrs. Dominiske Test., 8186; Mrs. Ed Tonner Affidavit, 7384; Mrs. Petrucci Test., 8194; William Snyder Test., 7371; Dr. Perry Jaffa Test., 7365; Mr. M. G. Low Test., 6853. Ometomica Covadle Affidavit; Leyor Fylor Affidavit; Mrs. James Fylor Affidavit; William Snyder Affidavit; Virginia Bertoloti Affidavit; Juanita Hernandez Affidavit; Mrs. Ed Tonner Affidavit. Affidavits found in CIR Papers, RG 174, NA. See also Mrs. Lee Champion speech, 21 May 1914, Washington, D.C., Meredith Papers.
107. O'Neal, 133–34.
108. Pearl Jolly's husband had left the strike region to look for work elsewhere. Several survivors contended that she and Tikas and were lovers. See Zeeze Papanikolas,

169–70; Pearl Jolly speech, 21 May 1914, Meredith Papers. See also Jolly Test., CIR Report, 6351.

109. Details of the massacre are taken from previously cited testimony and affidavits (see Note 106), with the following exceptions. All details regarding Mary Petrucci are taken from Petrucci Test., CIR Report, 8194; Lucy Huffaker interview with Mary Petrucci, undated Clip., File: "Colorado 4," Box 2, Mother Jones Papers. For the story of Tikas going back, see Pearl Jolly speech, 21 May 1914, Ellis Meredith Papers, CSHS. This account differs from that in Long, "Women of the Colorado Fuel and Iron Strike," in which I relied on Mary Thomas's version. Jolly's is more likely to be correct, because she was with Tikas at the time.

110. Memo, File 145, Box 7, CF&I Papers, CSHS.

111. An eyewitness account of the killing of Tikas is given in New York *World*, 5 May 1914.

112. McDonald Test., CIR Report, 6777.

113. Quoted in Zeese Papanikolas, 239.

114. E. L. Doyle to Jno P. White, 21 Apr. 1914, Env. 4, Doyle Papers.

115. L. M. Bowers to Rockefeller, 21 Apr. 1914, Item 101, Box 28, Bowers Papers.

116. "G. P. West Report on Ludlow," File: "West Report on Ludlow," Box: "Colorado Strike," CIR Papers, RG 174, NA.

117. "A Call to Rebellion," 22 Apr. 1914, Env. 17, Doyle Papers.

118. Author interview with Livodas, Denver, 1977.

119. Telegrams, Doyle to National Office, 22 and 23 Apr. 1914, Env. 4, Doyle Papers.

120. "1000 Strikers in Ambush for Troop Train," *New York Times*, 24 Apr. 1914, 10.

121. George P. West, "Report on Ludlow," 10 May 1914, 17, File: "West Report on Ludlow" Box: "Colorado Strike," CIR Papers, RG 174, NA.

122. Ibid., 3.

123. Rev. James McDonald Test., CIR Report, 6776.

CHAPTER 13: WE ARE UP AGAINST A STONE WALL

1. "Address by V. P. Hayes," Env. 17, Doyle Papers.

2. Ibid.

3. Cline, 160; Henry, 164.

4. Workman's Sick and Death Benefit Fund (New York) to Pres. Wilson, 23 Apr. 1914, File No. 168733-a, RG 60, NA. For numerous other telegrams, see File No. 168733-a, RG 60, NA, and File 41/10-E, Federal Mediation and Conciliation Papers.

5. United Garment Workers No. 139 to Pres. Wilson, 24 Apr. 1914, File No. 168733-a, RG 60, NA.

6. "1000 Strikers in Ambush for Troop Train," *New York Times*, 24 Apr. 1914, 10; McGovern, 309; Robinson, 247; "Arrangements Completed . . . ," *RMN*, 25 Apr. 1914, 4; "Determination of Mothers, Wives and Daughters . . . ," *RMN*, 26 Apr. 1914, 4; "Capital Ground Scene of Demonstrations," *RMN*, 27 Apr. 1914, 3.

7. "Citizens of Denver Scare Governor Ammons . . . ," *Appeal to Reason*, 2 May 1914, 3; McGovern, 418.

8. "Upton Sinclair Jailed," and "10,000 denounce Ludlow's Massacres," *Appeal to Reason*, 9 May 1914, 2, 16 and 23 May 1914; McGovern, 418.

9. *Appeal to Reason*, 30 May 1914; Upton Sinclair, "Chicago War Spreads . . . ," *Appeal to Reason*, 13 June 1914; Chicago *Tribune*, 1, 2, 3, and 4 May 1914.

10. UMWA Local 1802, Maryville, Ill., to Pres. Wilson, 27 Apr. 1914; Mass Meeting at Rock Springs, Wyo., to Pres. Wilson, 23 Apr. 1914; Arnot, Pa., UMWA Local 865, to Pres. Wilson, 28 Apr. 1914, File No. 168733-a, RG 60, NA.

11. J. C. Hudelson, V. P., First National Bank (Trinidad), to Pres. Wilson, 27 Apr. 1914, File No. 168733-a, RG 60, NA.
12. "Condensed Translation of Jack Corso," Johnston City, Ill., to Mr. Joseph Poggiani, File "D 15, 1914," UMWA Papers.
13. Boughton Test., CIR Report, 6363; "Ludlow, Being the Report of the Special Board of Officers . . . ," File 168733-a, RG 60, NA.
14. Jesse Welborn to McClement, 27 May 1914, File: "D 15, 1914," UMWA Papers; McGovern, 301.
15. "1000 strikers in Ambush," New York Times, 24 Apr. 1914, 10; Col. Edward Verdeckberg, "Report of the District Commander, Walsenburg, 28 Oct. 1913 to 5 May 1914," Verdeckberg Papers.
16. "Churchill Affidavit," 11 May 1914, CIR Report, 7383.
17. J. F. Welborn to Mr. McClement, 27 May 1914, File: "D 15, 1914," UMWA Papers; McGovern, 302–3.
18. "1,500 at Burial of Ludlow Dead," RMN, 25 Apr. 1914, 2.
19. Doyle Diary, 27 Apr. 1914, Env. 18, Doyle Papers,
20. "Striker Buried in Solemn Rite," RMN, 28 Apr. 1914, 3; "14 Dead in Day's Battle," RMN, 30 Apr. 1914, 1, 5.
21. UMWA, EBM, 5 May 1914, 8 May 1914; "UMWA Circular," 12 May 1914, Env. 6, Doyle Papers.
22. McGovern, 315–25.
23. E. L. Doyle to John P. White, 6 May 1914, Env. 4, Doyle Papers; McGovern, 318.
24. J. F. Welborn Test., CIR Report, 6722; Doyle Test., 6987. McGovern erred when he wrote: "Although there were several complaints from union spokesmen charging that the ban on strikebreakers had been violated, the charges were largely unsupported by the evidence." McGovern, 324.
25. Larsen, 74–82.
26. Clip., Lucy Huffaker interview with Mary Petrucci, File: "Colorado 4," Box 2, Mother Jones Papers.
27. L. M. Bowers to Rockefeller, 21 May 1914, Item 101, Box 28, Bowers Papers.
28. L. M. Bowers to Rockefeller, 16 Aug. 1914, Item 101, Box 28, Bowers Papers.
29. File: "CIR Correspondence and miscellaneous records relating to Ivy Lee," Box 7, CIR Papers.
30. George P. West to Frank Walsh, Denver, 27 June 1914, Box 10, CIR Papers.
31. D. 15 UMWA Special Convention, "Proceedings to Consider President Wilson's Proposition . . . Trinidad, Sept. 15–16, 1914," Western History Collection, Univ. of Colo., Boulder.
32. "Minutes, June 19 (Friday), June 22 (Monday), June 25 (Thursday), and Aug. 30 (Sunday) 1914, Env. 18, Doyle Papers.
33. "Minutes, Sept. 8, 1914," Ibid; UMWA, EBM, 1 Dec. 1914.
34. E. L. Doyle, "Minutes of a Meeting in Omaha with White, Green, Hayes . . . ," Env. 17, Doyle Papers; W. R. Fairley to W. B. Wilson, 8, 11, 14, and 16 Sept. 1914, File 41/10-A, Part IV, Box 89, Federal Mediation and Conciliation Service Papers.
35. D. 15 UMWA Special Convention, "Proceedings to Consider President Wilson's Proposition . . . Trinidad, Sept. 15–16, 1914," 4, 10, 11, 23, 42, Western History Collection, Univ. of Colo., Boulder.
36. McGovern, 349.
37. "Special Convention of UMWA, D. 15, 7 Dec. 1914," File 41/10-A, Federal Mediation and Conciliation Papers. As a face-saving procedure, the UMWA accepted the commission proposed in the Wilson proposal even though the operators

had rejected the proposal. For this commission's report, see Dispute Case Files, Box 92, RG 280, NA.
38. "Address by V. P. Hayes," Env. 17, Doyle Papers.

CHAPTER 14: ORGANIZING CONSENT

1. Eastman, "Class War in Colorado," 5.
2. Clip., L. M. Bowers, "Businessmen Waking Up," Leslie's Illustrated Weekly, Item 87, Box 27, Bowers Papers.
3. Bowers to Rockefeller, 18 Apr. 1914, Item 101, Box 28, Bowers Papers; CF&I, Twenty-Second Annual Report, 1914, 3, 13. Operations for 1913 had been quite profitable. Twenty-First Annual Report, 1913, 3.
4. "Colorado: America or Russia, Which?" File: "D 15," UMWA Papers.
5. Harper's Weekly 58 (May 1914), 4.
6. "J. C. H." to Mr. Thacher, 3 May 1914, Trinidad, File: "Colorado Strike Matters, T-Y," Box 20, RG 2, RFA, RAC.
7. Weinstein, The Decline of Socialism in America, 27, 87, 116–17.
8. Eastman, "The Nice People of Trinidad," 5–8; Eastman, "Class War in Colorado," 5–8.
9. Filler, Crusaders 98–99; Filler, "The Muckrakers in Middle America." Coal was an early theme of the muckrakers. See John Mitchell, "The Coal Strike," McClure's Magazine 20, No. 2 (Dec. 1902), 219–24; Ray Stannard Baker, "The Reign of Lawlessness; Anarchy and Despotism in Colorado," McClure's Magazine 23 (May–Oct. 1904), 43–56; Francis H. Nichols, "Children of the Coal Shadow," McClure's Magazine 20, No. 4 (Feb. 1903), 435–44.
10. Mellon, 419–20.
11. Fitch, "The Human Side of Large Outputs," 1706–1720; Theodore Roosevelt, "The Progressive Party," The Century Magazine 86, No. 6 (Oct. 1913), 833–34.
12. Clip., L. M. Bowers, "Businessmen Waking Up," Leslie's Illustrated Weekly, Item 87, Box 27, Bowers Papers.
13. Robinson, 246. See her obituary in the Denver Post, 10 July 1923, 1, 5.
14. "The Ludlow Camp Horror," New York Times, 23 Apr. 1914.
12. See also the Chicago Tribune and the New York World for April and May 1914.
15. "Strikers Veterans of Balkan War," New York Times, 25 Apr. 1914, 6; Wall Street Journal, 24 Apr. 1914, 6.
16. Jerome Greene to Rockefeller, 21 May 1914, File: "Mr. Greene," Box 20, Business Interests, RG 2, RFA, RAC.
17. Hiebert, 60, 64–69; Collier and Horowitz, 66; Seligmann, 14.
18. McGovern, 370–75.
19. L. M. Bowers to Mr. Chas. L. Sheldon, 29 Mar. 1913, Item 100, Box 28, Bowers Papers. See also Henry Cooper to L. M. Bowers, 2 June 1909, New York, Item 83, Box 27, Bowers Papers.
20. Quoted in Hiebert, 97.
21. Ivy Lee to Rockefeller, 16 Aug. 1914, File: "Colorado Strike Matter, Mr. Lee," Box 22, Business Interests, RG 2, RFA, RAC.
22. Quoted in McGovern, 407–10; Hiebert, 97–108. Copies of The Struggle in Colorado for Industrial Freedom may be found in File: "Colorado Strike, 1914," Box 95, Mary Heaton Vorse Papers, Labor History Archives, Wayne State Univ., Detroit.
23. Charles Eliot to Rockefeller, 1 July 1914, Asticou, Me., File: "Colo Strike," Box 19, Business Interests, RG 2, RFA, RAC.
24. Jerome D. Greene to Mr. Charles Loughridge, 27 July 1914, Folder 146, Box 19, Series 900, RG 3, RFA, RAC.

25. J. F. Welborn to Rockefeller, 8 Mar. 1915; Ivy Lee to Jerome D. Greene, 25 Sept. 1914; "Memorandum of Expenses, September 1–15, 1914, Statement A-4"; File: "Colorado Strike Matters, Mr. Lee," Box 22, Business Interests, RG 2, RFA, RAC.
26. Collier and Horowitz, 118.
27. Collier and Horowitz, 225–26; Herbert, 284–93 and 309–10.
28. Clip., L. M. Bowers, "Businessmen Waking Up," *Leslie's Illustrated Weekly*, Item 87, Box 27, Bowers Papers.
29. Filler, *Crusaders*, 330.
30. Ibid., 331–35, 339.
31. Ivy Lee to Rockefeller, 16 Aug. 1914, File: "Colorado Strike Matter, Mr. Lee," Box 22, Business Interests, RG 2, RFA, RAC.
32. Ivy Lee to Rockefeller, 16 Aug. 1914, File: "Colorado Strike Matter, Mr. Lee," Box 22, Business Interests, RG 2, RFA, RAC.
33. "Information Furnished by the Rockefeller Foundation in Response to Questionnaires Submitted by the United States Commission on Industrial Relations," Folder 130, Box 18, Series 900, RG 3, R Foundation Archives, RAC; Collier and Horowitz, 119–21; Scheinberg, 90–94.
34. Jerome Greene to W. L. MacKenzie King, 3 June 1914; Rockefeller, to W. L. MacKenzie King, 23 June 1914. Folder 149, Box 20, Series 900, RG 3, R Foundation Archives, RAC.
35. Rockefeller to W. L. MacKenzie King, 1 Aug. 1914; King to Rockefeller, 6 Aug. 1914. File: "CF&I, Welborn," Box 23, Business Interests, RG 2, RFA, RAC.
36. L. M. Bowers to Rockefeller, 16 Aug. 1914, File: "CF&I and Welborn," Box 23, Business Interests, RG 2, R Family Archives, RAC.
37. Jesse Welborn to Rockefeller, 20 Aug. 1914, File 2, Welborn Papers.
38. Quoted in Kolko, 206–7. See also Sklar.
39. Graebner, *Coal-Mining Safety*, 11–12, 34.
40. General works on the rise of corporate liberalism include Kolko; Weinstein, *The Corporate Ideal;* William Appleman Williams, *The Contours of American History* (Chicago, 1961, 1966), 345–478.
41. Weinstein, *The Corporate Ideal*, 182.
42. Adams, 146–75; Weinstein, *The Corporate Ideal*, 173.
43. Adams, 164. CIR Report, 7763–7897.
44. *New York Times*, 27 Jan. 1915, 6; *New York Times*, 29 Jan. 1915, 1; "Turns Upon John D. Jr." Washington *Post*, 29 Jan. 1915.
45. See, for example, "To Our Employees," 15 Dec. 1914, File: "CF&I Welborn," Box 23, Business Interests, RG 2, RFA, RAC.
46. Bowers was kept on a gradually shrinking retainer for several years, but did little more for the Rockefellers after Jan. 1915. Rockefeller to J. F. Welborn, 28 Dec. 1914; Bowers to J. F. Welborn, 29 Dec. 1914; Rockefeller to Bowers, 15 Jan. 1915 (Item 89, Box 27); Rockefeller to Bowers, 20 Jan. 1915; Rockefeller to Bowers, 29 Jan. 1915 (Item 89, Box 27); Bowers to Rockefeller, 21 Jan. 1915 (Item 102, Box 28), L. M. Bowers Papers. Bowers to Rockefeller, 17 Feb. 1915, File: "Bowers: 1912–1914," Box 21, Business Interests, RG 2, RFA, RAC.
47. CIR Report, 8592–8715; "Newsletters Say Mr. Rockefeller Ran War on Labor," New York *Herald*, 24 Apr. 1915; File: "Colorado Strike Statements to Press," Box 23, RG 2, RFA, RAC.
48. Many of these reforms were temporarily put into place during World War I. Weinstein, *The Corporate Ideal*, 207–13.
49. Harriman, 138.

50. Clip., "John D. Sleeps in Miners' Camp," File: "Colo 3," Box 2, Mother Jones Papers; "John D. Jr. Breaks Bread With Miners in Coal Camp," Denver *Post*, 20 Sept. 1915, 1.
51. The Rockefeller profit on the CF&I had amounted to nine million dollars in the previous twelve years. Graham, vii, liii.
52. Clip., "John D. Sleeps in Miners' Camp," File: "Colo 3," Box 2, Mother Jones Papers.
53. The plan is described in "Joint Representation of Employees and Management and Procedure in Industrial Relations . . . ," File 194, CF&I Papers; "Speech of Mr. David Griffiths, delivered Jan. 19, 1915," File: "CF&I, storage, F. C. #1," Box 13, RG 2, Business Interests, RFA, RAC. For previous company unions, see Perlman and Taft, 343–52.
54. "Mr. Welborn's Remarks at Pueblo meeting, Oct. 2, 1915," File 27, Welborn Papers.
55. Rockefeller to E. H. Weitzel, 13 Mar. 1916, File: "Welborn Corespondence, 1915–1922," Box 15, Business Interests, RG 2, Office of the Messrs. Rockefeller, RFA, RAC.
56. E. H. Weitzel to J. F. Welborn, 26 Jan. 1916, File: "J. F. Welborn Correspondence, 1915–1922." Box 15, Business Interests, RG 2, RFA, RAC.
57. Rockefeller to J. F. Welborn, 10 Feb. 1916, File: "J. F. Welborn Correspondence, 1915–1922," Box 15, Business Interests, RG 2, RFA, RAC.
58. E. H. Weitzel to J. F. Welborn, 1 Feb. 1916, File: "J. F. Welborn Correspondence, 1915–1922," Box 15, Business Interests, RG 2, RFA, RAC.
59. E. H. Weitzel to Jesse Welborn, 26 Jan. 1916, File: "J. F. Welborn Correspondence, 1915–1922," Box 15, Business Interests, RG 2, RFA, RAC.
60. E. H. Weitzel to Rockefeller, 19 Feb. 1916, File: "J. F. Welborn Correspondence, 1915–1922," Box 15, Business Interests, RG 2, RFA, RAC.
61. Rockefeller, to E. H. Weitzel, 13 Mar. 1916. Rockefeller's eight-page letter to Weitzel was tutored by W. L. MacKenzie King. King to Rockefeller, 25 Feb. 1916, File: "Welborn Correspondence, 1915–1922," Box 15, Business Interests, RG 2, Office of the Messrs. Rockefeller, RFA, RAC.
62. John A. Fitch, "Two Years of the Rockefeller Plan," 19.
63. Edward L. Doyle, "Miners Organize Their Own Unions," *American Federationist*, Nov. 1916.
64. "Circular District 15, Pueblo, Colo., 10 Feb. 1921," File: "UMWA 1915–1916," Box 16, Business Interests, RG 2, Office of the Messrs. Rockefeller, RFA, RAC.
65. Selekman and Van Kleeck, 232–65.
66. Brody, 55–59.
67. Selekman and Van Kleeck, 335, 303–49; Ware, *Labor in Modern Industrial Society*, 422–23.
68. "Offenses for Which an Employee May Be Suspended or Dismissed Without Further Notice, Effective May 1, 1921," File: "CF&I Co., 1927 Strike," Box 16, Business Interests, RG 2, Office of the Messrs. Rockefeller, RFA, RAC.
69. To refresh the reader's memory, the IWW was founded by the Western Federation of Miners in 1905.
70. "Open Letter," File: "CF&I Co. 1927 strike," Box 16, Business Interests, RG 2, Office of the Messrs. Rockefeller, RFA, RAC.
71. Dunn, 83; CF&I, *Thirtieth Annual Report, 1921*, 4; Selekman and Van Kleeck, 248–65.
72. Mother Jones speech, UMWA, *Porceedings of the Twenty-first Annual Convention, 1910*.

73. Gluck, 234–35.
74. The definitive biography is Dubofsky and Van Tine.
75. See John R. Shaefer and Henry Meyer, Auditors, 12 Sept. 1918, Doyle Papers; Dubofsky and Van Tine, 22–23.
76. Dubofsky and Van Tine, 25–31.
77. UMWA, EBM, 25 Oct. 1917.
78. Dubofsky and Van Tine, 31.
79. Duncan McDonald to Edward Doyle, 3 Jan. 1917, Springfield, Ill., Env. 2, Doyle Papers. John Walker submitted evidence of fraud to the executive board, but the board declined to investigate, because "the duties of the Board are confined to contests growing out of the report of the Tellers. . . ." UMWA, EBM, 9–20 Jan. 1917. See also John Walker to John R. Lawson, 1 Feb. 1917, Box 4, Walker Papers, Univ. of Illinois, Champaign/Urbana. Another example of shady doings was the Adolph Germer case. UMWA, EBM, 16–20 Nov. 1914, 2 Feb. 1915, 15 Sept. 1915.
80. Dubofsky and Van Tine, 30.
81. UMWA, *Proceedings of the Twenty-Fifth Consecutive and Second Biennial Convention, 1916*, Vol. 2, 756; Env. 18, Doyle Papers.
82. Dubofsky and Van Tine, 34.
83. Frank Hayes, Robert Harlin, J. M. Zimmerman, R. Watkins and William Green to the executive board, UMWA, 17 Jan. 1917, International Executive Board Documents and Circulars, File: "1917 F," UMWA Papers.
84. UMWA, EBM, 18, 19, and 20 Jan. 1917. As his membership was not in jeopardy, Lawson did not have the privilege of going over the head of the executive board to the ultimate decision-making body—the convention. John P. White to E. L. Doyle, 13 Feb. 1917, Env. 3, Doyle Papers.
85. Telegram, John P. White to E. L. Doyle, 12 Feb. 1917, Env. 3; John P. White et al., "To the Officers and Members of All Local Unions of District No. 15, UMWA," Env. 2, Doyle Papers.
86. John R. Lawson, "An Official Circular," 10 Feb. 1917, Env. 6, Doyle Papers. For the suspension of Districts 17 and 29, see UMWA, EBM, 2, 3, and 4 Feb. 1916.
87. Mother Jones to Doyle, 10 Apr. 1917, Env. 2, Doyle Papers.
88. Doyle to Mother Jones, 13 Nov. 1918, Env. 5, Doyle Papers.
89. John R. Lawson and E. L. Doyle, *An Election Steal*, File: 1.A2, Box 18, Colorado State Federation of Labor Papers, Western Historical Collection, University of Colorado Library, Boulder. "Mine Workers' President Dodges Investigations of the Election Steals in District 15," Env. 18, Doyle Papers; UMWA, EBM, 10 Apr. 1918, 11 July 1919. The formation of the dual union is reported in "Delegates Vote to Secede from UMW," *The Pueblo Chieftain*, 28 Feb. 1918, 5.
90. Dubofsky and Van Tine, 34, 35, 37; UMWA, EBM, 25 Oct. 1917.
91. UMWA, EBM, 25 Oct. 1917; Dubofsky and Van Tine, 37.
92. The board minutes of 14 Jan. 1919 state that this was the first board meeting since April 1917, an error given that the 25 Oct. 1917 board meeting approved Lewis's appointment as vice president. UMWA, EBM, 14 Jan. 1919.
93. UMWA, EBM, 10 and 11 July 1919; Frank J. Hayes to John L. Lewis, 1 Jan. 1920 (Denver) printed in UMWA, EBM, 6 Feb. 1920; James Lord to Adolph Germer, 14 Oct. 1951, File: "UMWA, 1900–1920," Box 13, Germer Papers.
94. Philip Murray to John L. Lewis, 15 Mar. 1923, File: "Misc. Alpha," UMWA Papers.
95. See U. S. Coal Commission, *Report*, five Parts. Senate Doc. 195, (Washington, D.C., 1925).
96. The first similar bill was introduced to the Senate by Senator William Kenyon, a

388 · WHERE THE SUN NEVER SHINES

Republican from the coal state, Iowa. The other was a proposal by the Secretary of Commerce. Herbert Hoover had been the world's best-known mining engineer, an early advocate of bargaining with labor in exchange for company-imposed efficiency in the production process. Jett Lauck was the UMWA economist who helped to draw up Lewis's bill. The Supreme Court declared the NRA unconstitutional in 1937. It was replaced by Robert Wagner's National Labor Relations Act, which made company unions unconstitutional. Seltzer, 51–55.

98. For Lewis's activities in the 1940s–1960s, see Dubofsky and Van Tine; Finley; Seltzer.
99. John D. Rockefeller, Jr., "Labor and Capital—Partners," *Atlantic Monthly* 117 (Jan. 1916), 12–21.
100. Weinstein, *The Corporate Ideal,* 237–38. For the suppression of dissenters, see William Preston, Jr., *Aliens and Dissenters: Federal Suppression of Radicals, 1903–1933,* (New York: 1963).
101. Mother Jones speech, UMWA, *Proceedings of the Twenty-Second Convention, 1911,* 259.
102. Philip Foner, ed., 297.

BIBLIOGRAPHY

BOOKS AND ARTICLES

Adams, Graham, Jr. *Age of Industrial Violence: 1910–1915: The Activities and Findings of the United States Commission on Industrial Relations.* New York: Columbia University Press, 1966.

Alden, H.M. "The Pennsylvania Coal Region." *Harper's Magazine* 27 (Sept. 1863).

Amsden, Jon, and Stephen Brier. "Coal Miners on Strike." *Journal of Interdisciplinary History* 7, No. 4 (Spring 1977), 583–616.

Ashton, T.S. *The Industrial Revolution: 1760–1830.* London and New York: Oxford University Press, 1948, 1969.

———. *Iron and Steel in the Industrial Revolution.* New York: Augustus M. Kelley, 1963, 1968.

Ashton, T.S., and Joseph Sykes. *The Coal Industry of the Eighteenth Century.* Manchester, England: Manchester University Press, 1929.

Aurand, Harold W. "The Workingman's Benevolent Association." *Labor History* 7, No. 1 (Winter 1966), 19–34.

———. "The Anthracite Strike of 1887–1888." *Pennsylvania History* 35 (April 1968), 189–95.

———. "Diversifying the Economy of the Anthracite Regions, 1880–1900." *The Pennsylvania Magazine of History and Biography* 94, No. 1 (Jan. 1970), 54–64.

———. *From the Molly Maguires to the United Mine Workers.* Philadelphia: Temple University Press, 1971.

Aurand, Harold W., and William Gudelunas, Jr. "The Mythical Qualities of the Molly Maguires." *Pennsylvania History* 49, No. 2, (Apr. 1982), 91–105.

Bailey, Kenneth R. "A Judicious Mixture: Negroes and Immigrants in the West Virginia Mines, 1880–1917." *West Virginia History* 34 (Jan. 1973), 141–61.

Baker, Ray Stannard. "The Reign of Lawlessness: Anarchy and Despotism in Colorado." *McClure's Magazine* 23 (May–Oct. 1904), 43–56.

Bancroft, Hubert Howard. *History of the Pacific States, Vol 20: History of Nevada, Colorado, and Wyoming.* San Francisco: The History Co., 1890.

Barendse, Michael A. *Social Expectations and Perception: the Case of the Slavic Anthracite Workers.* University Park, Pa.: Pennsylvania State University Press, 1981.

Barnum, Darold T. *The Negro in the Bituminous Coal Mining Industry* (Report No. 14 of *The Racial Policies of American Industries*). Philadelphia: University of Pennsylvania Press, 1970.

Beachley, Charles E. *History of the Consolidation Coal Co., 1864–1934.* New York: The Company, 1934.
" 'Befo de Wah' Mine Yields Curious Relics." *Coal Age* 25, No. 8 (21 Feb. 1924), 273–76.
Berg, Charles. "Process Innovation and Changes in Industrial Energy Use." *Energy II: Use, Conservation, and Supply.* Edited by Philip H. Abelson and Allen L. Hammond. Washington, D.C.: American Association for the Advancement of Science, 1978.
Berman, Edward. *Labor Disputes and the President of the United States.* New York: Columbia University Press, 1924.
Berthoff, Roland Tappan. *British Immigrants in Industrial America, 1790–1950.* Cambridge, Mass.: Harvard University Press, 1953.
———. "The Social Order of the Anthracite Region, 1825–1902." *The Pennsylvania Magazine of History and Biography* 89, No. 3 (July 1965), 261–91.
Beshoar, Barron. *Out of the Depths: The Story of John R. Lawson.* Denver: Golden Bell Press, n.d.
Billinger, Robert D. *Pennsylvania's Coal Industry.* Gettysburg: The Pennsylvania Historical Association, 1954.
Bimba, Anthony. *The Molly Maguires.* New York: International Publishers, 1932.
Binder, Frederick Moore. *Coal Age Empire: Pennsylvania Coal and its Utilization to 1860.* Harrisburg, Pa.: Pennsylvania Historical and Museum Commission, 1974.
Black, Isabella. "American Labor and Chinese Immigration." *Annals of Wyoming* 12 (Jan. 1940), 47–55, and (Apr. 1940), 153–61.
———. "The Blind Spot." *Coal Age* 25, No. 14 (3 Apr. 1924).
Bodnar, John. "Immigration and Modernization: The Case of Slavic Peasants in Industrial America." In *American Workingclass Culture,* edited by Milton Cantor. Westport, Conn.: Greenwood Press, 1979.
———. *Workers' World: Kinship, Community, and Protest in an Industrial Society, 1900–1940.* Baltimore: Johns Hopkins University Press, 1982.
———. *Anthracite People: Families, Unions and Work, 1900–1940.* Harrisburg, Pa.: Pennsylvania Historical and Museum Commission, 1983.
Boemeke, Manfred Franz. *The Wilson Administration, Organized Labor, and the Colorado Coal Strike.* Ann Arbor, Mich.: University Microfilms International, 1983.
Booth, Stephane Elise. *The Relationship Between Radicalism and Ethnicity in Southern Illinois Coal Fields, 1870–1940.* Ann Arbor, Mich.: University Microfilms International, 1983.
Boston, Ray. *British Chartists in America, 1839–1900.* Totawa, N.J.: Rowman and Littlefield, Inc., 1971.
Bowen, Eli. "Coal and the Coal-Mines of Pennsylvania," *Harper's New Monthly Magazine* 15, No. 88 (Sept. 1857).
Bowers, L.M. "The Great Strike in Colorado." *Leslie's Illustrated Weekly Newspaper,* 5 Feb. 1914, 127.
Boyd, R. Nelson. *Coal Pits and Pitmen.* London: Whitaker and Co., 1895.
Braverman, Harry. *Labor and Monopoly Capital.* New York: Monthly Review Press, 1974.
Brier, Stephen. "Interracial Organizing in the West Virginia Coal Industry." In *Essays in Southern Labor History,* edited by Gary M. Fink and Merl E. Reed. Westport, Conn.: Greenwood Press, 1977.
Briggs, Asa. *The Power of Steam.* Chicago: University of Chicago Press, 1982.
Brody, David. "The Rise and Decline of Welfare Capitalism." In *Workers in Industrial America,* edited by David Brody. New York: Oxford University Press, 1980.
Broehl, Wayne G. *The Molly Maguires.* Cambridge, Mass.: Harvard University Press, 1964.

Brophy, John. *A Miner's Life.* Madison, Wis.: University of Wisconsin Press, 1964.

Brundage, David. *The Making of Working Class Radicalism in the Mountain West: Denver, Colorado, 1880–1903.* Ann Arbor, Mich.: University Microfilms International, 1982.

Bryons, Bill. "Coal Mining in Twentieth Century Wyoming: A Brief History." *Journal of the West* 21 (Oct. 1982), 24–35.

Bunting, David G. *Statistical View of the Trusts.* Westport, Conn.: Greenwood Press, 1974.

Burgess, George H., and Miles C. Kennedy. *Centennial History of the Pennsylvania Railroad.* Philadelphia: The Pennsylvania Railroad Co., 1949.

Buttrick, John. "The Inside Contract System," *Journal of Economic History* 12, No. 3 (Summer 1952), 205–21.

Campbell, Robert A. "Blacks and the Coal Mines of Western Washington, 1888–1986." *Pacific Northwestern Quarterly* 73 (1982), 146–55.

Cardwell, D. S. L. *Turning Points in Western Technology.* New York: Science History Publications, 1972.

Chandler, Alfred, Jr. *The Railroads: the Nation's First Big Business.* New York: Harcourt, Brace and World, 1965.

———. "Anthracite Coal and the Beginning of the Industrial Revolution in the United States." *Business History Review* 46, No. 2 (Summer 1972), 141–81.

———. *The Visible Hand: The Managerial Revolution in American Business.* Cambridge, Mass.: Harvard University Press, 1977.

Cleland, Hugh G. "The Industrial Revolution." In *Encyclopedia Americana.* 1982. s.v.

Cline, Howard F. *The United States and Mexico.* New York: Atheneum, 1966.

Cochran, Thomas. *Railroad Leaders, 1845–1890.* Cambridge, Mass.: Harvard University Press, 1953.

Coleman, Walter J. *The Molly Maguire Riots.* New York: Arno and the New York Times, 1936, 1969.

Collier, Peter, and David Horowitz. *Rockefeller: An American Dynasty.* New York: Holt, Rinehart and Winston, 1976.

Combs, B. L. *Those Clouded Hills.* London: Cobbett Publishing, 1944.

Conway, Alan. *The Welsh in America: Letters From the Immigrants.* Minneapolis: University of Minnesota Press, 1961.

Corbin, David Alan. *Life, Work, and Rebellion in the Coal Fields: The Southern West Virginia Miners, 1880–1922.* Urbana: University of Illinois Press, 1981.

Crane, Paul, and Alfred Larson. "The Chinese Massacre." *Annals of Wyoming* 12 (Jan. 1940), 47–55, and (Apr. 1940), 153–61.

Daddow, Samual Harries, and Benjamin Bannan. *Coal, Iron, and Oil, or the Practical Miner.* Pottsville, Pa., 1866.

Davies, Edward J., II. *The Anthracite Aristocracy.* DeKalb, Ill.: Northern Illinois University Press, 1985.

Davis, Jerome. "Experimenting With the Human Mechanics of Industry." *Industrial Management* 15, No. 1, 23–26.

Deane, Phyllis. *The First Industrial Revolution.* Cambridge: Cambridge University Press, 1965, 1979.

Dekok, David. *Unseen Danger.* Philadelphia: University of Pennsylvania Press 1986.

Derry, T.K., and Trevor I. Williams. *A Short History of Technology From Earliest Times to A.D. 1900.* London and New York: Oxford University Press, 1961.

Dix, Keith. *Work Relations in the Coal Industry: The Hand Loading Era, 1880–1930.* Morgantown, W.Va.: West Virginia University Press, 1977.

Dobb, Maurice. *Studies in the Development of Capitalism.* New York: International Publishers, 1963.

Donovan, Arthur. "Carboniferous Capitalism: Excess Productive Capacity and Institutional Backwardness in the Coal Industry." In *Energy in American History,* edited by Arthur Donovan. New York: Pergamon Press, 1983.

Drury, Doris M. *A Study of the Literature on Accidents in the Coal Mines of the U.S. With Comparisons of Records in Other Coal Producing Countries.* Ann Arbor, Mich.: University Microfilms International, 1965.

Dubofsky, Melvyn. *We Shall Be All: A History of the IWW.* New York: Quadrangle/The New York Times Book Co., 1977.

———. *Industrialism and the American Worker, 1865–1920.* Arlington Heights, Ill.: AHM Publishing, 1975.

Dubofsky, Melvyn, and Warren Van Tine. *John L. Lewis.* New York: Quadrangle/The New York Times Book Co., 1977.

Dunn, Robert W. *Company Unions.* New York: Vanguard Press, 1927.

Easterlin, Richard A. "Population." In *Encyclopedia of American Economic History.* Vol. 1, edited by Glen Porter. New York: Charles Scribner's Sons, 1980.

Eastman, Max. "Class War in Colorado." *The Masses* 5, No. 9 (June 1913), 5–8.

———. "The Nice People of Trinidad." *The Masses* 5, No. 10 (July 1914), 5–8.

Eavenson, Howard Nicholas. *The First Century and a Quarter of the American Coal Industry.* Baltimore, Md.: Waverly Press, 1942.

Engels, Frederick. *The Condition of the Working Class in England.* (English Edition, 1892). Moscow: Progress Publishers, 1973.

Erickson, Charlotte. *American Industry and the European Immigrant, 1860–1885.* Cambridge, Mass.: Harvard University Press, 1957.

Evans, Chris. *The History of the United Mine Workers of America.* 2 Vols. Indianapolis, Ind.: United Mine Workers, 1914, 1918.

Featherling, Dale. *Mother Jones, the Miners' Angel.* Carbondale, Ill.: Southern Illinois University Press, 1974.

Filler, Louis. *Crusaders for American Liberalism,* New York: Collier Press, 1939, 1961.

———. "The Muckrakers in Middle America." In *Muckraking: Past, Present and Future,* edited by John M. Harrison and Harry H. Stein. University Park, Pa.: Pennsylvania State University Press, 1973.

Fink, Leon. *Workingmen's Democracy: The Knights of Labor and American Politics.* Urbana, Ill.: University of Illinois Press, 1983.

———. "The Uses of Political Power." In *Working-Class America,* edited by Frisch and Walkawitz. Urbana, Ill.: University of Illinois Press, 1983.

Finley, Joseph. *The Corrupt Kingdom: The Rise and Fall of the United Mine Workers.* New York: Simon and Schuster, 1972.

Finney, C. S., and John Mitchell. *A History of the Coke Industry in the United States.* Eastern Gas and Fuel Associates, 1961.

Fishback, Price V. *Employment Conditions of Blacks in the Coal Industry, 1900–1930.* Ann Arbor, Mich.: University Microfilms International, 1983.

Fisher, Chris. "The Free Miners of the Forest of Dean." In *Independent Collier: The Coal Miner as Archetypal Proletarian Reconsidered,* edited by Royden Harrison. Hassocks, Sussex: Harvester Press, 1978.

Fitch, John. "The Human Side of Large Outputs, V: The Steel Industry and the People in Colorado." *The Survey,* 3 Feb. 1912, 1706–20.

———. "Two Years of the Rockefeller Plan." *The Survey,* 6 Oct. 1917, 14–20.

Flinn, Michael W., with David Stoker. *The History of the British Coal Industry.* Vol. 2, *1700–1830: The Industrial Revolution.* Oxford: Oxford University Press, 1984.

Foner, Eric. "Class, Ethnicity, and Radicalism in the Gilded Age: The Land League and Irish-America." *Marxist Perspectives* 1 (Summer 1978), 6–55.

Foner, Philip S., ed. *Mother Jones Speaks: Collected Speeches and Writings.* New York: Monad Press, 1983.

Forbes, R. J. *Studies in Ancient Technology* 2d. ed. Vol. 1. Leiden, Netherlands: E.J. Brill, 1964.

Fraser, Marianne. "Warm Winters and White Rabbits: Folklore of Welsh and English Coal Miners." *Utah Historical Quarterly* 51 (1983), 246–258.

Fritz, Percy. *Colorado: the Centennial State.* New York: Prentice-Hall, Inc., 1941.

George, J. E. "The Coal Miners' Strike of 1897." *Quarterly Journal of Economics* 12 (Jan. 1898), 186–94.

Gibbons, P.E. "The Miners of Scranton, Pa." *Harper's New Monthly Magazine* 55 (Nov. 1877), 916–27.

Ginger, Ray. "Managerial Employees in Anthracite, 1902" *Journal of Economic History* 14 (1954), 146–157.

Gluck, Elsie. *John Mitchell: Labor's Bargain with the Gilded Age.* New York: John Day Co., 1929.

Goodrich, Carter. *The Miner's Freedom.* Boston: Marshal Jones, 1925.

Gordon, David M., Richard Edwards, and Michael Reich. *Segmented Work, Divided Workers: The Historical Transformation of Labor in the United States.* Cambridge: Cambridge University Press, 1982.

Gowaskie, Joe. "John Mitchell and the Anthracite Mine Workers: Leadership Conservatism and Rank and File Militancy." *Labor History* 27, No. 1 (Winter, 1985–1986), 54–83.

Graebner, William. "Great Expectations: The Search for Order in Bituminous Coal, 1890–1917." *Business History Review* 48 (Spring 1974), 49–72.

———. *Coal Mining Safety in the Progressive Period:* Lexington, Ky.: University Press of Kentucky. 1976.

Graham, John. Intro. to *The Coal War* by Upton Sinclair, vii–xcii. Boulder, Colo.: Colorado Associated University Press, 1976.

Green, Archie. *Only a Miner: Studies in Recorded Coal Mining Songs.* Urbana, Ill.: University of Illinois Press, 1972.

Green, James R. *Grass-Roots Socialism: Radical Movements in the Southwest, 1895–1943.* Baton Rouge, La.: Louisiana State University Press, 1978.

Greenberg, Dolores. "Energy Flow in a Changing Economy, 1815–1880." In *An Emerging Independent American Economy, 1815–1875*, edited by Joseph R. Frese, S.J. and Jacob Judd. Tarrytown, N.Y.: Sleepy Hollow Press, 1980.

———. "Reassessing the Power Patterns of the Industrial Revolution: An Anglo-American Comparison." *American Historical Review* 87 (1982), 1237–61.

Greene, Victor R. "A Study in Slavs, Strikes, and Unions: The Anthracite Strike of 1897." *Pennsylvania History* 31 (Apr. 1964), 199–215.

———. "The Poles and Anthracite Unions in Pennsylvania." *Polish American Studies* 22 (Jan.–July 1965), 10–16.

———. *The Slavic Community on Strike: Immigrant Labor in Pennsylvania Anthracite.* Notre Dame, Ind.: Notre Dame University Press, 1968.

Grob, Gerald N. *Workers and Utopia: A Study of Ideological Conflict in the American Labor Movement, 1865–1900.* Evanston, Ill.: Northwestern University Press, 1961.

Grogan, Dennis S. "Unionism in Boulder and Weld Countries to 1890." *The Colorado Magazine* 44 (Fall 1967), 324–41.

Gudelunas, William A., Jr. "Nativism and the Demise of Schuylkill County Whiggery: Anti-Slavery or Anti-Catholicism." *Pennsylvania History* 45, No. 3 (July 1978), 225–36.

Gudelunas, William A., Jr., and William G. Shade. *Before the Molly Maguires: the Emergence of the Ethno-Religious Factor in the Politics of the Lower Anthracite Region, 1844–1872.* New York: Arno Press, 1976.

Gutman, Herbert. "Five Letters of Immigrant Workers from Scotland to the United States." *Labor History* 9 (1968), 384–99.

———. "Black Coal Miners and the Greenback Labor Party in Redeemer, Alabama, 1878–1879." *Labor History* 10 (Summer 1969), 506–35.

———. "The Negro and the United Mine Workers of America; the Career and the Letters of Richard L. Davis." In *Workers in the Industrial Revolution,* edited by Peter N. Stearns and Daniel Walkawitz, 338–76. New Brunswick, N.J.: Transaction, 1974.

———. "The Buena Vista Affair, 1874–1875." In *Workers in the Industrial Revolution,* edited by Peter N. Stearns and Daniel Walkawitz, 194–231. New Brunswick N.J.: Transaction, 1974.

Hafen, LeRoy R. *Colorado: The Story of a Western Commonwealth.* Denver: Peerless Publ. Co., 1933.

Handlin, Oscar. *The Uprooted.* New York: Little Brown, 1951.

———. *A Pictorial History of Immigration.* New York: Crown, 1972.

Harriman, Mrs. J. Borden. *From Pinafores to Politics.* New York: Henry Holt, 1923.

Harrison, John M. and Harry H. Stein. *Muckraking: Past Present and Future.* University Park, Pa.: Pennsylvania State University Press, 1973.

Hartmann, Edward G. *Americans From Wales.* Boston: Christopher Publishing House, 1967.

Harvey, Katherine A. "The Civil War and the Maryland Coal Trade." *Maryland Historical Magazine* 62, No. 4 (Dec. 1967), 361–80.

———. "The Knights of Labor in the Maryland Coal Fields, 1878–1882." *Labor History* 10 (Fall 1969), 555–83.

———. *The Best Dressed Miners: Life and Labor in the Maryland Coal Regions, 1835–1910.* Ithaca, N.Y.: Cornell University Press, 1969.

Hershey, Robert D. Jr. "Coal—The Other Glut." *New York Times,* 16 May 1982, sec. F, 1.

Hiebert, Ray Eldon. *Courtier to the Crowd: The Story of Ivy Lee and the Development of Public Relations.* Ames, Iowa: Iowa State University Press, 1966.

Higham, John. *Strangers in the Land.* New Brunswick, N.J.: Rutgers University Press, 1955.

History of the Union Pacific Coal Mines, 1868–1940. Omaha, Nebr.: Colonial Press, 1940.

Hoffman, John N. *Anthracite in the Lehigh Region of Pennsylvania, 1820–1845.* Washington, D.C.: Smithsonian Institution Press, 1968.

———. *Girard Estate Coal Lands in Pennsylvania, 1801–1884.* Washington, D.C.: Smithsonian Institution Press, 1972.

Hourwich, Isaac A. *Immigration and Labor.* New York: C.P. Putnam's Sons, 1912.

Hudson Coal Company. *The Story of Anthracite.* New York: The Company, 1932.

Hunter, Louis C. "Waterpower in the Century of the Steam Engine." In *America's Wooden Age: Aspects of its Early Technology,* edited by Brooke Hindle. Tarrytown, N.Y.: Sleepy Hollow Restorations, 1975.

Ichioka, Uji. "Asian Immigrant Coal Miners and the United Mine Workers of America: Race and Class in Rock Springs, Wyoming, 1907." *Amerasia* 6:2 (1979), 1–23.

Jenks, Jeremiah W., and W. Jett Lauck. *The Immigration Problem.* New York: Funk and Wagnall's Co. 1917.

Jenks, Leland Hamilton. *The Migration of British Capital to 1875.* New York: Alfred A. Knopf, 1927.

Jensen, Mead L., and Alan Batemen. *Economic Mineral Deposits.* 3d ed., rev. New York: John Wiley and Sons, 1981.

John, Angela V. *By the Sweat of Their Brow: Women Workers at Victorian Coal Mines.* London: Croon Helm, 1980.

Johnson, James P. *The Politics of Soft Coal.* Urbana, Ill.: University of Illinois Press, 1979.

Jones, Eliot. *The Anthracite Coal Combination in the United States.* Cambridge, Mass.: Harvard University Press, 1914.

Jones, Mary Harris. *Autobiography.* Edited by Mary Field Parton. New York: Arno and the New York Times, 1925, 1969.

Kerr, George L. *Practical Coal Mining.* London: Charles Griffith, 1905.

Killeen, Charles E. *John Siney.* Ann Arbor, Mich.: University Microfilms International, 1942.

Klein, Maury. *The Life and Legend of Jay Gould.* Baltimore, Md.: Johns Hopkins University Press, 1986.

———. *Union Pacific: The Birth of a Railroad, 1862–1893.* New York: Doubleday, 1987.

Knight, Harold V. *Working in Colorado.* Boulder, Colo.: University of Colo. Center for Labor Education, 1971.

Kolko, Gabriel. *The Triumph of Conservatism.* Chicago: Quadrangle Books, 1963, 1967

Kornbluh, Joyce L. *Rebel Voices.* Ann Arbor, Mich.: University of Michigan Press, 1968.

Korson, George. *Minstrels of the Mine Patch.* Philadelphia: University of Pennsylvania Press, 1938.

———. *Coal Dust on the Fiddle.* Hatboro, Pa.: Folklore Associates, 1965.

LaGumina, Salvatore J. *The Immigrants Speak: Italian Americans Tell Their Story.* New York: Center for Migration Studies, 1979.

Lambie, Joseph T. *From Mine to Market: The History of Coal Transportation in the Norfolk and Western Railway.* New York: New York University Press, 1954.

Landes, David S. *The Unbound Prometheus: Technological Change and Industrial Development in Western Europe from 1750 to the Present.* Cambridge: Cambridge University Press, 1969.

———. "The Industrial Revolution." In *Encyclopaedia Britannica.* 1980. s.v.

Lane, Ann. "Recent Literature of the Molly Maguires." *Science and Society* 30 (Summer 1965), 309–17.

Larsen, Charles. *The Good Fight: The Life and Times of Ben B. Lindsey.* Chicago: Quadrangle 1972.

Laslett, John H. M. *Labor and the Left.* New York: Basic Books, 1970.

———. *Nature's Noblemen: The Fortunes of the Independent Collier in Scotland and the American Midwest, 1855–1899.* Los Angeles: Institute of Industrial Relations, UCLA, 1983.

Lee, Rose Hum. *The Growth and Decline of Chinese Communities in the Rocky Mountain Region.* New York: Arno Press, 1947, 1978.

Leifchild, J.P. *On Coal at Home and Abroad.* London: Longmans, Green and Co., 1873.

Leiserson, William, M. *Adjusting Immigrant and Industry.* New York: Harper and Bros., 1924.

LeKachman, Robert. "American Business—Then and Now." In *An Emerging Independent American Economy, 1815–1875,* edited by Joseph R. Frese, S.J. and Jacob Judd. Tarrytown, N.Y.: Sleepy Hollow Press, 1980.

Long, Priscilla. *Mother Jones: Woman Organizer, and Her Relations With Miner's Wives, Working Women and the Suffrage Movement.* (40 page monograph) Boston: South End Press, 1976.

———. "The Women of the Colorado Fuel and Iron Strike, 1913–14." In *Women,*

Work, and Protest: A Century of U.S. Women's Labor History, edited by Ruth Milkman. London and New York: Routledge and Kegan Paul, 1985.

Lonsdale, David, L. *The Movement for an Eight Hour Law in Colorado, 1893–1913.* Ann Arbor, Mich.: University Microfilms International, 1963.

Love, George H. *An Exciting Century of Coal! (1864–1964).* New York: Newcomen Society, 1955.

Lovejoy, Owen R. "The Extent of Child Labor in the Anthracite Coal Industry." *Annals of the American Academy of Political and Social Science* 29 (Jan. 1907), 26–34.

———. "Child Labor in the Soft Coal Mines." *Annals of the American Academy of Political and Social Science* 29 (Jan. 1907), 26–34.

MacFarlane, James. *The Coal Regions of America.* New York: D. Appleton and Co., 1875.

Maclean, Annie Marion. "Life in the Pennsylvania Coal Fields with Particular Reference to Women." *American Journal of Sociology* 14 (1909), 329–51.

Marcus, Irwin M. "Labor Discontent in Tioga Cy., Pa, 1865–1905: The Gutman Thesis: A Test Case." *Labor History* 14 (Summer 1973), 414–22

Marlatt, Gene Donald. *Joseph R. Buchanan.* Ann Arbor, Mich.: University Microfilms International, 1975.

Martin, Albro. "James J. Hill and the First Energy Revolution." *Business History Review* 50, No. 2 (Summer 1976), 179–97.

McAuliffe, Eugene. *History of the Union Pacific Coal Mines.* Omaha, Nebr.: Colonial Press, 1940.

McDonald, David, and Edward Lynch. *Coal and Unionism: A History of the American Coal Miners Unions.* Indianapolis, Ind.: Lynald Books, 1939.

McGovern, George S. *The Colorado Coal Strike, 1913–1914.* Ann Arbor, Mich.: University Microfilms International, 1953.

McGovern, George S., and Leonard F. Guttridge. *The Great Coalfield War.* Boston: Houghton Mifflin, 1972.

McHenry, J. Patrick. *A Short History of Mexico.* New York: Doubleday, 1970.

McLaurin, Melton A. "Knights of Labor." In *Essays in Southern Labor History,* edited by Fink and Reed. Westport, Conn.: Greenwood Press, 1976.

Mellon, Ben. "How Coal Owners Sacrifice Coal Workers." *Pearson's Magazine* 25, No. 4 (Apr. 1911), 419–29.

Mitchell, John. "The Coal Strike." *McClure's Magazine* 20, No. 2 (Dec. 1902), 219–24.

Montgomery, David. "Pennsylvania: An Eclipse of Ideology." In *Radical Republicans in the North: State Politics During Reconstruction* edited by James C. Mohr. Baltimore, Md.: Johns Hopkins University Press, 1976.

———. "The Irish and the American Labor Movement." In *America and Ireland, 1776–1976,* edited by David Noel Doyle and Owen Dudley Edwards. Westport, Conn.: Greenwood Press, 1980.

———. *Beyond Equality: Labor and the Radical Republicans, 1862–1872.* Urbana, Ill.: University of Illinois Press, 1967, 1980.

Moore, E.S. *Coal.* 2d ed. New York: John Wiley and Sons, 1940.

Morris, James O. "The Acquisitive Spirit of John Mitchell, UMW President, 1899–1908." *Labor History* 20 (1972), 5–43.

Mosgrove, Jed H. "Mine Gasses." In *Elements of Practical Coal Mining,* edited by Samuel M. Cassidy. New York: Society of Mining Engineers, 1973.

Nash, Michael. *Conflict and Accommodation: Some Aspects of the Political Behavior of America's Coal Miners and Steel Workers, 1890–1920.* Ann Arbor, Mich.: University Microfilms International, 1975.

National Coal Board, Scottish Division. *A Short History of the Scottish Coal Mining Industry.* Edinburgh and Glasgow: National Coal Board, 1958.

Nearing, Scott. *Anthracite.* Philadelphia: John Winston Co., 1915.

Nef, J.U. "Coal Mining and Its Utilization." In *A History of Technology.* Vol. 3. New York and Oxford: Oxford University Press, 1957.

Nicolls, Francis H. "Children of the Coal Shadow." *McClure's Magazine* 20, No. 4 (Feb. 1903), 435–44.

Nicolls, William Jasper. *The Story of American Coals.* Philadelphia: J.P. Lippincott, 1897.

Notarianni, Philip F. "Italian Involvement in the 1903–04 Strike in Southern Colorado and Utah." In *Pane E Lavoro: The Italian American Working Class,* edited by George E. Pozzetta. Toronto: Multicultural History Society of Ontario, 1980.

Novak, Michael. *The Guns of Lattimer.* New York: Basic Books, 1978.

Noyes, Robert, ed. *Coal Resources, Characteristics, and Ownership in the USA.* Park Ridge, N.J.: Noyes Data Corp., 1978.

Nyden, Paul. *Black Coal Miners in the United States.* New York: American Institute for Marxist Studies, 1974.

O'Higgins, Harvey. "John Lawson's Story." *Metropolitan Magazine* (Dec. 1916).

O'Neal, Mary Thomas. *Those Damn Foreigners.* Hollywood, Calif.: Minerva, 1971.

Palladino, Grace. *The Poor Man's Fight: Draft Resistance and Labor Organization in Schuylkill County, Pa., 1860–1865.* Ann Arbor, Mich.: University Microfilms International, 1983.

Papanikolas, Helen Z. "Utah's Coal Lands: A Vital Example of How America Became a Great Nation." *Utah Historical Quarterly* 43 (1975), 104–24.

Papanikolas, Zeese. *Buried Unsung: Louis Tikas and the Ludlow Massacre.* Salt Lake City: University of Utah Press, 1982.

Parton, Mary Field, ed. *The Autobiography of Mother Jones.* New York: Arno Press and the New York Times, 1925, 1969.

Paskoff, Paul F. *Industrial Evolution: Organization, Structure, and Growth of the Pennsylvania Iron Industry, 1760–1860.* Baltimore, Md.: Johns Hopkins University Press, 1983.

Patterson, Joseph F. "Old W. B. A. Days." *Historical Society of Schuylkill County Publications,* Vol. 2 (1910), 355–84.

———. "After the W.B.A.," *Historical Society of Schuylkill County Publications,* Vol. 4 (1912), 168–84.

———. "Reminiscences of John Maguire After Fifty Years of Mining." *Historical Society of Schuylkill County Publications,* Vol. 4 (1912), 305–26.

Perelmen, Michael. *Farming for Profit in a Hungry World.* Montclair, N.J.: Allanheld, Osmun and Co., 1977.

Perlman, Silig, and Philip Taft. *History of Labor in the United States, 1896–1932.* New York: Macmillan, 1935

Pierce, Harry H. "Anglo-American Investors and Investment in the New York Central Railroad." In *An Emerging Independent American Economy, 1815–1875,* edited by Joseph R. Frese and Jacob Judd. Tarrytown, N.Y.: Sleepy Hollow Press, 1980.

Pinkowski, Edward. *Lattimer Massacre.* Philadelphia: Sunshine Press, n.d.

Pittsburg-Buffalo Co. Pittsburgh, Pa.: The Company, 1911.

Plested, Dolores. "Where Coal Was Once King: A Journey Through the Life of a Small Coal Operator." *Colorado Heritage* No. 2 (1987), 30–43.

Pollard, Sydney. *The Genesis of Modern Management: A Study of the Industrial Revolution in Great Britain.* Cambridge, Mass.: Harvard University Press, 1965.

Porter, Eugene O. "The Colorado Coal Strike of 1913—An Interpretation." *The Historian* 12 (Autumn 1949), 3–27.

Powell, Allen Kent. "The 'Foreign Element' and the 1903–04 Carbon County Coal Miners' Strike." *Utah Historical Quarterly* 43 (Spring 1975), 125–54.

―――. *The Next Time We Strike: Labor in Utah's Coal Fields, 1900–1933.* Logan, Utah: Utah State University Press, 1985.

Pratt, Joseph A. "Natural Resources and Energy." In *Encyclopedia of Economic History.* Vol. 1, edited by Glen Porter. New York: Charles Scribner's Sons, 1980.

Pryde, George B. "The Union Pacific Coal Co." *Annals of Wyoming* 25 (1953), 190–205.

Raynes, J.R. *Coal and Its Conflicts.* London: Ernest Braun, 1928.

Rhodes, Steven, and Paulette Middleby, "Public Pressures, Technical Options: The Challenge of Controlling Acid Rain." *Environment* 25 (May 1983).

Ritson, J. A. S. "Metal and Coal Mining, 1750–1875." In *A History of Technology.* Vol. 4, edited by Singer, et al. New York and London: Oxford University Press, 1958.

Roberts, Peter. *The Anthracite Coal Industry.* New York: Macmillan, 1901.

―――. *Anthracite Coal Communities.* London and New York: Macmillan, 1904.

Robinson, Helen Ring. "The War in Colorado." *The Independent,* 11 May 1914, 245–49.

Rosenblum, Gerald. *Immigrant Workers, Their Impact on American Labor Radicalism.* New York: Basic Books, 1973.

Rosenburg, Nathan. "Technology." In *Encyclopedia of American Economic History.* Vol. 1, edited by Glenn Porter. New York: Charles Scribner's Sons, 1980.

Roy, Andrew. *The Coal Mines: A Description of the Various Systems of Working and Ventilating Mines.* Cleveland: Robinson and Savage and Co., 1876.

―――. *History of the Coal Miners in the United States.* Westport, Conn.: Greenwood Press, 1905, 1970.

Scamehorn, H. Lee. "John C. Osgood and the Western Steel Industry." *Arizona and the West* 15 (Summer 1973), 133–48.

―――. *Pioneer Steelmaker in the West: the Colorado Fuel and Iron Co., 1872–1903.* Boulder, Colo.: Pruett Publishing, 1976.

Schaefer, Donald Fred. *A Quantitative Description and Analysis of the Growth of the Pennsylvania Anthracite Coal Industry, 1820–1865.* New York: Arno Press, 1977.

Schallenberg, Richard H., and David A. Aulf. "Raw Materials Supply and Technological Change in the American Charcoal Iron Industry." *Technology and Culture* 18, No. 3 (July 1977), 436–66.

Scheinberg, Stephen J. *The Development of Corporate Labor Policy, 1900–1940.* Ann Arbor, Mich.: University Microfilms International, 1966.

Schlegel, Marvin Wilson. "The Workingman's Benevolent Association." *Pennsylvania History* 10 (Oct. 1943), 243–67.

―――. *Ruler of the Reading: The Life of Franklin B. Gowen, 1836–1889.* Harrisburg, Pa.: Archives Publishing, 1947.

Schmookler, Jacob. "The Bituminous Coal Industry." In *The Structure of American Industry.* Rev. ed. Edited by Walter Adams. New York: Macmillan, 1954.

Schurr, Sam H., and Bruce C. Netschert. *Energy in the American Economy.* Baltimore, Md.: Johns Hopkins University Press, 1960.

Schwieder, Dorothy. "Italian Americans in Iowa's Coal Mining Industry." *Annals of Iowa* 46 (Spring 1982), 263–78.

―――. *Black Diamonds: Life and Work in Iowa's Coal Mining Communities, 1895–1925.* Ames, Iowa: Iowa State University Press, 1983.

Selekman, Ben M., and Mary Van Kleek. *Employees Representation in Coal Mines.* New York: Russell Sage, 1924.

Seltzer, Curtis. *Fire in the Hole: Miners and Managers in the American Coal Industry.* Lexington, Ky.: University of Kentucky Press, 1985.

Shankman, Arnold. "Draft Resistance in Pennsylvania." *Pennsylvania Magazine of History and Biography* 101, No. 2 (Apr. 1977), 190–204.

Simon, Richard Mark. *The Development of Underdevelopment: The Coal Industry and Its Effects on the West Virginia Economy, 1880–1930.* Ann Arbor, Mich.: University Microfilms International, 1978.

Simonin, Louis. *Underground Life of Mines and Miners.* London: Chapman and Hall, 1969.

Sinclair, Upton. *King Coal.* Published by Author, 1921.

Sklar, Martin J. "Woodrow Wilson and the Political Economy of Modern United States Liberalism." *Studies on the Left* 1, No. 3 (1960), 17–47.

Smedley, Agnes. *Daughter of Earth.* Old Westbury, N.Y.: Feminist Press, 1973.

Stapleton, Darwin H. "The Diffusion of Anthracite Iron Technology: The Case of Lancaster County." *Pennsylvania History* 102 (April 1978), 147–57.

Steel, Edward M., ed. *The Correspondence of Mother Jones.* Pittsburgh, Pa.: University of Pittsburgh Press, 1985.

Straw, Richard Alan. *This Is Not a Strike, It Is Simply a Revolution: Birmingham Miners' Struggle for Power, 1894–1908.* Ann Arbor, Mich.: University Microfilms International, 1980.

Suggs, George G., Jr. *Colorado Conservatives Versus Organized Labor: A Study of the George Hamilton Peabody Administration, 1903–1905.* Ann Arbor, Mich.: University Microfilms International, 1964.

————. "Catalyst for Industrial Change: The WFM, 1893–1903." *The Colorado Magazine* 45 (Fall 1968), 322–37.

————. "The Colorado Coal Miners' Strike, 1903–04." *Journal of the West* 12 (Jan. 1973), 36–52.

Sullivan, Charles Kenneth. *Coal Men and Coal Towns: Development of the Smokeless Coal Fields of Southern West Virginia, 1873–1921.* Ann Arbor, Mich.: University Microfilms International, 1979.

Swank, James. *History of the Manufacture of Iron in All Ages.* Philadelphia: Published by Author, 1884.

Taylor, A. J. "The Sub-contract System in the British Coal Industry." In *Studies in the Industrial Revolution,* edited by L. S. Pressnell. London: University of London (Athone Press), 1960.

Taylor, Richard Cowling. *Statistics of Coal.* Philadelphia: J.W. Moore, 1848.

Te Brake, William H. "Air Pollution in Preindustrial London, 1250–1650." *Technology and Culture* 16, No. 3 (July 1975), 337–59.

Temin, Peter. *Iron and Steel in the Nineteenth Century.* Cambridge, Mass.: MIT Press, 1964.

Thompson, Alexander MacKenzie, III. *Technology, Labor, and Industrial Structure of the U.S. Coal Industry: An Historical Perspective.* New York: Garland Publishing, 1979.

Thompson, E. P. *The Making of the English Working Class.* New York: Vintage, 1963.

Trachtenberg, Alexander. *The History of Legislation for the Protection of Coal Miners in Pennsylvania, 1824–1915,* New York: International Publishers, 1942.

Ubbelohde, Carl, et al. *A Colorado History.* Boulder, Colo., Pruett Publishing Society, 1972.

Van Tine, Warren. *The Making of a Labor Bureaucrat: Union Leadership in the United States, 1870–1920.* Amherst, Mass.: University of Massachusetts Press, 1973.

Veblen, Thorstein. *The Theory of Business Enterprise.* New Brunswick, N.J.: Rutgers University Press, 1904, 1978.

Walker, Samuel. "Varieties of Workingclass Experience: The Workingmen of Scranton, Pa., 1855–1885." In *American Workingclass Culture,* edited by Milton Cantor. Westport, Conn.: Greenwood Press, 1979.

Wallace, Anthony F. C. *The Social Context of Innovation.* Princeton, N.J.: Princeton University Press, 1982.

————. *St. Clair.* New York: Alfred Knopf, 1987.

Ward, Robert David and William Warren Rogers. *Labor Revolt in Alabama: The Great Strike of 1894.* University, Ala.: University of Alabama Press, 1965.

Ware, Norman J. *The Labor Movement in the United States, 1860–1895.* Gloucester, Mass.: Peter Smith, 1929, 1959.

————. *Labor in Modern Industrial Society.* New York: Russell and Russell, 1935.

————. *The Slav Invasion and the Mine Workers.* Philadelphia: J.B. Lippincott, 1904.

Warne, Frank Julian. *The Immigrant Invasion.* New York: Dodd, Mead and Co., 1913.

Weinstein, James. *The Decline of Socialism in America.* New York: Monthly Review Press, 1967.

————. *The Corporate Ideal in the Liberal State, 1900–1918.* Boston: Beacon Press, 1968.

Weitz, Eric. "Class Formation and Labor Protest in the Mining Communities of Southern Illinois and the Ruhr, 1890–1925." *Labor History* 27, No. 1 (Winter 1985–1986), 85–105.

White, Everett M. "Fires and Explosions." *Elements of Practical Coal Mining.* 2d ed. by Crickmer and Zegeer. New York: Society of Mining Engineers, 1981.

Whitten, David O. *The Emergence of Giant Enterprise, 1860–1914.* Westport, Conn.: Greenwood Press, 1983.

Wieck, Edward A. *The American Miners' Association* New York: Russell Sage Foundation, 1940.

Williams, John Alexander. *West Virginia and the Captains of Industry.* Morgantown, W.Va.: West Virginia University Press, 1976.

Wiman, Erastus. "British Capital and American Industries." *North American Review* 150 (Feb. 1890), 220–34.

Wright, Carroll D. "An Historical Sketch of the Knights of Labor." *The Quarterly Journal of Economics* (Jan. 1887), 137–68.

Wrigley, E. A. "The Supply of Raw Materials in the Industrial Revolution." In *The Causes of the Industrial Revolution,* edited by R. M. Hartwell. London: Methuen and Co., 1967.

Yarrow, Michael. "The Labor Process in Coal Mining: Struggle for Control." In *Case Studies in the Labor Process,* edited by Andrew Zimbalist. New York: Monthly Review Press, 1979.

Yearly, Clifton K., Jr. *Enterprise and Anthracite.* Baltimore, Md.: Johns Hopkins University Press, 1961.

————. *Britons in American Labor.* Baltimore, Md.: Johns Hopkins University Press.

PERIODICALS

Camp and Plant

Trinidad Chronicle-News (1900–1904, 1913–1914)

The Irish World (1870–1876)

Journal of United Labor (1880–1886)

Denver Labor Enquirer (1884)

The Miners' Magazine (1900–1915)

National Labor Tribune (1884–1887)

Pueblo Courier and Pueblo Labor Advocate (1898–1904)

Rocky Mountain News (1900–1915)

Union Pacific Employees Magazine (1886–1894)

United Mine Workers' Journal (1891–1916)

The Workingman's Advocate (1871–1877)

MANUSCRIPT COLLECTIONS

Catholic University of America
  John Mitchell Papers
  Mother Jones Papers
  Terrence V. Powderly Papers

Colorado State Archives
  Gov. James H. Peabody Papers
  Government Records, Special Series, Colorado National Guard
  Gov. Elias Ammons Papers

Colorado State Historical Society
Ellis Meredith Papers
General William Jackson Palmer
Papers
Colorado Fuel and Iron Company
Papers
Jesse Floyd Welborn Papers
Edward Verdeckberg Papers
Denver Public Library, Western History
Dept.
John Chase Scrapbooks
Colorado National Guard Record
E. L. Doyle Papers
Frederick M. Farrar, Sr., Papers
Hildreth Frost Papers
George Minot Papers
Rocky Mountain Fuel Company
Papers
Harvard University, Baker Library
Colorado Fuel and Iron Company
Papers (Corporate Reports Dept.)
Buck Mountain Coal Company Papers
(Manuscript and Archives Dept.)
Library of Congress
Woodrow Wilson Papers
National Archives
Federal Mediation and Conciliation
Service (RG 280, Files 41/10-A;
41/10-E)
Records, Provost Marshal General's
Bureau (Entry 3050, R.G 110)
F.B.I. Files Dated 1909–1924 (RG 60,
Dept. of Justice)
File 168733-a (on Colorado Strike),
(RG 60, Dept. of Justice)
Commission on Industrial Relations
Papers (RG 174, Dept. of Labor)

Nebraska State History Society
Union Pacific Archives
New York Public Library
Frank Walsh Papers
Rockefeller Archive Center
Rockefeller Family Archives. Business
Interests (Colorado Fuel and Iron
Company)
Rocky Mountain Energy, Broomfield,
Colorado
Union Pacific Coal Company Papers
State Historical Society of Wisconsin
Adolph Germer Papers
SUNY, Binghamton
L.M. Bowers Papers
Union Pacific Railroad, Omaha,
Nebraska.
Union Pacific Coal Company Papers
United Mine Workers of America,
Washington, D.C.
UMWA Papers
University of Chicago
Thomas J. Morgan Papers
University of Colorado, Boulder,
Western History Collections
Colorado State Federation of Labor
Papers
Gov. Peabody Scrapbooks
Edward Keating Papers
Josephine Roche Papers
Western Federation of Miners Papers
University of North Carolina Library,
Southern Historical Collection,
Chapel Hill
Ethelbert Stewart Papers (Microfilm
Edition)

UNITED STATES GOVERNMENT DOCUMENTS

Averitt, Paul. "Coal." *United States Mineral Resources.* USGS Professional Paper 820. Washington, D.C., 1973.

Chamberlain, Rollin Thomas. *Notes on Explosive Gasses and Dusts.* USGS Bulletin 383. Washington, D.C., 1909.

Fay, Albert H. *Coal Mine Fatalities in the United States, 1870–1914.* Bureau of Mines Bulletin 115. Washington, D.C., 1910.

Hall, Clarence, and Walter O. Snelling. *Coal Mine Accidents: Their Causes and Prevention.* USGS Bulletin 333. Washington, D. C., 1907.

Henderson, Charles W. *Mining in Colorado: A History of Discovery, Development and Production,* USGS Professional Paper No. 138. Washington, D. C., 1926.

Humphrey, H. B. *Historical Summary of Coal Mine Explosions in the United States, 1810–1958.* Bureau of Mines Bulletin 586. Washington, D. C., 1960.

Interstate Commerce Commission. *Special Report No. 1: Intercorporate Relationships of Railways on the United States as of June 30, 1906.* Washington, D. C., 1908.

Interstate Commerce Commission. *Report of Investigation . . . Into Railroad Discriminations and Monopolies in Coal and Oil.* 28 April 1908, Senate Doc. 450, 60th Cong., 1st Session.

Machisak, John C., et al. *Injury Experience in Coal Mining, 1958.* Bureau of Mines Information Circular 8067. Washington, D. C., 1959.

Peterson, Florence. *Strikes in the United States, 1880–1936.* U.S. Bureau of Labor Statistics Bulletin No. 651. Washington, D. C., 1938.

Sheridan, Frank J. "Italian, Slavic, and Hungarian Unskilled Immigrant Laborers in the United States." *Bulletin of the Bureau of Labor No. 72.* 1907, 1971.

U.S. Census. *Reports.* 1840–1920.

———. 16th Census, 1940. *Population: Comparative Occupation Statistics for the United States, 1870–1940.* Washington, D. C., 1943.

U.S. Commission on Immigration. "Representative Community E." *Reports on Immigrants in Industries, Part 21: Diversified Industries, Vol. 1.* Washington D. C., 1911.

———. *Reports on Immigrants in Industries, Part 25: Japanese and Other Immigrant Races in the Pacific Coast and Rocky Mountain States.* Vol. 3. Washington D. C., 1911.

———. *Reports on Immigrants in Industries, Part I: Bituminous Coal Mining.* 2 vols. 61st. Cong., 2d Sess. Washington, D.C., 1911.

U.S. Commission on Industrial Relations. *Final Report and Testimony,* Vols. 7, 8, and 9. Senate Doc. 415, 64th Cong., 2d Sess. Washington, D.C., 1916.

U.S. Commissioner of Labor. *A Report on Labor Disturbances in the State of Colorado From 1880 to 1904, Inclusive,* Senate Doc. 122, 58th Cong., 2d Sess. Washington, D. C., 1905.

———. *Third Annual Report: Report of the Secretary of the Interior.* Vol. 5, 50th Cong., 1st. Sess. 1887.

———. *Sixth Annual Report.* 1890. H. R. 51st Cong., 2d. Sess., Exec. Doc. No. 265. Washington, D. C., 1891.

———. *Tenth Annual Report, 1894, Strikes and Lockouts.* Washington, D. C., 1896.

U.S. Congress, House Subcommittee on Mines and Mining. *Conditions in the Coal Mines of Colorado.* 2 vols. Pursuant to H. R. 387. 63d Cong., 2d. Sess. Washington, D.C., 1914.

U.S. Dept. of Labor, Women's Bureau. *Home Environment and Employment Opportunities of Women in Coal Mine Workers' Families.* Bulletin No. 45. Washington, D.C., 1925.

U. S. Geological Survey. *Mineral Resources of the United States.* Washington, D.C., 1882–1920.

———. *Twenty-Second Annual Report, 1900–1901, Vol. 2: Coal, Oil, and Cement.* Washington, D.C., 1902.

———. *The Prevention of Mine Explosions.* USGS Bulletin 369. Washington, D.C., 1980.

———. *Coal Resources of Colorado,* by E. R. Landis. USGS Bulletin 1072-C. Washington, D. C., 1959.

U.S. Industrial Commission. "The Foreign Born in the Coal Mines." *Reports of the Industrial Commission, Vol. 15: Immigration and Education.* Washington, D.C., 1901.

———. "National Labor Organizations in the United States," "The Coal Mine Industry," and "Labor Organizations of Mine Workers," *"Reports of the Industrial Commission, Vol. 17: On Labor Organizations, Labor Disputes and on Railway Labor."* Washington, D. C., 1901.

————. "The Organization of the Pittsburgh Coal Co." *Reports of the Industrial Commission, Viol. 2.* Washington, D. C., 1901.
Virtue, G. O. *The Anthracite Mine Laborers.* Bulletin of the Dept. of Labor No. 13 (Nov. 1897). Washington, D. C., 1897.

STATE DOCUMENTS
Colorado, Bureau of Labor Statistics. *Biennial Reports.* 1887–1915.
Colorado, Inspector of Coal Mines. *Reports.* 1884–1915.
Colorado General Assembly. "Report of Joint Committee to Investigate Conditions in Coal Mines." *Senate Journal 1901.* 536–73.
Illinois, Bureau of Labor Statistics. *Eighth Biennial Report, 1894.* Springfield, Ill., 1895.
————, Dept. of Mines and Minerals. *Twenty-Second Annual Coal Report, 1903.* Springfield, Ill., 1904.
————. Dept. of Mines and Minerals. *Compilation of the Reports of the Mining Industry of Illinois From the Earliest Records to the Close of the Year 1930.* Springfield, Ill., 1931.
Ohio. *Report of the Mining Commission, 1871.* Columbus, Ohio, 1872.
Pennsylvania Senate. "Report of the Committee on the Judiciary (General) . . . in Relation to the Coal Difficulties." *Pennsylvania Legislative Documents, 1971.* Document No. 39, 1515–1733.
Pennsylvania. *Industrial Statistics.* [Various titles]. 1872–1900.
Pennsylvania Geological Survey. "Mining Methods of the Pittsburgh Coal Region." *Annual Report for 1886.* Harrisburg, Pa., 1887.
Pennsylvania. Second Geological Survey. *A Special Report . . . Upon the Causes, Kinds and Amount of Waste in Mining Anthracite.* Vol. A2. Harrisburg, Pa., 1881.

UNPUBLISHED MANUSCRIPTS
Benson, Maxine Frances. "Labor and the Law in Colorado, 1915–1917." Master's Thesis, University of Colorado, 1962.
Castro, Jess Augustin. "Alexander Cameron Hunt: Colorado Territorial Governor, 1867–1869." Master's Thesis, University of Denver, 1957.
Fogel, Karen. "Mineworkers and Their Families in the Southern Colorado Coal Fields, 1913–1914." Honors Thesis, Harvard University, 1975.
McClurg, Donald Joseph. "Labor Organizations in the Coal Mines of Colorado, 1878–1933." PhD. Dissertation, University of California, 1959.

MINUTES, ANNUAL REPORTS, PROCEEDINGS
Colorado Coal and Iron Company. *Annual Reports.*
Colorado Fuel and Iron Company. *Annual Reports.*
Denver and Rio Grande Railway. *Annual Reports.*
UMWA. Executive Board Minutes. 1891–1921.
————. Proceedings of Annual and Biennial Conventions. 1891–1921.
————. "Proceedings of the Joint Convention of the N. P. U. and N. Y. A. 135, 1890." (In UMWA Papers).
————. "Constitution and Bylaws, 1890." (In UMWA Papers).
UMWA, District 15. "Proceedings of the Tenth Annual Convention of District No. 15 Held in Louisville, Colo., Feb. 18–22, 1913."
————. "Special Convention Proceedings Held at Trinidad, Colo., Sept. 15–16, 1914."
Union Pacific Coal Company. *Annual Reports.* 1891–1893, 1895, 1898, 1905–1906, 1912–1915.

INDEX

NOTE: Page numbers in italics refer to illustrations